CELL BIOLOGY AND GENETICS

CECIE STARR / RALPH TAGGART

Christine A. Evers

Lisa Starr

BIOLOGY
THE UNITY AND DIVERSITY OF LIFE
ELEVENTH EDITION

THOMSON

BROOKS/COLE

Australia • Brazil • Canada • Mexico • Singapore
Spain • United Kingdom • United States

PUBLISHER Jack C. Carey

VICE-PRESIDENT, EDITOR-IN-CHIEF Michelle Julet

SENIOR DEVELOPMENT EDITOR Peggy Williams

ASSOCIATE DEVELOPMENT EDITOR Suzannah Alexander

EDITORIAL ASSISTANT Chris Ziemba, Kristina Razmara

TECHNOLOGY PROJECT MANAGER Keli Amann

MARKETING MANAGER Ann Caven

MARKETING ASSISTANT Brian Smith

MARKETING COMMUNICATIONS MANAGER Nathaniel
  Bergson-Michelson

PROJECT MANAGER, EDITORIAL PRODUCTION Andy Marinkovich

CREATIVE DIRECTOR Rob Hugel

PRINT BUYER Karen Hunt

PERMISSIONS EDITOR Joohee Lee, Sarah Harkrader

PRODUCTION SERVICE Grace Davidson & Associates

TEXT AND COVER DESIGN Gary Head

PHOTO RESEARCHER Myrna Engler

COPY EDITOR Kathleen Deselle

ILLUSTRATORS Gary Head, ScEYEnce Studios, and Lisa Starr

COMPOSITOR Lachina Publishing Services

TEXT AND COVER PRINTER R.R. Donnelley/Willard

COVER IMAGE Ko'olau Mountains on the windward side of
Oahu, part of the Hawaiian Archipelago that is a natural
laboratory for the study of evolution.
Photographer Russ Lowgren

Printed in the United States of America
2 3 4 5 6 7 09 08 07 06

Library of Congress Control Number: 2005931272

ISBN 0-495-12578-4

For more information about our products, contact us at:
Thomson Learning Academic Resource Center
1-800-423-0563

For permission to use material from this text or product, submit
a request online at http://www.thomsonrights.com.
Any additional questions about permissions can be submitted
by e-mail to thomsonrights@thomson.com.

BOOKS IN THE BROOKS/COLE BIOLOGY SERIES

Biology: The Unity and Diversity of Life, Eleventh, Starr/Taggart
Engage Online for Biology: The Unity and Diversity of Life
Biology: Concepts and Applications, Sixth, Starr
Basic Concepts in Biology, Sixth, Starr
Biology Today and Tomorrow, Second, Starr
Biology, Seventh, Solomon/Berg/Martin
Human Biology, Sixth, Starr/McMillan
Biology: A Human Emphasis, Sixth, Starr
Human Physiology, Fifth, Sherwood
Fundamentals of Physiology, Second, Sherwood
Human Physiology, Fourth, Rhoades/Pflanzer

Laboratory Manual for Biology, Fourth, Perry/Morton/Perry
Laboratory Manual for Human Biology, Morton/Perry/Perry
Photo Atlas for Biology, Perry/Morton
Photo Atlas for Anatomy and Physiology, Morton/Perry
Photo Atlas for Botany, Perry/Morton
Virtual Biology Laboratory, Beneski/Waber
Introduction to Cell and Molecular Biology, Wolfe
Molecular and Cellular Biology, Wolfe
Biotechnology: An Introduction, Second, Barnum

Introduction to Microbiology, Third, Ingraham/Ingraham
Microbiology: An Introduction, Batzing
Genetics: The Continuity of Life, Fairbanks/Anderson
Human Heredity, Seventh, Cummings
Current Perspectives in Genetics, Second, Cummings
Gene Discovery Lab, Benfey

Animal Physiology, Sherwood, Kleindorf, Yarcey
Invertebrate Zoology, Seventh, Ruppert/Fox/Barnes
Mammalogy, Fourth, Vaughan/Ryan/Czaplewski
Biology of Fishes, Second, Bond
Vertebrate Dissection, Ninth, Homberger/Walker

Plant Biology, Second, Rost/Barbour/Stocking/Murphy
Plant Physiology, Fourth, Salisbury/Ross
Introductory Botany, Berg

General Ecology, Second, Krohne
Essentials of Ecology, Third, Miller
Terrestrial Ecosystems, Second, Aber/Melillo
Living in the Environment, Fourteenth, Miller
Environmental Science, Tenth, Miller
Sustaining the Earth, Seventh, Miller
Case Studies in Environmental Science, Second, Underwood
Environmental Ethics, Third, Des Jardins
Watersheds 3—Ten Cases in Environmental Ethics, Third,
Newton/Dillingham

Problem-Based Learning Activities for General Biology, Allen/Duch
The Pocket Guide to Critical Thinking, Second, Epstein

Thomson Higher Education
10 Davis Drive
Belmont, CA 94002-3098
USA

Asia (including India)
Thomson Learning
5 Shenton Way
#01-01 UIC Building
Singapore 068808

Australia/New Zealand
Thomson Learning Australia
102 Dodds Street
Southbank, Victoria 3006
Australia

Canada
Thomson Nelson
1120 Birchmount Road
Toronto, Ontario M1K 5G4

UK/Europe/Middle East/Africa
Thomson Learning
High Holborn House
50/51 Bedford Row
London WC1R 4LR
United Kingdom

CONTENTS IN BRIEF

These highlighted chapters are not included in Cell Biology and Genetics

DETAILED CONTENTS

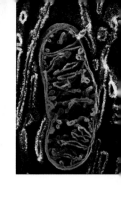

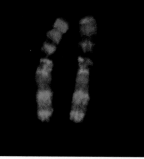

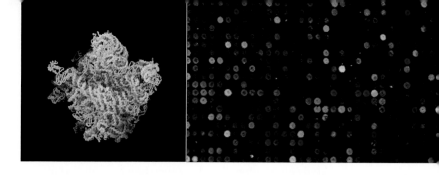

14 From DNA to Protein

15 Controls Over Genes

16 Studying and Manipulating Genomes

PREFACE

Biology's overriding paradigm is one of interpreting life's spectacular diversity as having evolved from simple molecular beginnings. For decades, researchers have been chipping away at the structural secrets of biological molecules. They are showing how different kinds are put together, how they function, and what happens when they mutate. They are casting light on how life originated, what happened over the past 3.8 billion years, and what the future may hold for humans and other organisms, individually and collectively.

This is profound stuff. *Yet introductory textbooks— including previous editions of this one—have not given students enough information to interpret for themselves the remarkable connection between molecular change, evolution, and their own lives.* We continue to rebuild this book in ways that clarify the connection. When students "get" the big picture of life, they become confident enough to think critically about the past, present, and future on their own. This is one of biology's greatest gifts.

MAKE THE DIRECT CONNECTION

New to this edition are connections that have direct and often controversial impact on our lives. What students learn today may help them make decisions tomorrow, in the voting booth as well as in their personal lives.

How Would You Vote?   Each chapter starts with an essay on a current issue that relates to its content (see list at right). Essays are expanded in custom *videoclips* and exercises on the student website and in the Power-Lecture, a one-stop PowerPoint tool for instructors.

We ask students in each chapter, *How would you vote on research or an application related to this issue?* Then the exercise invites them to read a balanced selection of articles, pro and con, before voting. Students throughout the country are already voting *online,* and are accessing campuswide, statewide, and nationwide tallies. This interactive approach to issues reinforces the premise that individual actions can make a difference.

HELP STUDENTS TRACK CONNECTIONS

Linking Concepts   Also new to this edition are links to concepts within and between chapters. Students have an easier time making connections when links guide them. Each chapter's opening spread has a section-by-section list of *key concepts,* each with a simple title. We repeat the titles at the top of appropriate text pages as reminders of the conceptual organization. A brief list of *links to earlier concepts* helps students assess concepts they should already know before they begin. For instance, before reading about neural function, they may wish to scan an earlier chapter's section on active transport. We repeat the linking icons in page margins to help students refer back to relevant sections.

Concept Spreads   Judging from extensive feedback, *concept spreads* are immensely popular. Students report that they can easily study and master parts of a chapter

IMPACTS/ISSUES ESSAYS

1   State of the world, including bioterrorism
2   Toxic chemicals in and around the body
3   Methane hydrates, energy, and extinctions
4   What is a cell and can researchers make one?
5   Membrane defects and cystic fibrosis
6   Student binge drinking and liver damage
7   Why your life depends on plants
8   Disorders arising from abnormal mitochondria
9   Henrietta Lacks, HeLa cells, and cervical cancer
10 Why females and males?
11 Rose breeding, a big business
12 Genes and mental illness
13 Cloning—reprogramming the DNA
14 The bioterror agent ricin affects ribosomes
15 Genes linked with breast cancer
16 Malnutrition, biotechnology, and Golden Rice
17 Asteroid impacts and human evolution
18 Natural selection and the rise of super rats
19 Hawaiian honeycreepers and humans
20 Bioprospecting in extreme environments
21 West Nile virus and Alexander the Great
22 Protists as architects and pathogens
23 Evolutionary perspective on deforestation
24 Reindeer, reindeer herders, and lichens
25 New neuroactive drugs from snail venom
26 How people interpret and misinterpret the past
27 A cautionary tale from Easter Island
28 Homeostasis and a lethal case of heat stroke
29 Plants and the rise and fall of civilizations
30 Phytoremediation at military waste sites
31 No more pollinators? No more chocolate
32 Plant hormones and our food supply
33 Superman and stem cell research
34 Dropping dead with Ecstasy
35 National security, sonar, and the whales
36 Endocrine disrupters and deformed frogs
37 Andro, creatine, and pumped-up athletes
38 From a bulldog to automated defibrillators
39 AIDS and a generation of African orphans
40 Cigarette smoking, biologically speaking
41 From early hominids to hefty humans
42 Urine tests, past and present
43 The point of mammalian (and human) sexual behavior
44 Fertility drugs and a plethora of babies
45 St. Matthews Island, the accidental experiment
46 What can stop the fire ants?
47 Global warming and the disappearing bayous
48 El Niños, super surfing, and suffering seals
49 Pheromones and the "killer bees"

when they have a small window of time rather than wading into an overwhelming mass of information. We place all text, art, and evidence in support of each concept in one section that consists of two facing pages, at most. Each section starts at a numbered tab and ends with a blue-boxed *on-page summary*, which students can use to check whether they understand the main points before turning to the next concept spread.

Concept spreads offer teachers flexibility in assigning topics to fit course requirements. Those who spend less time on, say, photosynthesis might bypass details on the properties of light and ATP formation. All spreads are part of the chapter topic, but some offer added depth.

Read-Me-First Art   *Read-me-first diagrams* are visual learning devices. Students can walk through them step-by-step as a preview of the text. The same art occurs in narrated, animated form online at BiologyNow and on the PowerLecture DVD. Repeated exposure to material in a variety of modalities accommodates diverse learning styles and reinforces knowledge of concepts.

Critical Thinking   Like all textbooks at this level, we walk students through examples of problem solving and experiments throughout the book. We integrate many simple experiments in the text proper. *Focus on Science* essays highlight the detailed ones. The entries *Experiment, examples*, and *Test, examples* in the main index list our selections.

Each chapter ends with a number of *Critical Thinking* questions, some with art, and some more challenging than others. They invite students to think outside the memorization box. We include an *annotated scientific paper* as an example of how scientists work and then report their findings (Appendix IX).

We selectively use some chapter introductions, even a few whole chapters, to emphasize scientific methods. However, introductory students are not scientists, so we do not shoehorn *all* material into an experimental context. We cannot expect them to learn the language and processes of science by intuition alone; we help them build these skills in a paced way.

Connections essays, flagged by vertical red bands on page edges, invite students to step back and connect the dots between chapters and units. An *Epilogue* invites them to make the most sweeping connection of all.

WRITE CLEARLY AND SELECTIVELY

Students often complain that their textbooks are dry and boring. We write to engage them without glossing over the science, even through such formidable topics as the laws of thermodynamics (Section 6.1).

To keep book length manageable, we were selective about which topics to include so that we could allocate enough space to explain them clearly. For instance, most nonmajors simply do not want to memorize each catalytic step of crassulacean acid metabolism. They do want to learn about the basis of sex, and many women would like to know what will go on inside their body if they become pregnant. Sufficient information will help them make informed decisions on many biology-related

issues, such as STDs, fertility drugs, prenatal diagnoses, gene therapies, and stem cell research. Over the years, students have written to tell us that they did read such material closely even when it was not assigned.

Our choices for which topics to condense, expand, or delete reflect more than three decades of feedback from teachers of millions of students around the world.

OFFER STUDENTS EASY-TO-USE MEDIA TOOLS

With this edition, logging in to online assets is simpler than before. Students register their free **1pass access code** at http://1pass.thomson.com and then log in to access all resources outlined below. An access code is packaged in each new copy of the book and also is available through e-commerce.

BiologyNow™   For each text chapter, *BiologyNow* provides *animated tutorials* as well as guidance to other resources that can help students master the content. Responses to *diagnostic pretest questions* yield a student-driven, *personalized learning plan*. Answer a question incorrectly, and the plan lists text sections, figures, and chapter videos. The plan also gives links to relevant animation.

The *How Do I Prepare* section has tutorials in math, chemistry, graphing, and basic study skills. *vMentor* is an online service with a live tutor who interacts with students through voice communication, whiteboard, and text messaging.

The *post-test* can be used as self-assessment tool or submitted to an instructor. BiologyNow is built in iLrn, Thomson Learning's course management system, so that answers and results can be fed directly to an electronic gradebook. BiologyNow also can be integrated with WebCT and Blackboard, so students may log in through these systems as well.

InfoTrac College Edition™   1pass also grants access to *InfoTrac College Edition*, an online database with more than 4,000 periodicals and close to 18 million articles. The articles are in full-text form and can be located easily and quickly with a *key word* search.

In BiologyNow, the *How Would You Vote* exercise for each chapter references specific InfoTrac articles and websites. As students vote on each issue, the site provides a running tally by campus, state, and the country. Instructors can assign the exercises from iLrn, or students can access them through the free website at:

http://biology.brookscole.com/starr11

This website also features more InfoTrac articles and web links, as well as *interactive flashcards* that define all of the book's boldface terms with pronunciation guides.

Also, an online *Issues and Resources Integrator* correlates chapter sections with applications, videos, InfoTrac articles, and websites. This guide is updated each semester.

Audiobook   New college editions of the textbook can also include free access to the *Audiobook*. Students can either listen to narration online or download mp3 files for use on a portable mp3 player.

Previous adopters of *Biology: The Unity and Diversity of Life* who continue to use the book may wish to scan this list of refinements and updates in chapter content. Again, all chapters start with new essays on current issues that are more likely to engage students.

A New Tree of Life Thanks to the wealth of information from biochemistry and molecular biology, the tree of life is coming into sharp focus. Working closely with Walter Judd at the University of Florida, we became confident enough to create a simple but powerful graphic of the evolutionary connections, including the main archaean and protist groups. We present the tree in Section 19.7 as a prelude to the diversity unit. We repeat parts of it in the diversity chapters, as regional road maps. The first chapter does have a simplified classification system in which six kingdoms are subsumed into three domains. It is all that students will require until they reach the later units on evolutionary principles and biodiversity.

We use informal names in the tree and in much of the text. Our advisors and focus-group participants agree that it is more important for introductory students to think about evolutionary connections than ongoing refinements in taxonomy. For reference purposes, Appendix I still provides taxonomic names for major groups.

Energy and Life's Organization Throughout the book, we strengthened the concept of energy flow as the basis of life's levels of organization. Students generally do not find bioenergetics an endearing topic, but we have two new, nonthreatening sections that might engage them through the use of real-life examples (6.1, 6.2). This conceptual overview may help them make sense of metabolism. The photosynthesis chapter (7) has reorganized, streamlined text, new art (including an update on chloroplast structure), and good bioenergetic and evolutionary threads.

Evolution We continue to apply evolutionary theory all through the book. The early, simple overview in Section 1.4 is enough to get students thinking about the concept without bogging them down in details. As examples, we invite them to think about why sexual reproduction evolved (the new Chapter 10 introduction), a possible evolutionary connection between mitosis and meiosis (10.6), and past and potential future threats of methane hydrate deposits on the seafloor. We continue to expand on the connections between mutation, evolution, and development, or "evo-devo" (15.3, 17.8, 17.9, 18.1, 25.1, 25.2, 25.8, 26.4, 26.14, 32.6, 34.1, 39.1, and 43.5 are some examples). The macroevolution chapter (18) has a new essay and text section on the adaptive radiation and current extinctions of the Hawaiian honeycreepers.

Cell Communication We strengthened the art and text presentations of signaling pathways. Sections 22.12 and 25.1 can get students thinking about the evolutionary origins of cell communication. Section 28.5 introduces a generalized graphic of signal reception, transduction, and response that we repeat throughout the animal and plant anatomy units. Section 28.5 also has a specific example: apoptosis in the development of the human hand. We build on this introduction in the sections on neural function (34.1–34.5), sensory function (35.1), hormone action in animals (36.2) and in plants (32.3), immune function (39.4, 39.6, 39.7), urine formation (42.4), and embryonic development (43.2, 43.5).

Unit I Chapter 1: reorganized, simplified, with a new section on experimental tests. Chapter 2: more accessible introduction to electrons and energy levels. Chapter 3: revised sections on levels of protein structure. Chapter 4: new art and text on cell structures. Chapter 5: sections on diffusion, transport mechanisms, and osmosis revised for clarity. Chapter 6: new art, more accessible writing. Chapter 7: extensively revised, updated. Chapter 8: new art for the Krebs cycle and the electron transfer system.

Unit II Chapter 9: new introduction and new essay on cancer. Chapter 10: stronger evolutionary framework for explaining meiosis. Chapter 11: section on impact of crossing over on gene linkage moved to this chapter; new genetics problems on rose and watermelon hybrid experiments. Chapter 12: reorganization of sections on autosomal and X-linked inheritance. Chapter 13: new, updated essay on how cloning methods manipulate information in DNA. Chapter 14: subtle rewrites for clarity, updated alternative mRNA processing. Chapter 15: reorganized, starts with overview of control points; introduces *ABC* flowering model; new essay on effect of mutant *Drosophila* master genes on development. Chapter 16: major reorganization and rewrite, new sections on genomics, organ factories, and other research that will have enormous impact on our lives.

Unit III Chapter 17: now starts with an essay on mass extinctions as one measure of geologic time; introduces evidence for evolution (fossils, radiometric dating, biogeography and plate tectonics, and comparative morphology, embryology, and biochemistry); has new examples of evolutionary constraints on developmental patterns for plants and animals. Chapter 18: some new examples, as in introductory essay; revised section on genetic drift; ends with essay on adaptation (adaptation to what?). Chapter 19: revised text and art, some new examples for reproductive isolating mechanisms and speciation models; updated sections on taxonomy and systematics; new tree of life.

Unit IV Chapter 20: now a less detailed introduction and timeline for the diversity chapters; updated sections on origin of life and of organelles. Chapter 21: essay on West Nile virus; rewrite and updates on bacterial and archaean lineages; updated essay on infectious disease, including SARS and prion diseases. Chapter 22: basically a new chapter on "protists" with clarified phylogenies; has new essay on amoebozoans and the origin of cell communication and multicellularity. Chapter 23: starts and ends with essays on deforestation; new overview of plant origins and evolution; charophyes now included in plant clade; more on flowering plants and pollinators to set the stage for Section 32.3. Chapter 24: new essay on lichens; new material on chytrids, microsporidians, *Neurospora* life cycle, and endophytic fungi.

Chapter 25: major new, updated sections on animal origins and characteristics, including the basis of large, internally complex bodies; reordering of placozoans, flatworms, roundworms, other major groups; essay on cephalopod evolution. Chapter 26: *Archaeopteryx* essay; revised sections on chordate family tree and vertebrate evolutionary trends; aquatic origin of tetrapods with new art; focus essay on vanishing amphibians; sections on dinosaurs now integrated in this chapter; revised text and art on rise of amniotes; new text and art on adaptive radiation of mammals; updated text and art on primate and hominid evolutionary trends.

Chapter 27: now includes a section on environmental impact of human population growth; Rachel Carson essay; conservation biology and sustaining biodiversity, with Monteverde Cloud Forest as an example; ends on a more optimistic note with examples of what might be accomplished by thinking outside the box.

Introduction to Units V and VI New to this edition. Chapter 28: presents common challenges facing plants and animals at increasingly complex levels structural organization; introduces concept of homeostasis and requirements for gas exchange, internal transport, and cell communication, as listed earlier; animal examples of homeostasis are temperature regulation and oxygen deficiency at high altitudes; plant examples are systemic acquired resistance and leaf folding in response to shifts in environmental conditions.

Unit V Chapter 29: reorganized, rewritten introduction to plant tissues with new graphics and micrographs. Chapter 31: incorporates new essay on coevolution of flowering plants and pollinators, including text and photograph of the hawkmoth and the Christmas Star; additional diagrams, improved classification of fruits. Chapter 32: heavily revised, updated; auxin signaling pathway used as new example of cell communication; ends the unit with a new connections essay on global protein deficiency and quinoa research (earlier edition's essay on pesticides is now in Chapter 47 as part of an expanded essay on biomagnification).

Unit VI Chapter 33: new essay on stem cell research; new micrographs; tissue descriptions rewritten, a bit more on muscle tissue; chapter builds on Section 25.1 introduction to compartmentalization (division of labor) as emergent property of multicellularity; new essay on vertebrate skin. Chapter 34: all sections simplified and reorganized; opens with Ecstasy essay; new overview of invertebrate and central/peripheral vertebrate nervous systems; rewritten sections on neurotransmitters and neuromodulators; new section on neuroglia; updated essay on psychoactive drugs. Chapter 35: new essay on oceanic noise pollution. Chapter 36: new essay and text

on hormone disruptors, amphibians. and human sperm counts; major text reorganization (gland by gland); new essay on diabetes and impact of stress on health.

Chapter 37: new section on origin of invertebrate and vertebrate skeletons. Chapter 38: new text, art on cardiac conduction system. Chapter 39: fully revised chapter; more accessible, yet reflects current research and up-to-date concepts, including evolutionary and functional interrelationship between adaptive and innate immunity; easier-to-follow art; update on HIV/AIDS. Chapter 40: rewrite on countercurrent flow in fish gills; new text on controls over respiration. Chapter 41: opens with essay on hormones and appetite; updated human nutritional requirements. Chapter 42: major reorganization and new step-by-step graphics and text on urine formation; expanded emphasis on kidney disorders.

Chapter 43: frog life cycle now integrated with the overview of stages of development; sections on cleavage, pattern formation, and constraints on development have been clarified; new Critical Thinking question on RNA interference for interested students (page 769); updates on aging hypotheses. Chapter 44: text tightened with some reorganization (sections on fertility control and STDs now located after human reproduction and before development); updates on STDs, pregnancy guidelines, fetal alcohol syndrome; oxytocin secretion during labor a new example of positive feedback; essay on bioethics of interventions in fertility now at end of chapter.

Unit VII Chapter 45: unit now opens with essay on reindeer overshooting carrying capacity on St. Matthews Island; revised sections on limiting factors; updates on human population growth; demographic transition model now qualified. Chapter 46: rewrites and new art for mutualism, competitive interactions, predation models; new examples of parasitism; revised community structure models; updates on *Codium* aquarium strain, kudzu, rabbits in Australia; some additions to section on biogeographic patterns; Critical Thinking question on Wallace's line (page 841). Note that Chapter 50 sections of the previous edition are now rolled into Chapters 47 and 48. Chapter 47: global warming and bayous; revised essays on pesticides and biomagnification, energy flow, watershed experiments; global water crisis integrated in this chapter; carbon cycle updates; global warming updates; revised nitrogen cycle. Chapter 48: text on solar energy and wind energy now integrated in section on global air circulation patterns; essay on air pollution (ozone thinning, smog, acid rain, particulates); better historical look at biogeographic realms; new section on differences in sunlight, soils, and moisture, with desert soils and moisture differences as case study; new essay on desertification, including a model for feedback loops between desertification, climate change, and losses in biodiversity; expansion of freshwater provinces with a section on water pollution; new section on north central Pacific gyre as garbage dump. Chapter 49: update on genetics of courtship behavior in fruit flies and pair bonds in voles; new sections on primate social behavior and on biological aspects of human behavior.

ACKNOWLEDGMENTS

Once again I am thinking about all of the advisors and reviewers who helped shape the content of this book in significant ways. Bits and pieces of earlier critiques still flit through my mind, and they are making me realize that educators are symbionts of the highest order. Why else would they give so much of themselves, year after year, to the cooperative enterprise called education? The demands are great and the money is not. Somehow they all have become hardwired to contribute to the greater good. Each year when I interact with them, some new manifestation of their commitment manages to take my breath away; which is why I am always uncomfortable with calling this book "the Starr book." It is our book.

Dan Fairbanks, John Jackson, Walt Judd, and Bill Wischusen have been fearless about approaching this established book with a sledgehammer. They are aware that research directions change, students change, and textbooks must change with them. They dig out flaws in my thinking with needles. I admire them immensely.

Biology: The Unity and Diversity of Life wins awards for subsuming design and art in the service of pedagogy instead of slapping them on as afterthoughts. All it takes is abiding passion for biology and a compulsive urge to assess trends in education, keep up with research, write, create art, and finely lay out workable concept spreads, all at the same time. For some time now, I have been mentoring Lisa Starr and Christine Evers in the odd science of being multitasking authors. Lisa is educated in biochemistry, biotechnology, and computer graphics; Chris in animal behavior, evolutionary biology, and media technologies. With their combined talents and dedication, they have become indispensable partners.

Starting with Susan Badger, Michael Johnson, and Sean Wakely, Thomson Learning proved once again with this edition that it is one of the world's foremost publishers. Michelle Julet, what would I do without your laser focus, arm twisting, intelligence, and wit? Peggy Williams is Empress of the Team. So many of the improvements in this book are direct outcomes of her extraordinary talent for organizing and interpreting the results of class tests, workshops, and focus groups. Andy Marinkovich and dear Grace Davidson continue to take my world-class compulsivity in stride and still get the book out on time. Amazing. Gary Head, aka The Rock, suffers through being my designer, year in, year out. Thank goodness for Myrna Engler's devotion to the team and Suzannah Alexander's quiet competence. Ann Caven, Kathie Head, Laura Argento, Keli Amann, Marlene Veach, Karen Hunt, Chris Ziemba, Kristina Razmara, the professionals at Lachina—the list goes on, but no listing conveys how this team interacts to create something extraordinary.

Jack Carey, how long ago was it that you signed me to do eighteen books to keep me from writing for anybody else in this lifetime? I never did get around to the other fourteen, but probably you forgive me. Through our long partnership, we helped move textbook publishing in new directions, to the benefit of students around the world. Never would have done it without you, partner.

CECIE STARR, *November 2005*

MAJOR ADVISORS AND REVIEWERS

DANIEL FAIRBANKS *Brigham Young University*
JOHN D. JACKSON *North Hennepin Community College*
WALTER JUDD *University of Florida*
E. WILLIAM WISCHUSEN *Louisiana State University*

JOHN ALCOCK *Arizona State University*
CHARLOTTE BORGESON *University of Nevada*
DEBORAH C. CLARK *Middle Tennessee University*
MELANIE DeVORE *Georgia College and State University*
TOM GARRISON *Orange Coast College*
DAVID GOODIN *The Scripps Research Institute*
CHRISTOPHER GREGG *Louisiana State University*
PAUL E. HERTZ *Barnard College*
TIMOTHY JOHNSTON *Murray State University*
EUGENE KOZLOFF *University of Washington*
ELIZABETH LANDECKER–MOORE *Rowan University*
KAREN MESSLEY *Rock Valley College*
THOMAS L. ROST *University of California, Davis*
LAURALEE SHERWOOD *West Virginia University*
STEPHEN L. WOLFE *University of California, Davis*

CONTRIBUTORS: 2004–2005 WORKSHOPS AND REVIEWS

ANDERSON, DIANNE *San Diego City College*
ASMUS, STEVE *Centre College*
BLAUSTEIN, ANDREW R. *Oregon State University*
BROSSAY, LAURENT *Brown University*
CAPORALE, DIANE *University of Wisconsin-Stevens Point*
CAWTHORN, J. MICHELLE *Georgia Southern University*
COLLIER, ALEXANDER *Armstrong Atlantic State University*
CONWAY, ARTHUR F. *Randolph-Macon College*
DAVIS, GEORGE T. *Bloomsburg University*
DeGRAUW, EDWARD *Portland Community College*
DELANEY, CYNTHIA LEIGH *University of South Alabama*
DeSAIX, JEAN *University of North Carolina*
DIERINGER, GREG *Northwest Missouri State University*
D'ORGEIX, CHRISTIAN *Virginia State University*
D'SILVA, JOSEPH G. *Norfolk State University*
FONG, APRIL *Portland Community College*
GARCIA, RIC A. *Clemson University*
HAIGH, GALE *McNeese State University*
HEARRON, MARY *Northeast Texas Community College*
HENDERSON, WILEY J. *Alabama A&M University*
HINTON, JULIANA *McNeese State University*
HUANG, SARA *Los Angeles Valley College*
HUFFMAN, DONNA *Calhoun Community College*
JONES, KEN *Dyersburg State Community College*
JULIAN, GLENNIS *University of Texas at Austin*
KROLL, WILLIAM *Loyola University, Chicago*
KURDZIEL, JOSEPHINE *University of Michigan*
LAMMERT, JOHN M. *Gustavus Adolphus College*
LAZOTTE, PAULINE *Valencia Community College*
MADTES, JR., PAUL *Mount Vernon Nazarene University*
MAJDI, BERNARD *Waycross College*
MATA, JUAN LUIS *University of Tennessee at Martin*
McKEAN, HEATHER R. *Eastern Washington University*
METZ, TIMOTHY *Campbell University*
MOSS, ANTHONY *Auburn University*
NEUFELD, HOWARD S. *Appalachian State University*
NOLD, STEPHEN *University of Wisconsin*
ORR, CLIFTON *University of Arkansas at Pine Bluff*
PETERS, JOHN *College of Charleston*
PLUNKETT, JENNIE *San Jacinto College*
POMARICO, STEVEN M. *Louisiana State University*
RAINES, KIRSTEN *San Jacinto College*
REED, ROBERT N. *Southern Utah University*
RIBEIRO, WENDA *Thomas Nelson Community College*
RICHEY, MARGARET G. *Centre College*
ROBERTS, LAUREL *University of Pittsburgh*

ROMANO, FRANK A., III,   *Jacksonville State University*
RUPPERT, ETTA   *Clemson University*
SHOFNER, MARCIA   *University of Maryland, College Park*
SIEVERT, GREG   *Emporia State University*
SIMS, THOMAS L.   *Northern Illinois University*
SONGER, STEPHANIE R.   *Concord University*
SPRENKLE, AMY B.   *Salem State College*
ST. CLAIR, LARRY   *Brigham Young University*
TEMPLET, ALICE   *Nicholls State University*
TURELL, MARSHA   *Houston Community College*
WALSH, PAT   *University of Delaware*
WILKINS, HEATHER DAWN   *University of Tennessee at Martin*
WINDELSPECHT, MICHAEL   *Appalachian State University*
WYGODA, MARK   *McNeese State University*
ZAHN, MARTIN D.   *Thomas Nelson Community College*
ZANIN, KATHY   *The Citadel*

CONTRIBUTORS: INFLUENTIAL CLASS TESTS AND REVIEWS

ADAMS, DARYL   *Minnesota State University, Mankato*
ANDERSON, DENNIS   *Oklahoma City Community College*
BENDER, KRISTEN   *California State University, Long Beach*
BOGGS, LISA   *Southwestern Oklahoma State University*
BORGESON, CHARLOTTE   *University of Nevada*
BOWER, SUSAN   *Pasadena City College*
BOYD, KIMBERLY   *Cabrini College*
BRICKMAN, PEGGY   *University of Georgia*
BROWN, EVERT   *Casper College*
BRYAN, DAVID W.   *Cincinnati State College*
BURNETT, STEPHEN   *Clayton College*
BUSS, WARREN   *University of Northern Colorado*
CARTWRIGHT, PAULYN   *University of Kansas*
CASE, TED   *University of California, San Diego*
COLAVITO, MARY   *Santa Monica College*
COOK, JERRY L.   *Sam Houston State University*
DAVIS, JERRY   *University of Wisconsin, LaCrosse*
DENGLER, NANCY   *University of California, Davis*
DESAIX, JEAN   *University of North Carolina*
DIBARTOLOMEIS, SUSAN   *Millersville University of Pennsylvania*
DIEHL, FRED   *University of Virginia*
DONALD-WHITNEY, CATHY   *Collin County Community College*
DUWEL, PHILIP   *University of South Carolina, Columbia*
EAKIN, DAVID   *Eastern Kentucky University*
EBBS, STEPHEN   *Southern Illinois University*
EDLIN, GORDON   *University of Hawaii, Manoa*
ENDLER, JOHN   *University of California, Santa Barbara*
ERWIN, CINDY   *City College of San Francisco*
FOREMAN, KATHERINE   *Moraine Valley Community College*
FOX, P. MICHAEL   *SUNY College at Brockport*
GIBLIN, TARA   *Stephens College*
GILLS, RICK   *University of Wisconsin, La Crosse*
GREENE, CURTIS   *Wayne State University*
GREGG, KATHERINE   *West Virginia Wesleyan College*
HARLEY, JOHN   *Eastern Kentucky University*
HARRIS, JAMES   *Utah Valley Community College*
HELGESON, JEAN   *Collin County Community College*
HESS, WILFORD M.   *Brigham Young University*
HOUTMAN, ANNE   *Cal State, Fullerton*
HUFFMAN, DAVID   *Southwestern Texas University*
HUFFMAN, DONNA   *Calhoun Community College*
INEICHER, GEORGIA   *Hinds Community College*
JOHNSTON, TAYLOR   *Michigan State University*
JUILLERAT, FLORENCE   *Indiana University, Purdue University*
KENDRICK, BRYCE   *University of Waterloo*
KETELES, KRISTEN   *University of Central Arkansas*
KIRKPATRICK, LEE A.   *Glendale Community College*
KREBS, CHARLES   *University of British Columbia*
LANZA, JANET   *University of Arkansas, Little Rock*
LEICHT, BRENDA   *University of Iowa*
LOHMEIER, LYNNE   *Mississippi Gulf Coast Community College*
LORING, DAVID   *Johnson County Community College*
MACKLIN, MONICA   *Northeastern State University*

MANN, ALAN   *University of Pennsylvania*
MARTIN, KATHY   *Central Connecticut State University*
MARTIN, TERRY   *Kishwaukee College*
MASON, ROY B.   *Mount San Jacinto College*
MATTHEWS, ROBERT   *University of Georgia*
MAXWELL, JOYCE   *California State University, Northridge*
McCLURE, JERRY   *Miami University*
McNABB, ANN   *Virginia Polytechnic Institute and State University*
MEIERS, SUSAN   *Western Illinois University*
MEYER, DWIGHT H.   *Queensborough Community College*
MICKLE, JAMES   *North Carolina State University*
MILLER, G. TYLER   *Wilmington, North Carolina*
MINOR, CHRISTINE V.   *Clemson University*
MITCHELL, DENNIS M.   *Troy University*
MONCAYO, ABELARDO C.   *Ohio Northern University*
MOORE, IGNACIO   *Virginia Tech*
MORRISON-SHETTLER, ALLISON   *Georgia State University*
MORTON, DAVID   *Frostburg State University*
NELSON, RILEY   *Brigham Young University*
NICKLES, JON R.   *University of Alaska, Anchorage*
NOLD, STEPHEN   *University of Wisconsin- Stout*
PADGETT, DONALD   *Bridgewater State College*
PENCOE, NANCY   *State University of West Georgia*
PERRY, JAMES   *University of Wisconsin, Center Fox Valley*
PITOCCHELLI, DR. JAY   *Saint Anselm College*
PLETT, HAROLD   *Fullerton College*
POLCYN, DAVID M.   *California State University, San Bernardino*
PURCELL, JERRY   *San Antonio College*
REID, BRUCE   *Kean College of New Jersey*
RENFROE, MICHAEL   *James Madison University*
REZNICK, DAVID   *California State University, Fullerton*
RICKETT, JOHN   *University of Arkansas, Little Rock*
ROHN, TROY   *Boise State University*
ROIG, MATTIE   *Broward Community College*
ROSE, GRIEG   *West Valley College*
SANDIFORD, SHAMILI A.   *College of Du Page*
SCHREIBER, FRED   *California State University, Fresno*
SELLERS, LARRY   *Louisiana Tech University*
SHAPIRO, HARRIET   *San Diego State University*
SHONTZ, NANCY   *Grand Valley State University*
SHOPPER, MARILYN   *Johnson County Community College*
SIEMENS, DAVID   *Black Hills State University*
SMITH, BRIAN   *Black Hills State University*
SMITH, JERRY   *St. Petersburg Junior College, Clearwater Campus*
STEINERT, KATHLEEN   *Bellevue Community College*
SUMMERS, GERALD   *University of Missouri*
SUNDBERG, MARSHALL D.   *Emporia State University*
SVENSSON, PETER   *West Valley College*
SWANSON, ROBERT   *North Hennepin Community College*
SWEET, SAMUEL   *University of California, Santa Barbara*
SZYMCZAK, LARRY J.   *Chicago State University*
TAYLOR, JANE   *Northern Virginia Community College*
TERHUNE, JERRY   *Jefferson Community College, University of Kentucky*
TIZARD, IAN   *Texas A&M University*
TRAYLER, BILL   *California State University at Fresno*
TROUT, RICHARD E.   *Oklahoma City Community College*
TURELL, MARSHA   *Houston Community College*
TYSER, ROBIN   *University of Wisconsin, LaCrosse*
VAJRAVELU, RANI   *University of Central Florida*
VANDERGAST, AMY   *San Diego State University*
VERHEY, STEVEN   *Central Washington University*
VICKERS, TANYA   *University of Utah*
VOGEL, THOMAS   *Western Illinois University*
WARNER, MARGARET   *Purdue University*
WEBB, JACQUELINE F.   *Villanova University*
WELCH, NICOLE TURRILL   *Middle Tennessee State University*
WELKIE, GEORGE W.   *Utah State University*
WENDEROTH, MARY PAT   *University of Washington*
WINICUR, SANDRA   *Indiana University, South Bend*
WOLFE, LORNE   *Georgia Southern University*
YONENAKA, SHANNA   *San Francisco State University*
ZAYAITZ, ANNE   *Kutztown University of Pennsylvania*

Introduction

Current configurations of the Earth's oceans and land masses—the geologic stage upon which life's drama continues to unfold. This composite satellite image reveals global energy use at night by the human population. Just as biological science does, it invites you to think more deeply about the world of life—and about our impact upon it.

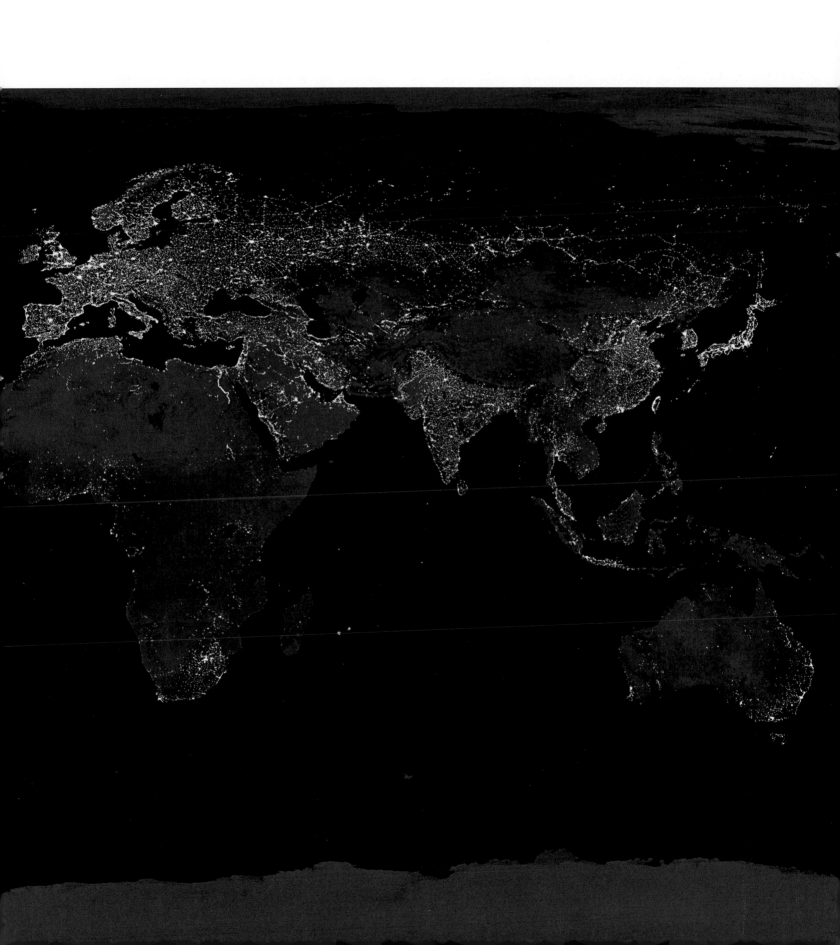

1 INVITATION TO BIOLOGY

What Am I Doing Here?

Leaf through a newspaper on any given Sunday and you might get an uneasy feeling that the world is spinning out of control. There is a lot about the Middle East, where great civilizations have come and gone. You will not find much on the spectacular coral reefs of the surrounding seas, especially at the northern end of the Red Sea. Now the news is about oil and politics, bioterrorists and war.

Think back on the 1991 Persian Gulf conflict, when an Iraqi dictator ordered 460 million gallons of crude oil to be dumped into the Gulf and oil fields to be set on fire. Thick smoke blocked out sunlight, and black rain fell. Today, the human population in neighboring Kuwait shows a spike in cancer rates, which may have been caused by dangerous particles in the smoke. Similarly, New Yorkers who breathed in the dense, noxious dust that billowed through the air

after the 2001 World Trade Center attack are still reporting chronic health problems.

Besides terrorists, nature itself seems to have it in for us. Cholera, the flu, and SARS pose global threats. A long-term AIDS pandemic is unraveling the fabric of African societies. Monstrous storms, droughts, heat waves, and fires batter the land (Figure 1.1). Once-vast glaciers and the polar ice caps are melting fast, the atmosphere is warming up, and living conditions may change just about everywhere.

It is enough to make you throw down the paper and go sit in a park. It is enough to make you wish you lived in the good old days, when things were so much simpler.

Of course, read up on the good old days and you will find they weren't so good. Bioterrorists were around in 1346, when soldiers catapulted the corpses of bubonic plague

Figure 1.1 Biology is a way of thinking critically about life, one that helps us understand nature and our place in it. It starts with the premise that any aspect of nature—including this forest fire in Montana—has one or more underlying causes. To the right, an oil field burning out of control during the Persian Gulf War, an example of human impact on nature.

victims into a walled city under siege. Infected people and rats fled the city and helped fuel the Black Death, a plague that left 25 million dead all across Europe. Later, in 1918, the Spanish flu raced around the world and left between 30 million and 40 million people dead. Like today, many felt helpless in a world that seemed out of control.

What it boils down to is this: For at least a couple of million years, we humans and our immediate ancestors have been trying to make sense of the natural world. We observe it, we come up with ideas, and we test the ideas. However, the more pieces of the puzzle we fit together, the bigger the puzzle gets. We are now smart enough to know that it is almost overwhelmingly big.

You might choose to walk away from the challenge and simply let others tell you what to think. Or you might choose to develop your own understanding of the puzzle. Maybe you are interested in the pieces that affect your health, the food you eat, or your home or family. Maybe you simply find organisms and their environment fascinating. Regardless of the focus, the scientific study of life—*biology*—can deepen your perspective on the world.

Throughout this book, you will come across examples of how organisms are constructed, how they function, where they live, and what they do. The examples support concepts which, when taken together, convey what "life" is. This chapter is an overview of the basic concepts. It also sets the stage for forthcoming descriptions of scientific observations, experiments, and tests that help show how you can develop, modify, and refine your views of life.

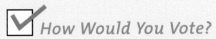

Watch the video online!

How Would You Vote?

The warm seas of the Middle East support some of the world's most spectacular coral reef ecosystems. Should the United States provide funding to help preserve these reefs? See BiologyNow for details, then vote online.

Key Concepts

LEVELS OF ORGANIZATION

The world of life has levels of organization that extend from atoms and molecules to the biosphere. The quality called "life" emerges at the level of cells.   Section 1.1

LIFE'S UNDERLYING UNITY

The world of life shows unity, for all organisms are alike in key respects. They require inputs of energy and materials to function and maintain their complex organization. They work to keep their internal operating conditions within a tolerable range. They all sense and respond to conditions inside and outside themselves. They inherit DNA from parents, which gives them a capacity to survive and reproduce.   Section 1.2

LIFE'S DIVERSITY

Millions of diverse kinds of organisms, or species, have appeared and disappeared over time. Each species is unique in some traits—that is, in some aspects of its body plan, functioning, and behavior.   Section 1.3

EXPLAINING UNITY IN DIVERSITY

Theories of evolution, especially a major theory of evolution by way of natural selection, help explain the link between life's unity and diversity. The theories are a useful foundation for research in all fields of biological inquiry.   Section 1.4

HOW WE KNOW

Biologists engage in systematic observations, hypotheses, predictions, and experimental tests in the outside world and in the laboratory. Well-designed tests can be repeated by others and yield the same results each time.
Sections 1.5–1.7

Links to Earlier Concepts

This book parallels nature's levels of organization, from atoms to the biosphere. Learning about the structure and function of atoms and molecules primes you to understand the structure of living cells. Learning about protein synthesis, active transport, and other processes that keep a single cell alive can help you understand how large organisms survive, because their trillions of living cells use the same processes. Knowing what it takes to survive can help you sense why and how organisms interact with one another and the environment.

At the start of each chapter, we will be reminding you of such structural and functional connections. Within chapters, you will come across keychain icons with cross-references that will link you to relevant sections in earlier chapters.

1.1 Levels of Organization in Nature

The world of life shows increasingly inclusive levels of organization. Take time to see how the levels connect to get a sense of how the topics of this book are organized and where they will take you.

MAKING SENSE OF THE WORLD

If we are interpreting the distant past correctly, the first humans lived in small bands that did not venture far from home. Safety, danger, and resources that did not extend beyond the immediate horizon comprised their world. Much later in time, human populations dispersed all around the globe. They soon had a lot more to observe, think about, and explain.

Today, even the far reaches of the known universe hold clues to our world. Scientists, clerics, farmers, astronauts, and anyone else who is of a mind to do so try to make sense of things. Interpretations differ, for no one person can be expert in everything learned so far or have foreknowledge of all that remains hidden. If you are reading this book, then you are starting to explore how a subset of scientists, the biologists, think

about things, what they found out, and what they are up to now. Alternative ways of explaining the world are available in books for nonscience classes, such as those dealing with philosophy and religion. As you read this book, keep an open mind. Doing so can help you make more enlightened decisions about which explanations work for you.

A PATTERN IN BIOLOGICAL ORGANIZATION

What do we call the external world in its entirety? *Nature.* Biologists are interested in its forms of life, past and present. They have studied life all the way down to interacting atoms and all the way up to the impacts of organisms on a global scale. In doing so, they discovered a great pattern of organization.

That pattern starts at the level of atoms, which are the smallest units of nature's fundamental substances (Figure 1.2*a*). At the next level, atoms have combined into larger units called molecules (Figure 1.2*b*).

The pattern reaches the threshold of life as certain molecules are assembled as cells (Figure 1.2*c*). Complex

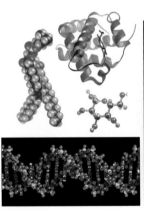

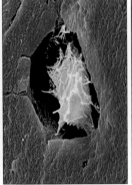
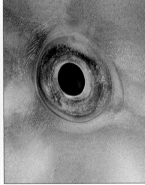
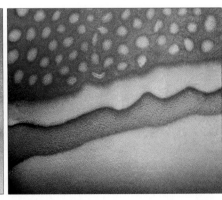

b molecule ────▶ **c cell** ────▶ **d tissue** ────▶ **e organ** ────▶ **f organ system** ──

Two or more joined atoms of the same or different elements. "Molecules of life" are complex carbohydrates, lipids, proteins, DNA, and RNA. Only living cells now make them.

Smallest unit that can live and reproduce on its own or as part of a multicelled organism. It has an outer membrane, DNA, and other components.

Organized array of cells and substances that are interacting in some task. Many cells (*white*) made this bone tissue from their own secretions.

Structural unit made of two or more tissues interacting in some task. A parrotfish eye is a sensory organ used in vision.

Organs interacting physically, chemically, or both in some task. Parrotfish skin is an integumentary system with tissue layers, organs such as glands, and other parts.

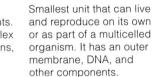

a atom

Elements are fundamental forms of matter. Atoms are the smallest units that retain an element's properties. Electrons, protons, and neutrons are its building blocks. This hydrogen atom's electron zips around a proton in a spherical volume of space.

Figure 1.2 *Animated!* Levels of organization in nature.

carbohydrates, complex fats and other lipids, proteins, DNA, and RNA—these are the "molecules of life." In nature, only living cells can make them. A **cell** is the smallest unit of life; it has a capacity to survive and reproduce on its own, given raw materials, an energy source, information encoded in its DNA, and suitable environmental conditions.

Each kind of organism, or **species**, consists of one or more cells. In multicelled species, cells form tissues, organs, and organ systems. Often there are trillions of specialized cells that interact directly or indirectly in the task of keeping the whole multicelled body alive. Figure 1.2d–g defines these levels of organization.

The population is the next level of organization. It is a group of the same species in some specified area. Fields of poppies in a valley or schools of fish in a lake are such groups (Figure 1.2h). The next level is the community. It includes all populations of all species in a specified area (Figure 1.2i). A community inside an underwater cave in the Red Sea, a forest in Argentina, or populations of tiny organisms that live, reproduce, and die quickly inside a flower are examples.

The next level of organization is the ecosystem, or a community interacting with its physical and chemical environment. The highest level, the biosphere, includes all regions of Earth's crust, waters, and atmosphere in which organisms live.

Bear in mind, life is more than the sum of its parts. At each successive level of organization, new properties emerge that are not inherent in any part by itself, as when living cells emerge from "lifeless" molecules. The interactions among parts generate **emergent properties**.

This book is a journey through the globe-spanning organization of life. Take a moment to study Figure 1.2. You can use it as a road map of where each part fits in the great scheme of nature.

Nature shows levels of organization, from the simple to the increasingly complex. Life's unique characteristics emerge as atoms and molecules interact and form cells. They extend from interactions among cells to populations, communities, ecosystems, and the biosphere.

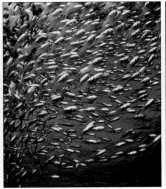

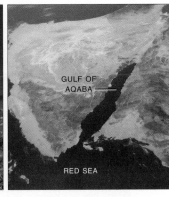

GULF OF AQABA

RED SEA

g multicelled organism

Individual made of different types of cells. Cells of most multicelled organisms, including this Red Sea parrotfish, are organized as tissues, organs, and organ systems.

h population

Group of single-celled or multicelled individuals of the same species occupying a specified area. This is a fish population in the Red Sea.

i community

All populations of all species occupying a specified area. This is part of a coral reef in the Gulf of Aqaba at the northern end of the Red Sea.

j ecosystem

A community that is interacting with its physical environment. It has inputs and outputs of energy and materials. Reef ecosystems flourish in warm, clear seawater throughout the Middle East.

k the biosphere

All regions of Earth's waters, crust, and atmosphere that hold organisms. In the vast universe, Earth is a rare planet. Without its abundance of free-flowing water, there would be no life.

1.2 Overview of Life's Unity

"Life" is not easy to define. It is just too big, and it has been changing for 3.8 billion years! Even so, you can characterize it in terms of its unity and diversity. Here's the unity part: All living things require inputs of energy and materials; they sense and respond to change, as when they adjust conditions inside their body; and they reproduce with the help of DNA. But they differ in the details of their traits. That's the diversity part—variation in traits.

ENERGY AND LIFE'S ORGANIZATION

Cells, remember, are the smallest units that are alive. To stay alive, they get energy from the environment and convert it to forms that help them do work, such as constructing and organizing molecules into cell parts. **Energy** is the capacity for doing work. Whether a cell is free-living or a tiny bit of a multicelled organism, its organization would end without continuous inputs of energy. Higher levels of organization—populations, communities, and ecosystems—also would fall apart in the absence of energy inputs from the environment.

Producers make their food from simple materials in the environment. Plants and other photosynthetic types use sunlight energy to construct sugars from carbon dioxide and water molecules. They use the sugars as packets of energy and as building blocks for making complex carbohydrates, fats, and proteins. Animals and decomposers are **consumers**. They cannot make food; they eat producers and other organisms. Decomposers break down the remains of organisms to simpler raw materials, some of which become cycled back to the producers.

We are outlining a series of energy transfers from the environment, through producers, then on through consumers, and back to the environment. It is a one-way flow of energy, because all the energy that enters the world of life in a given interval eventually leaves it. Why? At each transfer step, a small amount escapes as an unorganized form of energy: heat. All organisms are participants in this continuous, directional flow of energy (Figure 1.3).

ORGANISMS SENSE AND RESPOND TO CHANGE

All organisms are alike in another way. They sense changes in their surroundings and make controlled, compensatory responses to them. They do so with the assistance of **receptors**. Receptors are molecules and structures that detect stimuli, which are specific kinds of energy. Different receptors can respond to different stimuli. A stimulus may be sunlight energy, chemical energy, or the mechanical energy of a bite (Figure 1.4).

Activated receptors trigger changes in the activities of organisms. As a simple example, after you finish eating an apple, sugars leave your small intestine and enter blood. Think of blood and the fluid around cells as the body's *internal* environment. The composition and volume of that fluid must be kept within a range that your cells can tolerate. Too much or too little sugar changes the composition of blood, as happens with diabetes and other medical problems. Normally, when there is too much sugar, your pancreas secretes more insulin. Most living cells in your body have receptors for this hormone. It stimulates cells to take up sugar.

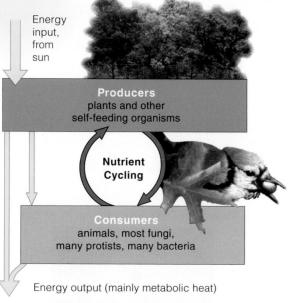

Energy input, from sun

Producers
plants and other self-feeding organisms

Nutrient Cycling

Consumers
animals, most fungi, many protists, many bacteria

Energy output (mainly metabolic heat)

Figure 1.3 *Animated!* The one-way flow of energy and cycling of materials in the world of life.

Figure 1.4 A roaring response to signals from pain receptors, activated by a lion cub flirting with disaster.

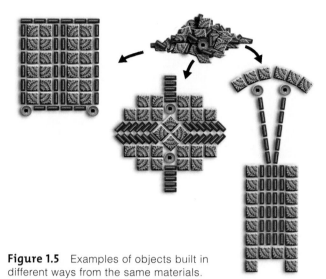

Figure 1.5 Examples of objects built in different ways from the same materials.

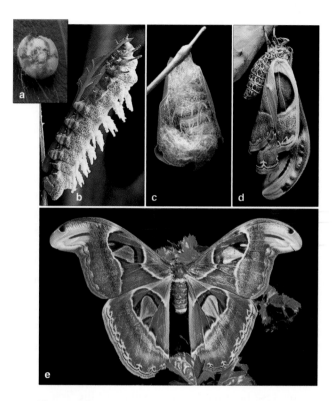

Figure 1.6 "The insect"—actually a series of stages of development guided largely by instructions in DNA. Here, a silkworm moth, from a fertilized egg (**a**), to a larval stage called a caterpillar (**b**), to a pupal stage (**c**), to the winged form of the adult (**d,e**).

When enough cells are doing so, the blood sugar level returns to the normal range.

As conditions in the internal environment change in potentially harmful ways, receptor-driven mechanisms kick in and return conditions to the state that cells can tolerate. **Homeostasis** is the name for this state, and it is a defining feature of life.

ORGANISMS GROW AND REPRODUCE

Organisms grow and reproduce based on information in **DNA**, a nucleic acid. DNA is *the* signature molecule of life. No chunk of granite or quartz has it.

DNA holds the information about building proteins from a few kinds of amino acids. Each protein has a particular amino acid sequence, which is the start of its particular shapes and properties.

By analogy, if you access suitable instructions and apply energy to the task, you might organize a pile of a few kinds of ceramic tiles into diverse patterns. Figure 1.5 has examples.

Protein-building information is vital for cell growth and reproduction. Proteins have many structural and functional roles. Important kinds function as enzymes, the cell's main worker molecules. With enzymes, cells build, split, and rearrange molecules exceedingly fast. Without enzymes, there could be no more complex carbohydrates, complex lipids, proteins, and nucleic acids. There could be no cells, no life.

In nature, an organism inherits DNA—the basis of its traits—from its parents. *Inheritance* is the acquisition of traits after parents transmit their DNA to offspring. Think about it. Why do baby storks look like storks

and not like pelicans? Because they inherited stork DNA, which is different from pelican DNA.

Reproduction refers to actual mechanisms by which parents transmit DNA to offspring. For trees, humans, and other large organisms, the information in DNA is used in ways that guide growth and *development*—the transformation of the first cell of the new individual through orderly stages. The outcome is a multicelled adult, typically with tissues and organs (Figure 1.6).

Life's levels of organization start with a one-way flow of energy from the environment, through producers and consumers, then back to the environment.

Organisms interact through this one-way flow of energy and through a cycling of raw materials.

Organisms maintain their organization by sensing and responding to change. Many responses return conditions in the body's internal environment to a range that cells can tolerate, a state called homeostasis.

Organisms grow and reproduce based on information encoded in DNA, which they inherit from their parents.

1.3   If So Much Unity, Why So Many Species?

Although unity pervades the world of life, so does diversity. Organisms differ enormously in body form, the functions of their body parts, and behavior.

Superimposed on life's unity is tremendous diversity. How many species are with us today? Estimates range as high as 100 million. And 99.9 percent of all species that ever lived are extinct. So far, we have named approximately 1.8 million species.

For centuries, many scholars have been organizing information about life's diversity. Carolus Linnaeus, a naturalist, came up with the strategy of giving each species a two-part name. The first part designates the **genus** (plural, genera). A genus is a grouping of one or more species characterized by certain traits, at least one of which is unique to them. The second part of the name designates a particular species within the genus that has at least one trait no other species has.

Scarus gibbus is the formal name for the humphead parrotfish shown in Figure 1.2g. A different species in its genus is named *S. coelestinus* (midnight parrotfish). This example also shows that you can abbreviate the genus designation after you spell it out the first time.

Later, ever more inclusive groupings were devised, such as phylum (plural, phyla), order, kingdom, and domain. These are rankings of classification systems, which simply are ways to organize knowledge about relationships among species. Observable traits are still markers, but so is a richly expanding base of molecular evidence of descent from a shared ancestor.

Most biologists now favor a classification system having three domains: Bacteria, Archaea, and Eukarya (Figure 1.7 and Table 1.1). Protists, plants, fungi, and animals make up domain Eukarya (Figure 1.8).

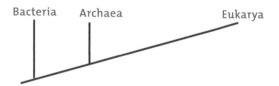

Figure 1.7   Three domains of life.

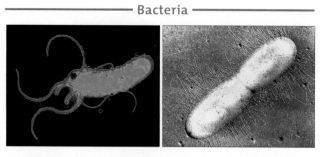

━━━━━━━━━━ Bacteria ━━━━━━━━━━

The most common prokaryotes; collectively, these single cells are the most metabolically diverse species on Earth.

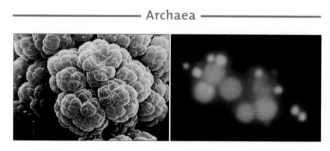

━━━━━━━━━━ Archaea ━━━━━━━━━━

These prokaryotes are evolutionarily closer to eukaryotes than to bacteria. *Left*, a colony of methane-producing cells. *Right*, two species from a hydrothermal vent on the seafloor.

Figure 1.8   A few representatives of life's diversity.

Bacteria (singular, bacterium) and **archaeans** are *prokaryotic* cells. These single-celled organisms do not have a nucleus, which is a membrane-bound sac that, in all other species, encloses DNA. Of all organisms, they show the greatest metabolic diversity. Different species are producers and consumers in near-boiling water, frozen desert rocks, sulfur-clogged lakes, and other exceptionally harsh environments. Experimental evidence suggests that the first cells on Earth faced similarly hostile challenges to survival.

Structurally, the **protists** are the simplest organisms that are *eukaryotic*, which means their cells contain a nucleus. Different kinds are producers or consumers. Many are single cells, larger and more complex than prokaryotes. Some are tree-size, multicelled seaweeds. Actually, the protists are so diverse that they are being reclassified into a number of separate major lineages.

Cells of species we call fungi, plants, and animals are eukaryotic. Most **fungi** are multicelled, and not all of them form mushrooms (the reproductive structures for species that grow mostly underground). Many are decomposers. They secrete enzymes that digest food outside their body, then individual cells absorb the bits. Nearly all **plants** are multicelled and photosynthetic.

Table 1.1	Comparison of Life's Three Domains
Bacteria	Single cells, prokaryotic (no nucleus). Most ancient lineage.
Archaea	Single cells, prokaryotic. Evolutionarily closer to eukaryotes.
Eukarya	Eukaryotic cells (with a nucleus). Single-celled and multicelled species categorized as protists, plants, fungi, and animals.

Protists Single-celled and multicelled eukaryotic species that range from the microscopic to giant seaweeds. Many biologists are now viewing the "protists" as many major lineages.

Plants Generally, photosynthetic, multicelled eukaryotes, many with roots, stems, and leaves. Plants are the primary producers for ecosystems on land. Redwoods and flowering plants are examples.

Fungi Single-celled and multicelled eukaryotes; different kinds are decomposers, parasites, or pathogens. Without decomposers, communities would become buried in their own wastes.

Animals Multicelled eukaryotes that ingest tissues or juices of other organisms. Like this basilisk lizard, most actively move about during at least part of their life.

They make all of their own food by using sunlight as an energy source, and atoms of carbon dioxide and water as building blocks.

All **animals** are multicelled consumers that ingest tissues or juices of other organisms. Herbivores are grazers, carnivores eat meat, scavengers eat almost anything edible, and parasites pilfer nutrients from a host's tissues. All animals grow and develop through a series of stages. Most of them actively move about during at least part of their lives.

Pulling this information together, are you getting a sense of what it means when someone says that life shows unity *and* diversity?

On the basis of observable traits and molecular evidence of shared ancestry, we rank species in ever more inclusive groupings. The largest groupings are domains: archaea, bacteria, and eukarya (protists, fungi, plants, and animals).

1.4 An Evolutionary View of Diversity

How can organisms be so much alike and still show tremendous diversity? A theory of evolution by way of natural selection is one explanation.

Individuals of a species share certain traits, which are aspects of their physical form, function, and behavior. Rarely are individuals exactly alike; they differ in the details. Except for identical twins, for instance, the 6.4 billion individuals of our species (*Homo sapiens*) vary in height, hair color, and other traits.

Variation in most traits arises through **mutations**, or changes in DNA, which offspring inherit from their parents. Most mutations have neutral or bad effects. Some cause a trait to change in a way that makes one individual of a population better adapted than others to prevailing conditions. That is, its bearer might have an easier time securing food, a mate, and so on—so it has a better chance of reproducing and passing on the mutation to offspring. What is the outcome? Consider how a naturalist, Charles Darwin, expressed it:

First, a natural population tends to increase in size until its individuals compete more and more for food, shelter, and other dwindling environmental resources.

Second, those individuals differ from one another in the details of their shared, heritable traits.

Third, bearers of adaptive forms of traits are more likely to survive and reproduce, so those forms tend to become more common over successive generations. This outcome is called **natural selection**.

Consider how pigeons vary in feather color, size, and other traits (Figure 1.9). Say that pigeon breeders prefer black, curly-tipped feathers. They select captive birds having the darkest, curliest-tipped feathers and

WILD ROCK DOVE

let only those birds mate. In time, no birds in their captive populations have light, uncurly feathers. By culling through many traits, breeders have developed well over 300 varieties of domesticated pigeons.

Pigeon breeding is a case of *artificial* selection. One form of a trait is favored over others in an artificial environment under contrived, manipulated conditions. Darwin saw that breeding practices could be an easily understood model for *natural* selection, a favoring of some forms of a given trait over others in nature.

Just as breeders are "selective agents" that promote reproduction of certain pigeons, different agents act on the range of variation in the wild. Among them are pigeon-eating peregrine falcons, as in Figure 1.9. The swifter or better camouflaged pigeons are more likely to avoid falcons and live long enough to reproduce, compared with not-so-swift or too-flashy pigeons.

When different forms of a trait are becoming more or less common over successive generations, evolution is under way. In biology, **evolution** simply means that heritable change is occurring in a line of descent. Later chapters get into actual mechanisms of evolution. For now, it is enough to remember these preview points.

> *Body form, function, and behavior are mostly heritable traits. Different forms of a trait arise through DNA mutations. One may be more adaptive than others to prevailing conditions.*
>
> *Natural selection is an outcome of differences in survival and reproduction among individuals of a population that vary in one or more heritable traits. Evolution, or change in lines of descent, gives rise to life's diversity.*

Figure 1.9 Outcome of artificial selection: just a few of the hundreds of varieties of domesticated pigeons, all descended from captive populations of wild rock doves. At right, peregrine falcons are agents of natural selection in the wild.

1.5 The Nature of Biological Inquiry

The preceding sections introduced some big concepts. Consider approaching this or any other collection of "facts" with a critical attitude. "Why should I accept that they have merit?" The answer requires a look at how biologists make inferences about observations, then test their inferences against actual experience.

OBSERVATIONS, HYPOTHESES, AND TESTS

To get a sense of "how to do science," you might start with practices that are common in scientific research:

1. Observe some aspect of nature and research what others have found out about it, then frame a question or identify a problem related to your observation.

2. Develop a **hypothesis**: a testable explanation of the observed phenomenon or process.

3. Using the hypothesis as a guide, make a **prediction** —a statement of what you should find in nature if you were to go looking for it. This is often called the "if–then" process. *If* gravity does not pull objects toward Earth, *then* it should be possible to observe an apple falling up, not down, from a tree.

4. Devise ways to **test** the accuracy of predictions, as by making systematic observations, building models, and conducting experiments. **Models** are theoretical, detailed descriptions or analogies that might help us visualize an object or event that has not been, or cannot be, directly observed.

5. If your tests do not confirm the prediction, check to see what might have gone wrong. It may be that you overlooked a factor that had impact on the results. Or maybe the hypothesis is not a good one.

6. Repeat the tests or devise new ones—the more the better, because hypotheses that withstand many tests have a higher probability of being useful.

7. Objectively analyze and report the test results, as well as the conclusions you drew from them.

You might hear someone refer to these practices as "the scientific method," as if all scientists march to the drumbeat of an absolute, fixed procedure. They do not. Many observe and describe some aspect of nature and leave the hypothesizing to others. A few are lucky; they stumble onto information that they are not even looking for. Of course, it isn't always a matter of luck. Chance seems to favor a mind that has already been prepared, by education and experience, to recognize what the information might mean.

However, scientists do have something in common. It is a critical attitude about testing ideas in rigorous ways that are designed to disprove them.

Careful observations are a logical way to test the predictions that flow from a hypothesis. So are **experiments**, or tests carried out under controlled conditions that researchers manipulate. Such tests are carried out in nature and in laboratories, and they remove irrelevant factors that might skew the results. You will find two examples in the section to follow.

ABOUT THE WORD "THEORY"

Suppose a hypothesis has not been disproved after years of rigorous tests. Scientists use it to interpret more data or observations, which often involve more hypotheses. When a hypothesis meets these criteria, it may become accepted as a **scientific theory**.

You may hear people apply the word "theory" to a speculative idea, as in the phrase, "It's just a theory." However, a scientific theory differs from speculation in a big way. *After testing a scientific theory's predictive power many times and in many ways in the natural world, researchers have yet to find evidence that disproves it.* That is why the theory of evolution by natural selection is respected. It has been used successfully to explain a diverse number of questions about the natural world, such as how life diversified, how river dams can alter ecosystems, and why antibiotics can stop working.

Perhaps a well-tested theory is as close to the truth as we can get. For instance, after more than a century of many thousands of tests, Darwin's theory holds, with only minor modification. Yet we cannot prove it holds under all possible conditions, because doing so would take an infinite number of tests. We *can* say that a theory has a high probability of not being wrong. Even then, biologists keep on looking for information and devising tests that may disprove its premises. This willingness to modify even an entrenched theory is a strength of science, not a weakness.

Scientific inquiry into nature involves asking questions, formulating hypotheses, making predictions, testing predictions, and objectively reporting the results.

A scientific theory is a time-tested intellectual framework that is used to interpret a broad range of observations and data. Scientific theories remain open to rigorous tests, revision, and tentative acceptance or rejection.

1.6 The Power of Experimental Tests

Experiments are tests that can simplify observation in nature, because conditions under which observations are made can be controlled. Well-designed experiments test predictions about what you will find in nature when a hypothesis is correct—or won't find if it is wrong.

AN ASSUMPTION OF CAUSE AND EFFECT

A scientific experiment starts with this premise: *Any aspect of nature has an underlying material cause that can be tested through controlled experiments.* This premise sets science apart from faith in the supernatural ("beyond nature"). It means a hypothesis must be testable in the natural world in ways that might well disprove it.

Most aspects of nature are outcomes of interacting variables. A **variable** is a feature of an object or event that may differ over time or among the representatives of that object or event.

Researchers design experiments to test one variable at a time. They establish a **control group**, a standard used for comparison against one or more **experimental groups**. One type of control group is *identical* with an experimental group, but the test conditions are not. Those conditions differ by one variable. Another type of control group *differs* from an experimental group in one variable, but the test conditions are identical for both groups.

EXAMPLE OF AN EXPERIMENTAL DESIGN

In 1996 the FDA approved Olestra®, a type of synthetic fat replacement made from sugar and vegetable oil, as a food additive. Potato chips were the first Olestra-laced food product on the market in the United States. Controversy raged. After eating the chips, some people complained of bad cramps. Two years later, researchers at Johns Hopkins University designed an experiment. If Olestra causes gastrointestinal cramps, then people who eat food that contains Olestra are more likely to get cramps than people who do not.

To test this prediction, they used a Chicago theater as the "laboratory." They asked more than 1,100 people between ages thirteen and thirty-eight to watch a movie and eat their fill of potato chips. Each person got an unmarked bag that held thirteen ounces of chips. The individuals who received a bag of Olestra-laced potato chips were the experimental group. Individuals who got a bag of regular chips were the control group.

Afterward, researchers contacted the subjects and tabulated reports of gastrointestinal cramps. Of 563 people in the experimental group, 89 (or 15.8 percent) complained of problems. But so did 93 of 529 people

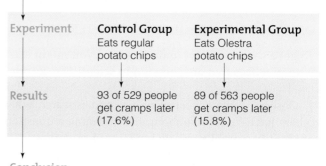

Hypothesis
Olestra® causes intestinal cramps.

Prediction
People who eat potato chips made with Olestra will be more likely to get intestinal cramps than those who eat potato chips made without Olestra.

| Experiment | Control Group
Eats regular potato chips | Experimental Group
Eats Olestra potato chips |
|---|---|---|
| Results | 93 of 529 people get cramps later (17.6%) | 89 of 563 people get cramps later (15.8%) |

Conclusion
Percentages are about equal. People who eat potato chips made with Olestra are just as likely to get intestinal cramps as those who eat potato chips made without Olestra. These results do not support the hypothesis.

Figure 1.10 *Animated!* Example of a typical sequence of steps taken in a scientific experiment.

(17.6 percent) in the control group, who had munched on regular chips! The experiment yielded no evidence that eating Olestra-laced potato chips, at least in one sitting, causes gastrointestinal problems (Figure 1.10).

EXAMPLE OF A FIELD EXPERIMENT

Many toxic or unpalatable species are vividly colored and often have distinct patterning. Predators learn to avoid individuals that display these visual cues after eating a few of them and getting sick.

In some cases, two bad-tasting species of butterflies resemble each other. **Mimicry** is a case of looking like something else and confusing predators (or prey). The naturalist Fritz Müller wondered why such a visual similarity persists. As he hypothesized, it may benefit both species. He knew that young birds catch and eat butterflies before learning to avoid eating bad-tasting ones. If both of the butterfly species are sampled, then each would lose fewer individuals to predatory birds.

Durrell Kapan, an evolutionary biologist, tested the hypothesis with *Heliconius cydno* in a rain forest. There are two forms of these bad-tasting butterflies. One has yellow markings on its wings, and the other does not (Figure 1.11*a*). Moreover, the yellow-marked butterfly resembles *H. eleuchia*, another bad-tasting species in a different part of the forest. The wings of *H. eleuchia* also have yellow markings (Figure 1.11*b*).

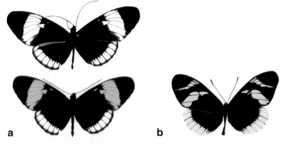

Figure 1.11 *Heliconius* butterflies. (**a**) Wing markings of the two forms of *H. cydno*. (**b**) Wing markings of *H. eleuchia*.

(**c**) Kapan's experiment with *Heliconius* butterflies in Ecuador. *H. cydno* butterflies with white or yellow wing markings were captured and transferred to habitats of *H. eleuchia*, a species that resembles yellow-marked *H. cydno*. Predatory birds (**d**), familiar with bad-tasting *H. eleuchia*, avoided yellow-marked *H. cydno* butterflies but ate most of those with yellow-free wings.

Experimental Group
34 yellow
H. cydno butterflies

Control Group
46 white
H. cydno butterflies

Experiment
Both groups are introduced into isolated habitats of yellow *H. eleuchia*. Resighted individuals are counted every day for two weeks.

Results
Control group (white *H. cydno*) is selected against: 37 of 46 (80%) disappear, compared with 20 of 34 (59%) of the experimental butterfly group.

Kapan predicted: If birds have learned to avoid *H. eleuchia* in their part of the forest, then they also will avoid imports of yellow-marked *H. cydno* butterflies. For the control group, Kapan used captured *H. cydno* butterflies with yellow-free wings. *H. cydno* butterflies with yellow-marked wings made up the experimental group. He released both of the groups into isolated *H. eleuchia* habitats. For two weeks, the approximate life span of the butterflies, he counted the survivors.

Kapan found that butterflies of the experimental group were less likely to survive in the new forest habitat. Because they did not display the visual cue recognized by the local birds—yellow markings—they were more likely to be eaten. The control group did better, as the test results in Figure 1.11c indicate. We can expect that predatory birds recognized this batch of imports as bad tasting and avoided them. Besides confirming Kapan's prediction, these test results also provide evidence of natural selection in action.

BIAS IN REPORTING RESULTS

Experimenters run a risk of interpreting data in terms of what they wish to prove or dismiss. That is why they prefer *quantitative* reports, with actual counts or some other precise measurements. With such reports, researchers get a chance to repeat an experimental test and check conclusions. Appendix IX gives an example.

This last point gets us back to the value of thinking critically. Scientists must keep asking themselves: *Will my observations or experiments show that a hypothesis is false?* They are expected to put aside pride or bias by testing ideas in ways that might prove them wrong. Even if someone won't do so, others will, for science works as a community that is both cooperative and competitive. Ideally, individuals will share their ideas, knowing that it is as useful to expose errors as it is to applaud insights. They can and often do change their mind when shown contradictory evidence.

Experiments simplify observations in nature by restricting a researcher's focus to one variable at a time. A variable is any feature of an object or event that may differ over time or among representatives of the object or event.

Tests are based on the premise that any aspect of nature has one or more underlying causes. Scientific hypotheses can be tested in ways that might disprove them.

1.7 The Limits of Science

Beyond the realm of scientific inquiry, some events are unexplained. Why do we exist, for what purpose? Why do we have to die at a particular moment? Such questions lead to *subjective* answers, which come from within, as an outcome of all the personal experiences and mental connections that shape our consciousness. People differ enormously in this regard, which is why subjective answers do not readily lend themselves to scientific analysis and experiments.

This is not to say subjective answers are without value. No human society can function for long unless its individuals share a commitment to standards for making judgments, even if they are subjective. Moral, aesthetic, philosophical, and economic standards vary from one society to the next. But they all guide people in deciding what is important and good, and what is not. All attempt to give meaning to what we do.

Every so often, scientists stir up controversy when they explain something that was thought to be beyond natural explanation, or belonging to the supernatural. This is often the case when a society's moral codes are interwoven with religious interpretations of the past. Exploring a long-standing view of the natural world from a scientific perspective might be misinterpreted as questioning morality, even though the two are not the same thing.

As one example, centuries ago in Europe, Nikolaus Copernicus studied the planets and decided that Earth circles the sun. Today this seems obvious. Back then, it was heresy. The prevailing belief was that the Creator made Earth—and, by extension, humans—as the fixed center of the universe. Galileo Galilei, another scholar, thought the Copernican model of the solar system was a good one and said so. He was forced to retract his statement, on his knees, and put Earth back as the fixed center of things. Word has it that he muttered, "Even so, it *does* move." Later on, Darwin's theory of evolution also ran up against prevailing beliefs.

Today, as then, society has sets of standards. Those standards might be questioned when a new, natural explanation runs counter to supernatural beliefs. This does not mean that scientists who raise questions are less moral, less lawful, less sensitive, or less caring than anyone else. It means a specific standard guides their work: *Explanations about nature must be testable in the external world, in ways that others can repeat.*

Science does not address subjective questions. The external world, not internal conviction, is the testing ground for the theories generated in science.

Summary

Section 1.1 Nature has increasingly inclusive levels of organization, with life emerging at the cellular level. All organisms consist of one or more cells. In most multicelled species, the cells are organized as tissues, organs, and organ systems. Individuals of the same species in a specified area form a population, and all populations in the same area form a community. An ecosystem is a community and its environment. The biosphere includes all regions of Earth's atmosphere, waters, and land that hold systems of life.

Distinctive properties emerge at each successive level of organization. It takes interactions among the parts to generate these emergent properties of life.

Biology⊜Now
Explore levels of biological organization with the interaction on BiologyNow.

Section 1.2 Life shows unity. All organisms require energy and raw materials from the environment to grow, maintain their organization, and reproduce, based on information encoded in DNA. All sense and respond to change. They work to counter changes in their internal environment so that conditions remain tolerable for cell activities, a state called homeostasis (Table 1.2).

Biology⊜Now
Using instructions with the animation on BiologyNow, view how different objects are assembled from the same materials. Also view energy flow and materials cycling.

Section 1.3 As in the past, species now show great diversity. Each species has unique aspects of body form, function, and behavior. Species are ranked in ever more inclusive groupings in classification systems, starting with a two-part name (genus and species name). One classification system assigns all species to three domains: Bacteria, Archaea, and Eukarya. Protists, plants, fungi, and animals make up Domain Eukarya.

Biology⊜Now
Explore the characteristics of the three domains of life with the interaction on BiologyNow.

Section 1.4 Life's diversity arises through mutation: change in the structure of DNA molecules. Mutations are the basis for variation in heritable traits, which are the traits that parents bestow on offspring. Such traits include most details of body form and function.

Individuals of a population differ in the details of their shared heritable traits. Variant forms of traits may affect the ability to survive and reproduce. The adaptive forms give their bearers a competitive edge, so they tend to become more common among successive generations; less adaptive traits become less common or are lost. Thus the traits that help define the population (and species) may change over successive generations; that is, the population may evolve. The outcome of differences in reproduction among individuals that differ in one or more heritable traits is called natural selection.

Learn more about natural selection and evolution with InfoTrac readings on BiologyNow.

Read the InfoTrac article "Will We Keep Evolving?" Ian Tattersall, Time, *April 2000.*

Section 1.5 Scientific methods differ, but they all are based on the premise that any aspect of nature has one or more underlying causes. Researchers observe some object or event, form hypotheses (testable explanations about it), make predictions about what they can expect to find if the hypothesis is not wrong, and then test the predictions. Their tests may involve making more observations, building models, or doing experiments.

Scientists analyze and share test results. A hypothesis that does not hold up under repeated testing is modified or discarded. A scientific theory is a set of hypotheses that can be used to explain a broad range of observations and data. Many diverse tests have supported it.

Sections 1.6, 1.7 Supernatural explanations cannot be tested. Science deals only with aspects of nature that lend themselves to systematic observation, hypotheses, predictions, and experimental tests. Most aspects are outcomes of many interacting variables that differ among individuals and over time. A scientific experiment can simplify observations in nature and in the laboratory because the variables can be precisely manipulated and controlled. A scientist changes one variable at a time and observes what happens. A typical experiment is designed so that one or more experimental groups can be compared with a control group.

Self-Quiz

Answers in Appendix II

1. The smallest unit of life is the _____ .

2. _____ is required to maintain levels of biological organization, from cells to populations, communities, and even entire ecosystems.

3. _____ is a state in which the internal environment is being maintained within a tolerable range.

4. Researchers assign all species to one of three _____ .

5. DNA _____ .
   a. contains instructions for building proteins
   b. undergoes mutation
   c. is transmitted from parents to offspring
   d. all of the above

6. _____ is the acquisition of traits from parents who transmit their DNA to offspring.
   a. Reproduction   c. Homeostasis
   b. Development   d. Inheritance

7. Differences in heritable traits arise through _____ .

8. A trait is _____ if it improves an organism's ability to survive and reproduce in the prevailing environment.

9. A control group is _____ .
   a. the standard against which experimental groups can be compared
   b. the experiment that gives conclusive results

Table 1.2	Summary of Life's Characteristics

Shared characteristics that reflect life's unity

1. Life emerges at the level of cells. All organisms consist of one or more cells.

2. In nature, only organisms make complex carbohydrates and lipids, proteins, and nucleic acids. They all use these molecules of life as building blocks and energy sources, and for the preservation of heritable information.

3. Organisms require ongoing inputs of energy to maintain their complex organization. All obtain energy from the environment and convert it to forms that can be used for growth, survival, and reproduction.

4. Organisms sense and make controlled responses to conditions in their external and internal environments.

5. Organisms grow and reproduce based on heritable information in DNA.

6. The traits that define a population of organisms can change over the generations; the population can evolve.

Foundations for life's diversity

1. Mutations (heritable changes in the structure of DNA) give rise to variation in heritable traits, which are most details of body form, function, and behavior.

2. Diversity is the sum total of variations that accumulated in different lines of descent over the past 3.8 billion years, as by natural selection and other processes of evolution.

---

10. Match the terms with the most suitable description.

    ____ emergent   a. statement of what you expect to
         properties   find in nature based on hypotheses
    ____ natural   b. testable explanation about what
         selection   causes an event or aspect of nature
    ____ scientific   c. requires interaction of parts that
         theory   make up a new level of organization
    ____ hypothesis   d. a time-tested, related set of
    ____ prediction   hypotheses that explains a broad
                               range of observations and data
                            e. outcome of differences in survival
                               and reproduction among individuals
                               of a population that differ in the
                               details of one or more traits

Additional questions are available on **Biology ⨍ Now™**

Critical Thinking

1. Assess your bedroom. Is the bed made? Are the sheets clean? Are socks and underwear folded and put away? Are clothes strewn all over the floor? Now explain what the bedroom has in common with a living cell.

2. It is often said that only living things respond to the environment. Yet even a rock shows responsiveness, as when it yields to gravity's force and tumbles down a hill or changes its shape slowly under the repeated batterings of wind, rain, or tides. So how do living things differ from rocks in their responsiveness?

3. Witnesses in a court of law are asked to "swear to tell the truth, the whole truth, and nothing but the truth." What are some of the problems inherent in the question? Can you think of a better alternative?

a Natalie, blindfolded, randomly plucks a jelly bean from a jar of 120 green and 280 black jelly beans; a ratio of 30 to 70 percent.

b The jar is hidden before she removes her blindfold. She observes a single green jelly bean in her hand and assumes the jar holds only green jelly beans.

c Still blindfolded, Natalie randomly picks 50 jelly beans from the jar and ends up with 10 green and 40 black ones.

d The larger sample leads her to assume one-fifth of the jar's jelly beans are green and four-fifths are black (a ratio of 20 to 80). Her larger sample more closely approximates the jar's green-to-black ratio. The more times Natalie repeats the sampling, the greater the chance she will come close to knowing the actual ratio.

Figure 1.12 A simple demonstration of sampling error.

4. The Olestra potato chip experiment in Section 1.6 was a *double-blind* study: Neither the subjects of the experiment nor the researchers who made the follow-up phone calls knew which potato chips were in which bag. What are some of the challenges a researcher must consider when performing a double-blind study?

5. Suppose an outcome of some event has been observed to happen with great regularity. Can we predict that the same thing will always happen again? Not really, because there is no way for us to account for all of the possible variables that might affect the outcome. To illustrate this point, Garvin McCain and Erwin Segal offer a parable:

> Once there was a highly intelligent turkey. The turkey lived in a pen, attended by a kind, thoughtful master. It had nothing to do but reflect on the world's wonders and regularities. It observed some major regularities.
> Morning always started out with the sky turning light, followed by the clop, clop, clop of the master's footsteps, which was always followed by the appearance of delicious food. Other things varied—sometimes the morning was warm and sometimes cold—but food always followed footsteps. The sequence of events was so predictable that it eventually became the basis of the turkey's theory about the goodness of the world.
> One morning, after more than 100 confirmations of the goodness theory, the turkey listened for the clop, clop, clop, heard it, and had its head chopped off.

Scientists understand that all well-tested theories about nature have a high probability of not being wrong. They realize, however, that any theory is subject to modification if and when contradictory information becomes available. The absence of absolute certainty has led some people to conclude that "facts are irrelevant—facts change." If that is so, should we just stop doing scientific research? Why or why not?

6. Many magazines are loaded with articles on exercise, diet, and many other health-related topics. Some authors recommend a specific diet or dietary supplement. What kinds of evidence should the articles include so that you can decide whether to accept the recommendations?

7. Rarely can experimenters observe all individuals of a group. They select subsets or samples of populations, events, and other aspects of nature. However, they must try to avoid bias, which means risking a test by using subsets that are not really representative of the whole. *Sampling error* can occur when estimates are based on a limited sample rather than the whole population (Figure 1.12). Test results are less likely to be distorted when a sampling is large and the test is repeated. Explain how sampling error could have affected results of the potato chip experiment described in Section 1.6 if the experimenters had not been careful.

8. In 1988 Dr. Randolph Byrd and his colleagues started a study of 393 patients admitted to the San Francisco General Hospital Coronary Care Unit. In the experiment, born-again Christian volunteers were asked to pray daily for a patient's rapid recovery and for prevention of complications and death.

None of the patients knew if he or she was being prayed for. None of the volunteers or patients knew each other. Byrd categorized how each patient fared in the hospital as "good," "intermediate," or "bad." He determined that patients who had been prayed for fared a little better than those who had not. His was the first experiment that had documented statistically significant results that seemed to support the prediction that prayer might have beneficial effects for seriously ill patients.

His published results engendered a storm of criticism, mostly from scientists who cited bias in the experimental design. For instance, Byrd had categorized the patients after the experiment was over. Think about how bias might play a role in interpreting medical data. Why do you suppose the experiment generated a heated response from many in the scientific community?

I Principles of Cellular Life

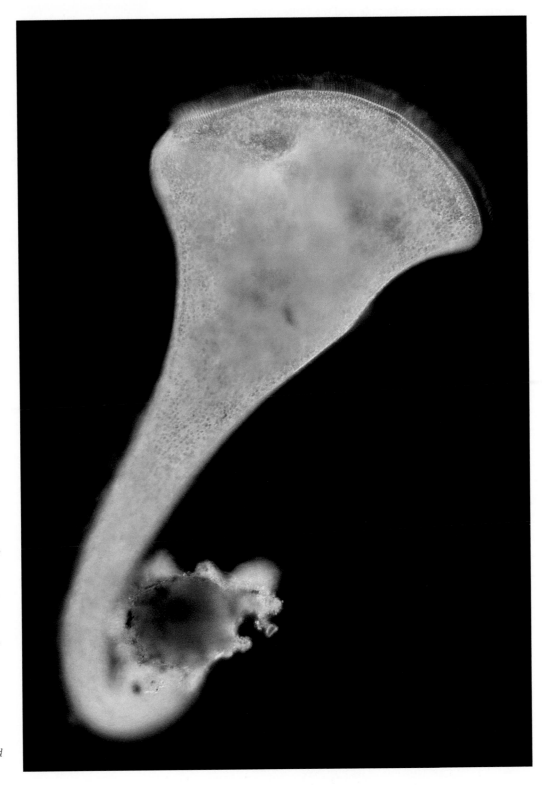

Staying alive means securing energy and raw materials from the environment. Shown here, a living cell of the genus Stentor. This protist has hairlike projections around the opening to a cavity in its body, which is about two millimeters long. Its "hairs" of fused-together cilia beat the surrounding water. They create a current that wafts food into the cavity.

2 LIFE'S CHEMICAL BASIS

What Are You Worth?

Hollywood thinks Leonardo DiCaprio is worth $20 million per movie, the Yankees think shortstop Alex Rodriguez is worth $217 million per decade, and the United States thinks the average teacher is worth $44,367 per year. Chemically, though, how much is the human body really worth (Figure 2.1a)?

Each of us is a collection of **elements**, or fundamental substances that each consist of only one kind of atom. An **atom** is the smallest unit of an element that still retains the element's properties. It occupies space, has mass, and cannot be broken down into something else, at least by everyday means.

Oxygen, carbon, hydrogen, nitrogen, and calcium are the main elements in organisms. Next are phosphorus, potassium, sulfur, sodium, and chlorine. There are a lot of *trace* elements, each making up less than 0.01 percent of the body's weight. Selenium and lead are examples.

Wait a minute! Selenium, lead, mercury, arsenic, and many other elements are toxic, right? So how can they be part of the collection? We're finding that trace amounts

Figure 2.1 (a) What are you worth, chemically speaking? (b) Proportions of the most common elements in a human body, Earth's crust, and seawater. How are they similar? How do they differ?

a

Mass of Elements in a 70-Kilogram Human Body		Cost (Retail)
Oxygen	43.00 kilograms (kg)	$0.021739
Carbon	16.00 kg	6.400000
Hydrogen	7.00 kg	0.028315
Nitrogen	1.80 kg	9.706929
Calcium	1.00 kg	15.500000
Phosphorus	780.00 grams (g)	68.198594
Potassium	140.00 g	4.098737
Sulfur	140.00 g	0.011623
Sodium	100.00 g	2.287748
Chlorine	95.00 g	1.409496
Magnesium	19.00 g	0.444909
Iron	4.20 g	0.054600
Fluorine	2.60 g	7.917263
Zinc	2.30 g	0.088090
Silicon	1.00 g	0.370000
Rubidium	0.68 g	1.087153
Strontium	0.32 g	0.177237
Bromine	0.26 g	0.012858
Lead	0.12 g	0.003960
Copper	72.00 milligrams (mg)	0.012961
Aluminum	60.00 mg	0.246804
Cadmium	50.00 mg	0.010136
Cerium	40.00 mg	0.043120
Barium	22.00 mg	0.028776
Iodine	20.00 mg	0.094184
Tin	20.00 mg	0.005387
Titanium	20.00 mg	0.010920
Boron	18.00 mg	0.002172
Nickel	15.00 mg	0.031320
Selenium	15.00 mg	0.037949
Chromium	14.00 mg	0.003402
Manganese	12.00 mg	0.001526
Arsenic	7.00 mg	0.023576
Lithium	7.00 mg	0.024233
Cesium	6.00 mg	0.000016
Mercury	6.00 mg	0.004718
Germanium	5.00 mg	0.130435
Molybdenum	5.00 mg	0.001260
Cobalt	3.00 mg	0.001509
Antimony	2.00 mg	0.000243
Silver	2.00 mg	0.013600
Niobium	1.50 mg	0.000624
Zirconium	1.00 mg	0.000830
Lanthanum	0.80 mg	0.000566
Gallium	0.70 mg	0.003367
Tellurium	0.70 mg	0.000722
Yttrium	0.60 mg	0.005232
Bismuth	0.50 mg	0.000119
Thallium	0.50 mg	0.000894
Indium	0.40 mg	0.000600
Gold	0.20 mg	0.001975
Scandium	0.20 mg	0.058160
Tantalum	0.20 mg	0.001631
Vanadium	0.11 mg	0.000322
Thorium	0.10 mg	0.004948
Uranium	0.10 mg	0.000103
Samarium	50.00 micrograms (µg)	0.000118
Beryllium	36.00 µg	0.000218
Tungsten	20.00 µg	0.000007
Grand Total		**$118.63**

Human		Earth's Crust		Seawater	
Oxygen	61.0%	Oxygen	46.0%	Oxygen	85.7%
Carbon	23.0	Silicon	27.0	Hydrogen	10.8
Hydrogen	10.0	Aluminum	8.2	Chlorine	2.0
Nitrogen	2.6	Iron	6.3	Sodium	1.1
Calcium	1.4	Calcium	5.0	Magnesium	0.1
Phosphorus	1.1	Magnesium	2.9	Sulfur	0.1
Potassium	0.2	Sodium	2.3	Calcium	0.04
Sulfur	0.2	Potassium	1.5	Potassium	0.03

b

of at least some of them have vital functions. For instance, even a little selenium is toxic, but *too* little can cause heart problems and thyroid disorders.

Superficially, then, the human body can be viewed as a balanced collection of elements. The amounts are worth no more than $118.63, and the kinds are not even unique; they occur in Earth's crust and even seawater (Figure 2.1*b*). However, the *proportions* of elements in humans and other organisms are unique relative to nonliving things. Look at all of that carbon, for instance! Also, you will never find a clod of dirt or a volume of seawater that comes close to the *structural and functional organization* of a living body. Assembling that collection of elements into an organized, operational body takes a fabulous molecular library (DNA), enzymes and other metabolic workers, and large, ongoing inputs of energy (just ask any pregnant woman).

Remember this when someone tries to say "chemistry" has nothing to do with you. It has everything to do with you. People, toothpaste, turkeys, refrigerators, jet fuel, health, disease, corsages, acid rain, nerve gas, old-growth forests—name any living or nonliving bit of the universe, and chemistry is part of it.

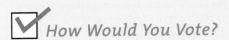

Watch the video online!

How Would You Vote?

Fluoride helps prevent tooth decay. But too much wrecks bones and teeth, and causes birth defects. A lot can kill you. Many communities in the United States add fluoride to their supply of drinking water. Do you want it in yours? See BiologyNow for details, then vote online.

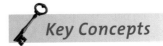
Key Concepts

ATOMS AND ELEMENTS

An element is a fundamental substance made of one type of atom. The atom is the smallest unit of an element that still retains the element's properties, and its building blocks are protons, electrons, and neutrons. Isotopes are atoms of an element that vary in the number of neutrons. Sections 2.1, 2.2

WHY ELECTRONS MATTER

Atoms acquire, share, and give up electrons. Whether one atom will bond with others depends on the number and arrangement of its electrons. Section 2.3

ATOMS BOND

The bonding behavior of biological molecules starts with the number and arrangement of electrons in each type of atom. Ionic, covalent, and hydrogen bonds are the main categories of bonds between atoms in biological molecules. Section 2.4

NO WATER, NO LIFE

Life originated in water and is adapted to its properties. Water has temperature-stabilizing effects. Many kinds of substances dissolve easily in it. Water also shows cohesion. Section 2.5

HYDROGEN IONS RULE

Life depends on precise controls over the formation, use, and buffering of hydrogen ions. Section 2.6

Links to Earlier Concepts

With this chapter, we start at the base of life's levels of organization, so take a moment to review the simple chart in Section 1.1. It all starts with atoms and energy. Life's organization requires tapping into a great one-way flow of energy and storing it in bonds between atoms (1.2).

The chapter also has a simple example of how the body's built-in mechanisms help return the internal environment to a homeostatic state when conditions shift beyond ranges that cells can tolerate (1.2).

2.1 Start With Atoms

LINK TO
SECTION
1.1

Know a bit about protons, neutrons, and electrons, and you have a clue to why the elements that make up the body behave as they do. Each element's unique properties start with the number of protons in its atoms.

An element, again, is a fundamental substance made of only one kind of atom. Atoms are built from three kinds of subatomic particles: protons, electrons, and neutrons. Each **proton** carries a positive *charge*, which is a defined amount of electricity. You can symbolize a proton as p^+. An atom's nucleus, or core region, holds one or more protons. Except for the hydrogen atom, it also holds **neutrons**, which carry no charge. Moving around the atomic nucleus are one or more **electrons**, which carry a negative charge (e^-). Figure 2.2 shows a few simple models for atomic structure.

The positive charge of one proton and the negative charge of one electron balance each other. Therefore, an atom that has the same number of electrons and protons has no net electrical charge.

Each element has a unique *atomic number*, which is the number of protons in the nucleus of its atoms. A hydrogen atom has one proton, so the atomic number is 1. For carbon, with six protons, it is 6.

Protons and neutrons contribute to an atom's mass. (Electrons are too tiny to do so.) We can assign each element a *mass number*, or the total number of protons *and* neutrons in the atomic nucleus. For carbon, with six protons and six neutrons, the mass number is 12.

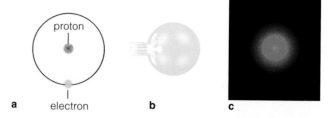

Figure 2.2 Different ways to represent atoms, using hydrogen (H) as the example. (**a**) The shell model, good for showing the number of electrons and their organization around the nucleus. (**b**) Ball models show the sizes of atoms relative to one another. (**c**) Electron density clouds are best at conveying the distribution of electrons around the nucleus.

Why bother with the number of electrons, protons, and neutrons? Knowing them can help you predict how each kind of element will behave under a variety of conditions inside and outside the body.

Elements were being classified in terms of chemical similarities long before their subatomic particles were discovered. In 1869, Dmitry Mendeleev, known more for his extravagant hair than his discoveries (he cut it only once a year), arranged the known elements in a repeating pattern, based on their chemical properties. By using gaps in this **periodic table of the elements**, Mendeleev correctly predicted the existence of many elements that had not yet been discovered.

Elements fall into order in the table according to their atomic number (Figure 2.3). All elements in each vertical column have the same number of electrons that are available for interaction with other atoms. As a result, they behave in similar ways. For example, helium, neon, radon, and other gases in the farthest right column of the periodic table are *inert* elements. Not one of the electrons in their atoms is available for chemical interactions. Such elements rarely do much; they occur mostly as solitary atoms.

You won't find all of the elements in nature. Those after atomic number 92 are extremely unstable. Some have been formed in exceedingly small quantities in laboratories—sometimes no more than a single atom —and they wink out of existence fast.

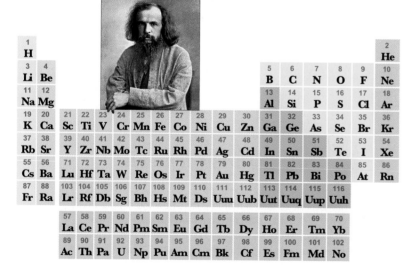

Figure 2.3 Periodic table of the elements and Dmitry Mendeleev, who created it. Some symbols for elements are abbreviations for their Latin names. For instance, Pb (lead) is short for *plumbum;* the word "plumbing" is related, because ancient Romans used lead to make their water pipes.

An element is a fundamental substance consisting of only one kind of atom. Atoms are the smallest units that retain an element's properties.

An atom consists of one or more positively charged protons, negatively charged electrons, and (except for hydrogen) neutrons. Whether any given atom will interact with others depends on how many electrons it has.

2.2 Putting Radioisotopes To Use

*All elements are defined by the number of protons in
their atoms—but an element's atoms can differ in their
number of neutrons. We call such atoms **isotopes** of the
same element. Some are radioactive.*

In 1896, Henri Becquerel made a chance discovery. He had
placed some uranium crystals in a desk drawer, next to a
coin and metal screen on top of some sheets of opaque black
paper. Underneath that paper was a photographic plate.
A day later, the physicist used the film and developed it.
Oddly, a negative image of the coin and the metal screen
showed up. Becquerel hypothesized that energy radiating
from the uranium salts had passed through the paper—
which was impenetrable to light—and exposed the film
around both metal objects.

As we now know, uranium has isotopes—fifteen of them.
Most naturally occurring elements do. Carbon has three
isotopes, nitrogen has two, and so on. A superscript number
to the left of an element's symbol is the isotope's mass
number. For instance, carbon's three natural isotopes are
^{12}C (carbon 12, the most common form, with six protons,
six neutrons), ^{13}C (with six protons, seven neutrons), and
^{14}C (with six protons, eight neutrons).

Some isotopes are unstable, or radioactive. A radioactive
isotope, or **radioisotope**, spontaneously emits energy in the
form of subatomic particles and x-rays when its nucleus
disintegrates. This process is called **radioactive decay**, and
it can transform one element into another. As an example,
^{13}C and ^{14}C are radioisotopes of carbon. Each predictably
decays with a particular amount of energy into a more
stable product. After 5,700 years, about half of the atoms in
a sample of ^{14}C will have turned into ^{13}N (nitrogen) atoms.
Researchers use radioactive decay to estimate the age of
rocks and biological remains, as Section 17.5 explains.

The different isotopes of an element are still the same
element. For the most part, carbon is carbon, regardless of
how many neutrons it has. Living systems use ^{12}C the same
way as ^{14}C. Knowing this, researchers or clinicians who want
to track a particular substance construct a **tracer**. Tracers
are molecules in which a radioisotope has been substituted
for a more stable isotope. They can be delivered into a cell or
multicelled body, even into populations used in laboratory
experiments. The energy from radioactive decay is like a
shipping label. It helps researchers track the pathway or
destination of a substance of interest with the help of
radioactivity-detecting instruments.

For example, Melvin Calvin and his colleagues used a
tracer to discover specific reaction steps of photosynthesis.
They let growing plants take up a radioactive gas (carbon
dioxide made with ^{14}C). By using radioactivity-detecting
instruments, they tracked the carbon radioisotope through
steps by which plants produce simple sugars and starches.

Radioisotopes also are used in medicine. *PET* (short for
Positron-Emission Tomography) uses radioisotopes to study
metabolism. Clinicians attach a radioisotope to glucose or
another sugar. They inject this tracer into a patient, who
is moved into a PET scanner (Figure 2.4*a*). Cells in different
parts of the body absorb the tracer at different rates. The
scanner detects radiation caused by energy from the decay
of the radioisotope. That radiation is used to form an image
on a monitor, as in Figure 2.4. Such images reveal variations
and abnormalities in metabolic activity.

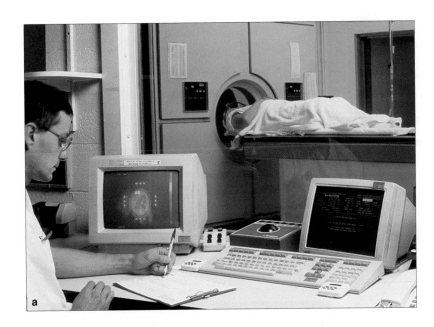

a

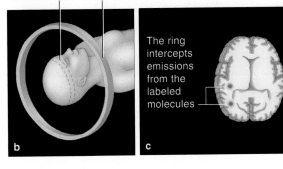

portion of the
patient's body
being scanned

detector ring
inside the
PET scanner

b

The ring
intercepts
emissions
from the
labeled
molecules

c

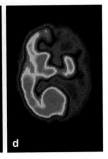

d

Figure 2.4 *Animated!* (**a**) Patient whose brain is being
probed in a PET scanner. (**b**,**c**) A ring of detectors intercepts
radioactive emissions from tracers that had been injected into
the patient. The body region of interest is scanned. Computers
analyze and color-code the number of emissions from each
location in the scanned region. Results are converted into
digital images and displayed on computer screens.

(**d**) Different colors in a scan signify differences in metabolic
activity. Cells in the left half of this brain absorbed and used
labeled molecules at expected rates. Cells in the right half
showed little activity. This patient has a neurological disorder.

2.3　What Happens When Atom Bonds With Atom?

LINK TO
SECTION
1.1

Atoms acquire, share, and donate electrons. The atoms of some elements do this quite easily; others do not. Why is this so? To come up with an answer, look to the number and arrangement of electrons in atoms.

ELECTRONS AND ENERGY LEVELS

In our world, simple physics explains the motion of an apple falling from a tree. Tiny electrons belong to a strange world where everyday physics does not apply. (If electrons were as big as apples, you would be 3.5 times taller than our solar system is wide.) Different forces bring about the motion of electrons, which can get from here to there without going in between!

We can calculate where an electron is, although not exactly. The best we can do is say that it is somewhere in a fuzzy cloud of probability density. Where it can go in the cloud depends on how many other electrons belong to the atom. The electrons become arranged in orbitals, or volumes of space around the atomic nucleus. Many orbitals, each with a characteristic three-dimensional shape, are possible.

An atom has the same number of electrons as protons. Most atoms have many electrons. How are they all arranged, given that electrons repel each other? Think of each atom as a multilevel apartment building with lots of vacant rooms to rent to electrons, and a nucleus in the basement. Each "room" is one orbital, and it rents out to two electrons at most. An orbital holding one electron only has a vacancy; another electron can move in.

Each floor in that atomic apartment building corresponds to an energy level. There is only one room on the first floor (one orbital at the lowest energy level, closest to the nucleus). It fills first. For hydrogen, the simplest atom, a lone electron occupies the room (Figure 2.5). Helium, with its two electrons, has no vacancies at the first (lowest) energy level. In larger atoms, more electrons rent the second-floor rooms. If the second floor is filled, then more electrons rent third-floor rooms, and so on. *Electrons fill orbitals at successively higher energy levels.*

The farther an electron is from the basement (the nucleus), the greater its energy. An electron in a first-floor room can't move to the second or third floor, let alone the penthouse, unless a boost of energy puts it there. Suppose the electron absorbs just enough energy from, say, sunlight, to get excited about moving up. Move it does. If nothing fills that lower room, though, the electron will quickly return to it, emitting extra energy as it does. Later on, you will see how cells in plants and in your eyes harness and use that energy.

FROM ATOMS TO MOLECULES

In shell models for atoms, nested "shells" correspond to energy levels. They give us an easy way to check for electron vacancies, as in Figure 2.6. Bear in mind, atoms do not look like these flat diagrams. The shells are not three-dimensional volumes of space, and they certainly don't show the electron orbitals.

The atoms with vacancies in their outermost shell tend to give up, acquire, or share electrons. Actually, what we call **chemical bonds** are just a case of atoms sharing their electrons with one another. An atom with no vacancies rarely bonds with others. But the most common atoms in organisms—such as oxygen, carbon, hydrogen, nitrogen, and calcium—do have vacancies in orbitals at their outermost energy level. They tend to participate in bonds.

A **molecule** is simply two or more atoms of the same or different elements joined in a chemical bond.

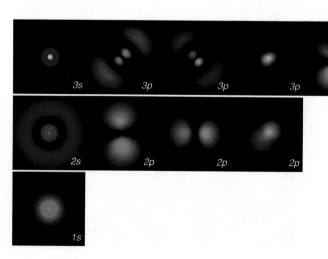

third
energy level
(second floor)

second
energy level
(first floor)

first
energy level
(closest to the
basement)

Figure 2.5 Models for the first, second, and third levels of the atomic apartment building. Each model is a three-dimensional approximation of an electron orbital. Colors are most intense in locations where electrons are most likely to be in any given instant. Orbitals farthest from the nucleus have greater energy and are more complex.

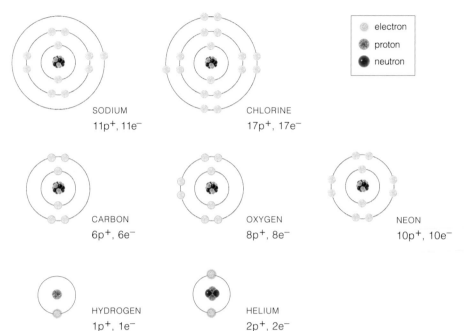

c **Third shell** This shell corresponds to the third energy level. It has nine orbitals (one *s*, three *p*, and five *d* orbitals), or room for eighteen electrons. Sodium has one electron in the third shell of orbitals, and chlorine has seven. Both have vacancies, so they both are receptive to chemical bonding.

b **Second shell** This shell, which corresponds to the second energy level, has one *s* orbital and three *p* orbitals—room for a total of eight electrons. Carbon has six electrons, two in the first shell and four in the second shell. It has four vacancies. Oxygen has two vacancies. Both carbon and oxygen form chemical bonds.

a **First shell** A single shell corresponds to the first energy level, which has a single orbital (*1s*) that can hold two electrons. Hydrogen has only one electron in this shell and gives it up easily. A helium atom has two electrons (no vacancies) and usually does not enter into chemical bonds.

SODIUM
$11p^+, 11e^-$

CHLORINE
$17p^+, 17e^-$

CARBON
$6p^+, 6e^-$

OXYGEN
$8p^+, 8e^-$

NEON
$10p^+, 10e^-$

HYDROGEN
$1p^+, 1e^-$

HELIUM
$2p^+, 2e^-$

electron
proton
neutron

Figure 2.6 *Animated!* Shell models, which help us visualize vacancies in an atom's outermost orbitals. Each circle, or shell, represents all orbitals at one energy level. Larger circles correspond to higher energy levels. Such models are highly simplified. A more realistic rendering would show electrons as fuzzy clouds of probability density about 10,000 times larger than the nucleus.

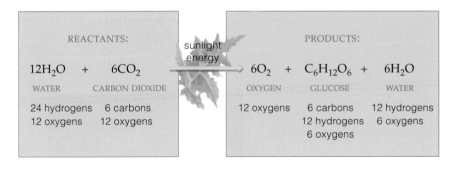

REACTANTS:

$12H_2O$ + $6CO_2$

WATER   CARBON DIOXIDE

24 hydrogens   6 carbons
12 oxygens   12 oxygens

sunlight energy

PRODUCTS:

$6O_2$ + $C_6H_{12}O_6$ + $6H_2O$

OXYGEN   GLUCOSE   WATER

12 oxygens   6 carbons   12 hydrogens
             12 hydrogens   6 oxygens
             6 oxygens

Figure 2.7 Chemical bookkeeping. We use formulas when writing out chemical equations, which represent reactions between atoms and molecules. Substances entering a reaction (reactants) are written to the left of a reaction arrow, and products to the right. How many molecules (or atoms) enter as reactants or form as products are indicated by a number that precedes their formula. *The same number of atoms that enter a reaction must be there at the end.* The atoms get shuffled around, but they never vanish. To be sure you wrote an equation correctly, count the atoms.

You can write a molecule's chemical composition as a formula, which uses symbols for the elements present and subscripts for the number of atoms of each kind of element (Figure 2.7). For example, one molecule of water has the chemical formula H_2O. The subscript number shows that there are two hydrogen (H) atoms for each oxygen (O) atom. If you have six molecules of water, then you would write $6H_2O$.

Compounds are molecules that consist of two or more different elements in proportions that never do vary. Water is an example. All water molecules have one oxygen atom bonded to two hydrogen atoms. The ones in rain clouds, the seas, a Siberian lake, flower petals, your bathtub, or anywhere else have twice as many hydrogen atoms as oxygen atoms. In a **mixture**, two or more substances intermingle without bonding.

For example, when you swirl sugar into water, you make a mixture. The proportions of elements in this mixture, or any other kind of mixture, can vary.

Electrons occupy orbitals, or defined volumes of space around an atom's nucleus. Successive orbitals correspond to levels of energy, which become higher with distance from the atomic nucleus.

One or at most two electrons can occupy any orbital. The atoms with vacancies in orbitals at their highest level tend to interact and form bonds with other atoms.

A molecule is two or more atoms joined in a chemical bond. In compounds, atoms of two or more elements are bonded together. A mixture consists of intermingled substances.

2.4 Major Bonds in Biological Molecules

Electrons of one type of atom interact with electrons of others in specific ways. Those interactions give rise to the distinctive properties of biological molecules.

ION FORMATION AND IONIC BONDING

An electron, recall, has a negative charge equal to a proton's positive charge. When an atom contains as many electrons as protons, these charges balance each other, so the atom has a net charge of zero. When an atom *gains* an extra electron, it acquires a net negative charge. When an atom *loses* an electron, it acquires a net positive charge. Either way, it has become an **ion**.

Consider: A chlorine atom has seven protons. It has seven electrons (one vacancy) in the third orbital level—which is most stable when filled with eight. This atom tends to attract an electron from someplace else. With that extra electron, it becomes a chloride ion (Cl^-), with a net negative charge.

Also consider: A sodium atom has eleven protons and eleven electrons. Its second orbital level is full of electrons, and only one electron is in the third orbital level. Giving up the one electron is easier than getting seven more. When it does so, the atom still has eleven protons. But now it has ten electrons. It has become a sodium ion, with a net positive charge (Na^+).

Remember that opposite charges attract each other. When a positively charged ion encounters a negatively charged ion, the two may associate closely with each other. A close association of ions is an **ionic bond**. For example, Figure 2.8*a* shows a crystal of table salt, or NaCl. In such crystals, ionic bonds hold ions of sodium and chloride in an orderly, cubic arrangement.

COVALENT BONDING

In an ionic bond, an atom that has lost one or more electrons associates with an atom that gained one or more electrons. What if both atoms have room for an extra electron? They can *share* one in a hybrid orbital that spans both atomic nuclei. The vacancy in each atom becomes filled with the shared electrons.

When atoms share two electrons, they are joined in a single **covalent bond** (Figure 2.8*b*). Such bonds are stable and are much stronger than ionic bonds.

We can represent covalent bonds as single lines in structural formulas, which show how the atoms of a molecule are physically arranged. A line between two atoms represents a pair of electrons that are being shared in a single covalent bond. To give examples of this bonding pattern, molecular hydrogen (H_2) has one covalent bond and can be written as H—H. Two

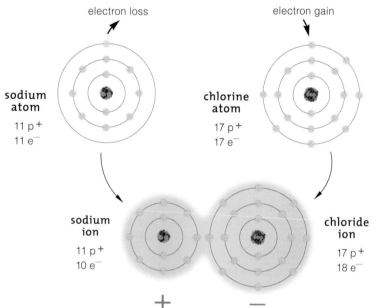

electron loss   electron gain

sodium
atom
11 p$^+$
11 e$^-$

chlorine
atom
17 p$^+$
17 e$^-$

sodium
ion
11 p$^+$
10 e$^-$

chloride
ion
17 p$^+$
18 e$^-$

+   —

1 mm

a Example of ongoing interactions called ionic bonding. In each crystal of table salt, or NaCl, many sodium ions and chloride ions are staying close together because of the mutual attraction of their opposite charges.

Figure 2.8 *Animated!* Important bonds in biological molecules.

atoms share two electron pairs in a *double* covalent bond. Molecular oxygen (O=O) is like this. Others share three electron pairs in a *triple* covalent bond, as in molecular nitrogen (N≡N). Each time you take a breath, O_2 and N_2 molecules flow toward your lungs.

In a *nonpolar* covalent bond, two atoms are sharing electrons equally, so the molecule shows no difference in charge between the two "ends" of the bond. We find such bonds in molecular hydrogen (H_2), oxygen (O_2), and nitrogen (N_2).

In a *polar* covalent bond, two atoms do not share electrons equally. Why not? The atoms are of different elements, and one has more protons than the other. The one with the most protons exerts more of a pull on the electrons, so its end of the bond ends up with a slight negative charge. We say it is "electronegative." The atom at the other end of the bond ends up with a slight positive charge. For instance, a water molecule (H—O—H) has two polar covalent bonds. The oxygen atom carries a slight negative charge, and each of its two hydrogen atoms carries a slight positive charge.

HYDROGEN BONDING

A **hydrogen bond** is a weak attraction that has formed between a covalently bound hydrogen atom and an electronegative atom in a different molecule or in a different region of the same molecule.

Because hydrogen bonds are weak, they form and break easily. Collectively, however, many hydrogen bonds contribute to the properties of liquid water, as you will see next.

Hydrogen bonds also play important roles in the structure and function of biological molecules. They often form between different parts of large molecules that have folded over on themselves and hold them in particular shapes. Many of these bonds hold DNA's two nucleotide strands together. Figure 2.8c hints at the number of these interactions in DNA.

Ions form when atoms acquire a net charge by gaining or losing electrons. Two ions of opposite charge attract each other. They can associate in an ionic bond.

In a covalent bond, atoms share a pair of electrons. When atoms share the electrons equally, the bond is nonpolar. When the sharing is not equal, the bond is polar—slightly positive at one end, slightly negative at the other.

In a hydrogen bond, a covalently bound hydrogen atom attracts a small, negatively charged atom in a different molecule or in a different region of the same molecule.

Two hydrogen atoms, each with one proton, share two electrons in a single nonpolar covalent bond.

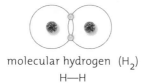

molecular hydrogen (H_2)
H—H

Two oxygen atoms, each with eight protons, share four electrons in a nonpolar double covalent bond.

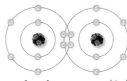

molecular oxygen (O_2)
O=O

Oxygen has vacancies for two electrons in its highest energy level orbitals. Two hydrogen atoms can each share an electron with oxygen. The resulting two polar covalent bonds form a water molecule.

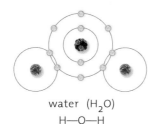

water (H_2O)
H—O—H

b Covalent bonding. Each atom becomes more stable by sharing electron pairs in hybrid orbitals.

Two molecules interacting weakly in one H bond, which can form and break easily.

hydrogen bond

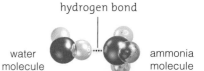

water molecule   ammonia molecule

H bonds helping to hold part of two large molecules together.

Many H bonds hold DNA's two strands together along their length. Individually each one is weak, but collectively they can stabilize DNA's large structure.

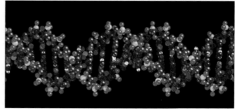

c Hydrogen bonds. Such bonds can form at a hydrogen atom that is already covalently bonded in a molecule. The atom's slight positive charge weakly attracts an atom with a slight negative charge that is already covalently bonded to something else. As shown, this can happen between one of the hydrogen atoms of a water molecule and the nitrogen atom of an ammonia molecule.

2.5 Water's Life-Giving Properties

No sprint through basic chemistry is complete unless it leads to the collection of molecules called water. Life originated in water. Organisms still live in it or they cart water around with them inside cells and tissue spaces. Many metabolic reactions use water. A cell's structure and shape absolutely depend on it.

POLARITY OF THE WATER MOLECULE

Figure 2.9*a* shows the structure of a water molecule. Two atoms of hydrogen have formed polar covalent bonds with an oxygen atom. The molecule has no net charge. Even so, the oxygen pulls the shared electrons more than the hydrogen atoms do. Thus, the molecule of water has a slightly negative "end" that is balanced out by its slightly positive "end."

The water molecule's polarity attracts other water molecules. Also, the polarity is so attractive to sugars and other polar molecules that hydrogen bonds form easily between them. That is why polar molecules are known as **hydrophilic** (water-loving) substances.

That same polarity repels oils and other nonpolar molecules, which are **hydrophobic** (water-dreading) substances. Shake a bottle filled with water and salad oil, then set it on a table. Soon, new hydrogen bonds replace the ones that the shaking broke. The reunited water molecules push out oil molecules, which cluster as oil droplets or as an oily film at the water's surface.

The same kinds of interactions proceed at the thin, oily membrane between the water inside and outside cells. Membrane organization—and life itself—starts with such hydrophilic and hydrophobic interactions. You will read about membrane structure in Chapter 5.

WATER'S TEMPERATURE-STABILIZING EFFECTS

Cells are mostly water, and they also release a lot of metabolic heat. The many hydrogen bonds in water keep cells from cooking in their own juices. How? All bonds vibrate nonstop, and they move more as they absorb heat. **Temperature** is a measure of molecular motion. Compared to most other fluids, water absorbs more heat energy before it gets measurably hotter. So water serves as a heat reservoir, and its temperature remains relatively stable. Over time, increases in heat step up the motion within water molecules. Before that happens, however, much of the heat will go into disrupting hydrogen bonds between molecules.

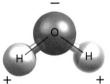

slight negative charge
on the oxygen atom

The + and – ends balance each other; the whole molecule carries no net charge, overall.

slight positive charge
on the hydrogen atoms

a

Figure 2.9 *Animated!* Water, a substance essential for life.

(**a**) Polarity of an individual water molecule.

(**b**) Hydrogen bonding pattern among water molecules in liquid water. Dashed lines signify hydrogen bonds, which break and re-form rapidly.

(**c**) Hydrogen bonding in ice. Below 0°C, every water molecule hydrogen-bonds with four others, in a rigid three-dimensional lattice. The molecules are farther apart, or less densely packed, than they are in liquid water. As a result, ice floats on water.

Thanks partly to rising levels of methane and other greenhouse gases that are contributing to global warming, the Arctic ice cap is melting. At current rates, it will be gone in fifty years. So will the polar bears. Already their season for hunting seals is shorter, bears are thinner, and they are giving birth to fewer cubs.

Figure 2.10 Spheres of hydration around two ions.

Figure 2.11 Examples of water's cohesion. (**a**) When a pebble hits liquid water and forces molecules away from the surface, the individual water molecules do not fly every which way. They stay together in droplets. Why? Countless hydrogen bonds exert a continuous inward pull on individual molecules at the surface.

(**b**) And just how does water rise to the very top of trees? Cohesion, and evaporation from leaves, pulls it upward.

a

b

When water temperature is stable, hydrogen bonds form as fast as they break. When water gets hotter, the increase in molecular motion can keep the bonds broken, so individual molecules at the water's surface can escape into air. By this process, **evaporation**, heat energy converts liquid water to gaseous form. The increased energy has overcome the attraction between water molecules, which break free. Water's surface temperature decreases during evaporation.

Evaporative water loss helps you and some other mammals cool off when you sweat on hot, dry days. Sweat, about 99 percent water, evaporates from skin.

Below 0°C, water molecules do not move enough to break their hydrogen bonds, so they become locked in the latticelike bonding pattern of ice (Figure 2.9c). Ice is less dense than water. During winter freezes, ice sheets may form near the surface of ponds, lakes, and streams. The ice "blanket" insulates the liquid water beneath it and helps protect many fishes, frogs, and other aquatic organisms against freezing.

WATER'S SOLVENT PROPERTIES

Water is an excellent *solvent*, meaning ions and polar molecules easily dissolve in it. A dissolved substance is known as a **solute**. In general, a substance is said to be *dissolved* after water molecules cluster around ions or molecules of it and keep them dispersed in fluid.

A clustering of water molecules around a solute is called a *sphere of hydration*. Such spheres form around any solute in cellular fluids, tree sap, blood, the fluid in your gut, and every other fluid associated with life. Watch it happen after you pour table salt (NaCl) into a cup of water. In time, the crystals of salt separate

into ions of sodium (Na^+) and chloride (Cl^-). Each Na^+ attracts the negative end of some water molecules even as Cl^- attracts the positive end of others (Figure 2.10). Spheres of hydration formed this way keep the ions dispersed in fluid.

WATER'S COHESION

Still another life-sustaining property of water is its cohesion. **Cohesion** means something is showing a capacity to resist rupturing when it is stretched, or placed under tension. You see its effect when a tossed pebble breaks the surface of a lake, a pond, or some other body of liquid water (Figure 2.11a). At or near the surface, uncountable numbers of hydrogen bonds are exerting a continuous, inward pull on individual molecules. Bonding creates a high surface tension.

Cohesion is working inside organisms, too. Plants, for example, absorb nutrient-laden water when they grow. Columns of liquid water rise inside pipelines of vascular tissues, which extend from roots to leaves. Water evaporates from leaves when molecules break free and diffuse into air (Figure 2.11b). The cohesive force of hydrogen bonds pulls replacements into the leaf cells, in ways explained in Section 30.3.

Being polar, water molecules hydrogen-bond to one another and to other polar (hydrophilic) substances. They tend to repel nonpolar (hydrophobic) substances.

The unique properties of liquid water make life possible. Water has temperature-stabilizing effects, cohesion, and a capacity to dissolve many substances easily.

2.6 Acids and Bases

LINK TO
SECTION
1.2

Ions dissolved in fluids inside and outside each living cell influence its structure and function. Among the most influential are hydrogen ions. They have far-reaching effects largely because they are chemically active and because there are so many of them.

THE pH SCALE

At any instant in liquid water, some water molecules split into ions of hydrogen (H^+) and hydroxide (OH^-). These ions are the basis of the **pH scale**. The scale is a way to measure the concentration of hydrogen ions in solutions such as seawater, blood, or sap. The greater the H^+ concentration, the lower the pH. Pure water (not rainwater or tap water) always has as many H^+ as OH^- ions. This state is neutrality, or pH 7.0 (Figure 2.12).

A decrease in pH by just one unit from neutrality corresponds to a tenfold increase in H^+ concentration, and an increase by one unit corresponds to a tenfold decrease in H^+ concentration. One way to get a sense of the range is to taste dissolved baking soda (pH 9), water (pH 7), and lemon juice (pH 2).

HOW DO ACIDS AND BASES DIFFER?

When dissolved in water, substances called **acids** *donate* hydrogen ions and **bases** *accept* hydrogen ions. *Acidic* solutions, such as lemon juice, gastric fluid, and coffee, release H^+; their pH is below 7. *Basic* solutions, such as seawater and egg white, easily combine with H^+. Basic solutions, which also are known as alkaline solutions, have a pH above 7.

Nearly all of life's chemistry occurs near pH 7. Most of your body's internal environment (tissue fluids and blood) is between pH 7.3 and 7.5. Seawater is more basic than body fluids of the organisms living in it.

Acids and bases can be weak or strong. The weak acids, such as carbonic acid (H_2CO_3), are stingy H^+ donors. Strong acids readily give up H^+ in water. An example is the hydrochloric acid that dissociates into H^+ and Cl^- inside your stomach. The H^+ makes your gastric fluid far more acidic, which in turn activates protein-digesting enzymes.

Too much HCl can cause an *acid stomach*. Antacids taken for this condition, including milk of magnesia, release OH^- ions that combine with H^+ to reduce the pH of stomach contents.

Actually, strong acids and bases can cause severe chemical burns. That is why we are supposed to read the labels on containers of ammonia, drain cleaner, and many other common household products. That is why we are not supposed to let a car battery's sulfuric acid drip on skin.

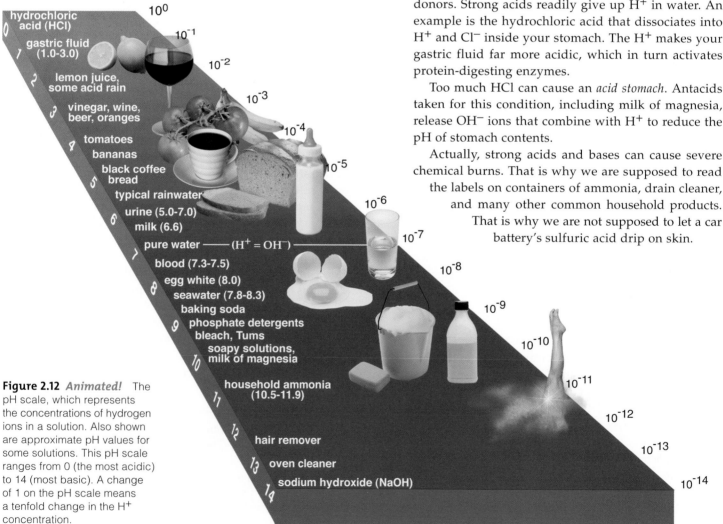

Figure 2.12 Animated! The pH scale, which represents the concentrations of hydrogen ions in a solution. Also shown are approximate pH values for some solutions. This pH scale ranges from 0 (the most acidic) to 14 (most basic). A change of 1 on the pH scale means a tenfold change in the H^+ concentration.

hydrochloric acid (HCl) — 0
gastric fluid (1.0–3.0) — 1
lemon juice, some acid rain — 2
vinegar, wine, beer, oranges — 3
tomatoes bananas — 4
black coffee bread — 5
typical rainwater
urine (5.0–7.0)
milk (6.6) — 6
pure water ——— ($H^+ = OH^-$) — 7
blood (7.3–7.5)
egg white (8.0) — 8
seawater (7.8–8.3)
baking soda
phosphate detergents
bleach, Tums — 9
soapy solutions, milk of magnesia — 10
household ammonia (10.5–11.9) — 11
hair remover — 12
oven cleaner — 13
sodium hydroxide (NaOH) — 14

10^0, 10^{-1}, 10^{-2}, 10^{-3}, 10^{-4}, 10^{-5}, 10^{-6}, 10^{-7}, 10^{-8}, 10^{-9}, 10^{-10}, 10^{-11}, 10^{-12}, 10^{-13}, 10^{-14}

Figure 2.13 Emissions of sulfur dioxide from a coal-burning power plant. Airborne pollutants such as sulfur dioxide dissolve in water vapor to form acidic solutions. They are a component of acid rain.

At high concentrations, strong acids or bases that enter an ecosystem can kill organisms. For instance, fossil fuel burning and nitrogen-containing fertilizers release strong acids that lower the pH of rainwater (Figure 2.13). Some regions are sensitive to this *acid rain*. Alterations in the chemical composition of soil and water harm fishes and other organisms in these regions. We return to this topic in Section 48.2.

SALTS AND WATER

A **salt** is any compound that dissolves easily in water and releases ions *other than* H^+ and OH^-. It commonly forms when an acid interacts with a base. For example:

$$HCl\ (acid)\ +\ NaOH\ (base)\ \rightleftharpoons\ NaCl\ (salt)\ +\ H_2O$$

HYDROCHLORIC   SODIUM   SODIUM
ACID   HYDROXIDE   CHLORIDE

NaCl, the salt product of this reaction, dissociates into sodium ions (H^+) and chloride ions (Cl^-) when it is dissolved in water. Many of the ions that are released when salts dissolve in fluid are important components of cellular processes. For example, ions of sodium, potassium, and calcium are essential for nerve and muscle cell functions. They also help plant cells take up water from soil.

BUFFERS AGAINST SHIFTS IN pH

Cells must respond fast to even slight shifts in pH, because excess H^+ or OH^- can alter how biological molecules function. Responses are rapid with **buffer systems**. Think of such a system as a dynamic chemical partnership between a weak acid and its salt. These two related chemicals work in equilibrium to counter slight shifts in pH. For example, if a small amount of a strong base enters a buffered fluid, the weak acid partner can neutralize excess OH^- ions by donating some H^+ ions to the solution.

Most body fluids are buffered. Why? Enzymes, receptors, and all other essential biological molecules work most efficiently within a narrow range of pH. Deviation from the range disrupts cellular processes.

Carbon dioxide, a by-product of many reactions, becomes part of a buffer system as it combines with water to form carbonic acid and bicarbonate. When the pH of human blood rises slightly, carbonic acid can neutralize the excess OH^- by releasing hydrogen ions, which combine with OH^- to form water:

$$OH^-\ +\ H_2CO_3\ \longrightarrow\ HCO_3^-\ +\ H_2O$$

CARBONIC   BICARBONATE   WATER
ACID   (salt)

When blood becomes more acidic, bicarbonate mops up excess H^+ and thus shifts the balance of the buffer system toward the acid:

$$HCO_3^-\ +\ H^+\ \longrightarrow\ H_2CO_3$$

BICARBONATE   CARBONIC
                  ACID

Buffer systems can neutralize only so many excess ions. With even a slight excess above that point, the pH swings widely. When the blood pH (7.3–7.5) falls even to 7, buffering fails, and the consequences can be severe. An individual may fall into a *coma*, an often irreversible state of unconsciousness. This happens in *respiratory acidosis*. Carbon dioxide accumulates, too much carbonic acid forms, and blood pH plummets. By contrast, when the blood pH increases even to 7.8, *tetany* may occur; skeletal muscles cannot be released from contraction. *Alkalosis* is a rise in blood pH that, if not reversed by medical treatment can be lethal.

Ions dissolved in fluids on the inside and outside of cells have key roles in cell function. Acidic substances release hydrogen ions, and basic substances accept them. Salts are compounds that release ions other than H^+ and OH^-.

Acid–base interactions help maintain pH, which is the H^+ concentration in a fluid. Buffer systems help maintain the body's acid–base balance at levels suitable for life.

Summary

Introduction   Chemistry can help us understand the composition and behavior of the substances that make up cells, organisms, and all components of the biosphere. Table 2.1 summarizes some key chemical terms that you will encounter throughout this book.

Section 2.1   All substances consist of one or more elements. Atoms are the smallest units that still retain the element's properties. An uncharged atom consists of one or more positively charged protons, an equal number of negatively charged electrons, and (except for hydrogen) one or more neutrons, which carry no charge. Protons and neutrons occupy an atom's core region, or nucleus, and essentially account for its mass.

Section 2.2   Atoms of an element typically differ in the number of neutrons; they are isotopes. Radioisotopes are unstable, and their nucleus spontaneously decays.

Biology⑧Now
Learn about how radioisotopes are used in a PET scan with the animation on BiologyNow.

Section 2.3   Whether one atom will interact with others depends on the number and arrangement of its electrons. Electrons occupy orbitals (volumes of space) around the atomic nucleus. The shell model for atomic structure is a diagram with successively larger circles, or shells, that keep track of all electrons in the orbitals at a given energy level.

When an atom has one or more vacancies in orbitals, it interacts with other atoms by donating, accepting, or sharing electrons (forming chemical bonds).

Biology⑧Now
Use the animation and interaction on BiologyNow to investigate electron distribution and the shell model.

Section 2.4   Each chemical bond is an interaction between the electron structures of atoms. The main types are called ionic, covalent, and hydrogen bonds.

When an atom loses or gains one or more electrons, it becomes an ion, with a positive or a negative charge. In an ionic bond, a positive ion and a negative ion stay together by mutual attraction of their opposite charges.

Atoms often fill vacancies in their outermost orbitals by sharing one or more pairs of electrons. Two atoms share electrons equally in a *nonpolar* covalent bond. The sharing is unequal in a *polar* covalent bond, so the bond has a slight negative charge at one end and a slight positive charge at the other. The charges balance, so the participating atoms carry no net charge, overall.

In a hydrogen bond, a covalently bound hydrogen atom weakly attracts an electronegative atom that is bound in a different molecule or a different region of the same molecule.

Biology⑧Now
Compare the types of chemical bonds in biological molecules using the animation on BiologyNow.

Section 2.5   Polar covalent bonds join three atoms in a water molecule (two hydrogen atoms and one oxygen). The polarity of the water molecule invites extensive hydrogen bonding between molecules in bodies of water. The polarity is the basis of hydrogen bonding, which gives liquid water a notable ability to resist temperature changes, to show internal cohesion, and to easily dissolve diverse polar or ionic substances. These properties of water help make life possible.

Biology⑧Now
Explore the structure and properties of water with the animation on BiologyNow.

Table 2.1	Summary of Important Players in the Chemical Basis of Life
Element	Fundamental substance consisting of one kind of atom
Atom	Smallest unit of an element that still retains element's properties. Occupies space, has mass, and cannot be broken apart by ordinary physical or chemical means.
Proton (p^+)	Positively charged particle of the atomic nucleus
Electron (e^-)	Negatively charged particle that can occupy a volume of space (orbital) around the nucleus
Neutron	Uncharged particle of the atomic nucleus
Isotope	One of two or more forms of an element's atoms that differ in the number of neutrons in the nucleus
Radioisotope	An unstable isotope that emits particles and energy; has an unstable combination of protons and neutrons
Tracer	Molecule that incorporates one or more atoms of a radioisotope. Used with tracking devices to identify the movement or destination of the molecule or atom in a metabolic pathway, the body, or some other system
Ion	An atom that has gained or lost an electron and carries a positive or negative charge. A proton without an electron zipping around it is a hydrogen ion (H^+)
Molecule	Unit of matter in which two or more atoms of the same element, or different ones, are bonded together
Compound	Molecule of two or more different elements in unvarying proportions (e.g., water)
Mixture	Intermingling of two or more elements or compounds in proportions that usually vary
Solute	Any molecule or ion dissolved in some solvent
Hydrophilic substance	Polar molecule or molecular region that can readily dissolve in water
Hydrophobic substance	Nonpolar molecule or molecular region that strongly resists dissolving in water
Acid	Substance that releases H^+ when dissolved in water
Base	Substance that accepts H^+ when dissolved in water
Salt	Compound that releases ions other than H^+ or OH^- when dissolved in water

Section 2.6 The pH scale is used to measure the hydrogen ion (H^+) concentration of a solution. A typical pH range is from 0 (highest H^+ concentration; most acidic) to 14 (lowest H^+ concentration; the most basic or alkaline). At pH 7, or neutrality, H^+ and OH^- concentrations are equal.

Salts are compounds that dissolve easily in water and release ions other than H^+ and OH^-. Acids release H^+ ions in water. Bases combine with them. A buffer system is a dynamic chemical partnership between a weak acid or base and its salt. The two go back and forth donating and accepting ions to counter slight shifts in pH and thus maintain a favorable pH. Most biological processes operate within a narrow pH range.

Biology⊜Now
Investigate the pH of common solutions with the interaction on BiologyNow.

Figure 2.14 Laboratory of a typical alchemist.

Self-Quiz

Answers in Appendix II

1. Is this statement false: Every type of atom consists of protons, neutrons, and electrons.

2. Electrons carry a _____ charge.
   a. positive   b. negative   c. zero

3. A(n) _____ is any molecule to which a radioisotope has been attached for research or diagnostic purposes.
   a. ion   b. isotope   c. element   d. tracer

4. Atoms share electrons unequally in a(n) _____ bond.
   a. ionic   c. polar covalent
   b. hydrogen   d. nonpolar covalent

5. In a hydrogen bond, a covalently bound hydrogen atom weakly attracts an _____ atom in a different molecule or a different region of the same molecule.
   a. electronegative   b. electropositive

6. Liquid water shows _____ .
   a. polarity   d. cohesion
   b. hydrogen-bonding capacity   e. b through d
   c. notable heat resistance   f. all of the above

7. Hydrogen ions (H^+) are _____ .
   a. the basis of pH values   d. dissolved in blood
   b. unbound protons   e. both a and b
   c. targets of certain buffers   f. a through d

8. When dissolved in water, a(n) _____ donates H^+, and a(n) _____ accepts H^+.

9. A(n) _____ is a dynamic chemical partnership between a weak acid and its salt.
   a. ionic bond   c. buffer system
   b. solute   d. solvent

10. Match the terms with their most suitable description.
    ____ trace element   a. atomic nucleus components
    ____ salt   b. two atoms sharing electrons
    ____ covalent   c. any polar molecule that readily
         bond   dissolves in water
    ____ hydrophilic   d. releases ions other than H^+ and
         substance OH^- when dissolved in water
    ____ protons,   e. makes up less than 0.001
         neutrons   percent of body weight

Additional questions are available on **Biology⊜Now™**

Critical Thinking

1. Some molecules consist of atoms of a single element, but others are compounds. Explain which type of molecule you would expect to be more abundant in living things.

2. *Ozone* is a chemically active form of oxygen gas. High in Earth's atmosphere, a vast layer of it absorbs about 98 percent of the sun's harmful rays. Normally, oxygen gas consists of two oxygen atoms joined in a double nonpolar covalent bond: O=O. Ozone has three covalent bonds in this arrangement: O=O—O. It is highly reactive with a variety of substances, and it gives up an oxygen atom and releases gaseous oxygen (O=O). Using what you know about chemistry, explain why you think it is so reactive.

3. Some undiluted acids are less corrosive than when diluted with a little water. In fact, lab workers are told to wipe off splashes with a towel before washing. Explain.

4. Medieval scientists and philosophers called alchemists were predecessors of modern-day chemists (Figure 2.14). Many tried to transform lead (atomic number 82) into gold (atomic number 79). Why didn't they succeed?

5. David, an inquisitive three-year-old, poked his fingers into warm water in a metal pan on the stove and didn't sense anything hot. Then he touched the pan itself and got a nasty burn. Explain why water in a metal pan heats up far more slowly than the pan itself.

6. Why can water striders (Figure 2.15) and the basilisk lizard shown in Figure 1.8 walk on water?

7. Why do you think H^+ is often written as H_3O^+?

Figure 2.15 Water strider, not sinking.

3 MOLECULES OF LIFE

Science or the Supernatural?

About 2,000 years ago in the mountains of Greece, the oracle of Delphi made rambling, cryptic prophecies after inhaling sweet-smelling fumes that had collected in the sunken floor of her temple. She actually was babbling in a hydrocarbon-induced trance. We now know her temple was perched on intersecting, earthquake-prone faults. When the faults slipped, methane, ethane, and ethylene seeped out from the depths. All three gases are colorless hallucinogens.

Ancient Greeks thought Apollo spoke to them through the oracle; they believed in the supernatural. Scientists looked for a natural explanation, and they found carbon compounds behind her words. Why is their explanation more compelling? It started with tested information about the structure and effects of natural substances, and it was based on analysis of gaseous substances at the site.

All three gases consist only of carbon and hydrogen atoms; hence the name, hydrocarbons. Thanks to scientific inquiry, we now know a lot about them. Consider methane. It was present when Earth first formed. It is released when volcanoes erupt, when we burn wood or peat or fossil fuels, and when termites and cattle pass gas. Methane collects in the atmosphere and in ocean depths along the continental shelves. Methane also is one of the greenhouse gases and a contributing factor in global warming.

And methane all by itself may be big trouble. Long ago, organic remains of marine organisms sank to the bottom of the ocean. Today, a few kilometers below the sediments that slowly accumulated on top of them, the remains have become food for methane-producing archaeans. Collectively, their metabolic activity produces tremendous quantities of methane. All of that gas bubbles upward and seeps from the seafloor (Figure 3.1b). At these methane seeps, the low temperature and high water pressure "freeze" methane into icy methane hydrate.

There may be a thousand billion tons of frozen methane hydrate on the seafloor. It is the world's largest reservoir of natural gas, but we do not have a safe, efficient way to retrieve it. Why not? The icy crystals are unstable. They instantaneously fall apart into methane gas and liquid water as soon as the temperature goes up or the pressure goes down. It does not take much, only a few degrees.

Methane hydrate can disintegrate explosively. It can cause an irreversible chain reaction that may vaporize neighboring deposits on the seafloor. We see plenty of evidence of small methane hydrate explosions in the past that pockmarked the ocean floor. Immense explosions have caused underwater landslides that stretched, almost unbelievably, from one continent to another.

methane

Figure 3.1 *Left*, ruins of the Temple of Apollo, where hydrocarbon gases seep out from the ground. *Right*, microorganisms and bubbles of methane gas almost 230 meters (750 feet) below sea level in the Black Sea. The methane is produced by archaeans far beneath the seafloor, then seeps into deep ocean water.

Watch the video online!

Also consider this: The greatest of all mass extinctions occurred 250 million years ago and marked the end of the Permian period. All but about 5 percent of the species in the seas and about 70 percent of the known plants, insects, and other species on the land abruptly vanished. Scientists, who are not given to hyperbole, call it The Great Dying.

Chemical clues locked in fossils dating from that time point to a sharp spike in the atmospheric concentration of carbon dioxide—not just any carbon dioxide, but molecules that had been assembled by living things. Methane hydrate disintegrated abruptly, and in a gargantuan burp, millions of tons of methane exploded from the seafloor. Methane-eating bacteria converted nearly all of it to carbon dioxide —which displaced most of the oxygen in the seas and sky.

Too much carbon dioxide, too little oxygen. Imagine being transported abruptly to the top of Mount Everest and trying to jog in the "thin air," with its lower oxygen concentration. You would pass out and die. Before The Great Dying, free oxygen made up about 35 percent of the atmosphere. After the burp, its concentration plummeted to 12 percent. We can expect that most animals on land and in the seas suffocated.

The methane problem is closer than you might think. Not long ago, researchers found vast methane hydrate deposits 96 kilometers (60 miles) or so off the coast of Newport, Oregon, and off the Atlantic seaboard. What is to become of us if there is another methane burp?

In short, knowledge about lifeless substances can tell you a lot about life, including your own. It will serve you well when you turn your mind to just about any topic concerning the past, present, and future—from ancient myths, to health or disease, to forests, to physical and chemical conditions that affect life everywhere.

 How Would You Vote?

Should we work toward developing the vast undersea methane deposits as an energy source, given that the environmental costs and risks to life are unknown? See BiologyNow for details, then vote online.

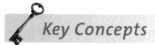 **Key Concepts**

NO CARBON, NO LIFE

We define cells partly by their capacity to assemble the organic compounds called complex carbohydrates and lipids, proteins, and nucleic acids. These large molecules of life have a backbone of carbon atoms, and functional groups attached to the backbone influence their properties. All are assembled from cellular pools of simple sugars, fatty acids, amino acids, and nucleotides. Sections 3.1, 3.2

CARBOHYDRATES

Carbohydrates are the most abundant biological molecules in nature. The simple sugars function as quick energy sources or transportable forms of energy. Complex carbohydrates are structural materials or energy reservoirs. Section 3.3

LIPIDS

Some kinds of complex lipids function as the body's energy reservoirs, others as structural components of cell membranes, as waterproofing or lubricating substances, and as signaling molecules. Section 3.4

PROTEINS

Structurally and functionally, proteins are the most diverse molecules of life. They include enzymes, structural materials, signaling molecules, and transporters. Sections 3.5, 3.6

NUCLEOTIDES AND NUCLEIC ACIDS

Both DNA and RNA are nucleic acids made of a few kinds of nucleotide subunits. They interact as the cell's system of storing, retrieving, and translating heritable information about building all the proteins necessary for life. Section 3.7

 Links to Earlier Concepts

You are about to enter the next level of organization in nature, as represented by the molecules of life. Keep the big picture in mind by quickly scanning Section 1.1 once again.

You will be building on your understanding of how electrons are arranged in atoms (2.3) as well as the nature of covalent bonding and hydrogen bonding (2.4). Here again, you will be considering one of the consequences of mutation in DNA (1.4), this time with sickle-cell anemia as the example.

3.1 Molecules of Life—From Structure to Function

LINK TO
SECTIONS
1.1, 2.4

Under present-day conditions in nature, only living cells make complex carbohydrates, lipids, proteins, and nucleic acids. Different classes of these biological molecules are a cell's instant energy sources, structural materials, metabolic workers, cell-to-cell signals, and libraries and translators of hereditary information.

WHAT IS AN ORGANIC COMPOUND?

The molecules of life are **organic compounds**, which are defined as containing the element carbon and at least one hydrogen atom. The term is a holdover from a time when chemists thought "organic" substances were the ones made naturally in living organisms only, as opposed to the "inorganic" substances that formed abiotically. The term persists, although scientists now synthesize organic compounds in laboratories and have reason to believe that organic compounds were present on Earth before organisms were.

The **hydrocarbons** consist only of hydrogen atoms covalently bonded to carbon. Examples are gasoline and other fossil fuels. Like other organic compounds, each has a specific number of atoms that are arranged in specific ways. Each organic compound has one or more **functional groups**, which are particular atoms or clusters of atoms covalently bonded to carbon.

In this book we use the following color code for the main atoms of organic compounds:

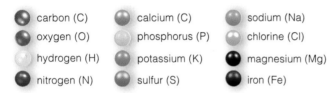

- carbon (C)
- oxygen (O)
- hydrogen (H)
- nitrogen (N)
- calcium (C)
- phosphorus (P)
- potassium (K)
- sulfur (S)
- sodium (Na)
- chlorine (Cl)
- magnesium (Mg)
- iron (Fe)

START WITH CARBON'S BONDING BEHAVIOR

Living things consist mainly of oxygen, hydrogen, and carbon (Figure 2.1). Their oxygen and hydrogen are primarily in the form of water. Put water aside, and carbon makes up more than half of what is left.

Carbon's importance to life starts with its versatile bonding behavior. *Each carbon atom can covalently bond with as many as four other atoms.* Such bonds, in which two atoms share one, two, or three pairs of electrons, are relatively stable. They join carbon atoms together as a backbone to which hydrogen, oxygen, and other elements are attached. In those configurations—in the arrangement of atoms and the distribution of electric charge—we find clues to how the different molecules of life will function and what their three-dimensional shapes will be.

WAYS TO REPRESENT ORGANIC COMPOUNDS

Methane is the simplest organic compound to think about. This colorless, odorless gas is present in the atmosphere, sea sediments, termite colonies, stagnant swamps, and stockyards. Its four hydrogen atoms are covalently bonded to one carbon atom (CH_4). You can use a ball-and-stick model to depict bond angles and show how the mass of this molecule or any other is distributed (in atomic nuclei). A space-filling model is better at conveying a molecule's size and surfaces:

structural formula for methane   ball-and-stick model   space-filling model

Let's use a ball-and-stick model to depict an organic compound with six covalently bonded carbon atoms from which hydrogen *and* oxygen atoms project:

ball-and-stick model for the linear structure of glucose

This type of carbon backbone sometimes forms chains inside cells. But most of the time it coils back on itself, and its two ends connect to form a ring structure:

six-carbon ring structure of glucose that usually forms inside cells

We typically depict carbon ring structures in simpler ways. A flat structural model may show the carbons but not other atoms bonded to them. If an icon for the ring shows no atoms at all, it is understood that one carbon atom occupies each "corner" of the ring:

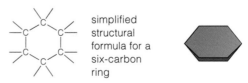

simplified structural formula for a six-carbon ring   icon for a six-carbon ring

Figure 3.2 shows ways to represent hemoglobin, a much larger molecule. You and all other vertebrates make this protein, which transports oxygen to tissues throughout your body. The ball-and-stick and space-filling models can give you an idea of this molecule's mass and structural complexity. But neither will tell you much about its oxygen-transporting function.

Now look at Figure 3.2c. This ribbon model shows how a hemoglobin molecule consists of four chains. As you will see later, each chain is a string of subunits called amino acids. Different regions of each chain are straight, folded, and coiled. For now, it is enough to know hemoglobin's three-dimensional shape includes four pockets, each containing a small cluster of atoms called a heme group. A heme group binds or releases oxygen in different body regions in response to how concentrated this gas is in different tissues.

More sophisticated models are now in use. Certain computer models, for instance, show local differences in electric charge across molecular surfaces. The areas color-coded, say, red on one molecule's surface might be attractive to a blue surface on another part of the same molecule or a different one (Figure 3.3).

Ultimately, such insights into the three-dimensional structure of molecules help us understand how cells and multicelled organisms function. For instance, virus particles can infect a cell when they dock at specific proteins located at the cell surface. Like Lego blocks, the proteins have ridges, clefts, and charged regions at their surface that can fit precisely into ridges, clefts, and charged regions of a protein at the surface of the virus. If a researcher can design a drug molecule that matches up with a viral protein and figure out how to deliver enough copies of it into a patient, then a lot of virus particles may be tricked into binding with the decoys instead of infecting body cells.

You will come across different kinds of molecular models throughout this book. In each case, the model selected gives you a glimpse into the structure and function of the molecule being described.

Carbohydrates, lipids, proteins, and nucleic acids are the main biological molecules—the organic compounds that only living cells assemble under present-day conditions in nature.

Organic compounds have diverse, three-dimensional shapes and functions that start with a carbon backbone and the bonding arrangements that arise from it.

Insights into the structure of molecules ultimately help us understand how cells, and multicelled organisms, function.

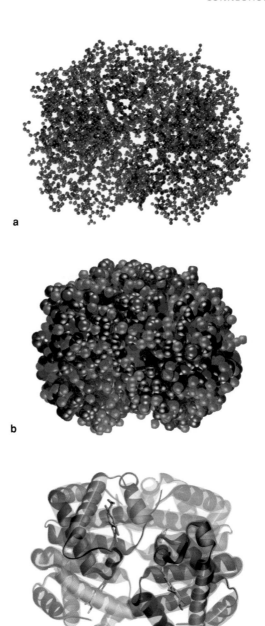

a

b

c

Figure 3.2 Visualizing the structure of hemoglobin, the oxygen-transporting molecule in red blood cells. (**a**) Ball-and-stick model, (**b**) space-filling model, and (**c**) ribbon model, with four heme groups (*red-orange*). Unlike the color coding for atoms, colors used for ribbon models and simple icons for complex molecules vary, depending on the context.

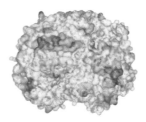

Figure 3.3 Model for the charged surface regions of a hemoglobin molecule. In this case, *blue* indicates positive charge and *red* indicates negative charge.

3.2 How Do Cells Build Organic Compounds?

LINK TO
SECTION 2.3

Before taking a run through the characteristics of the main biological molecules, get acquainted with their building blocks and how they are put together.

FOUR FAMILIES OF BUILDING BLOCKS

What is your favorite flower? Cells in the plant that made it turned carbon (from carbon dioxide), water, and the sun's energy into small organic compounds. The four main families of these small compounds are called simple sugars, fatty acids, amino acids, and nucleotides. Many kinds of molecules in each family contain two to thirty-six carbon atoms, at most.

Cells maintain and replenish pools of small organic compounds, which collectively account for about 10 percent of all organic material in a cell. They use up some molecules as ongoing sources of energy. They use others as individual subunits, or **monomers**, of the larger molecules necessary for their structure and functioning. The larger molecules—**polymers**—consist of three to millions of subunits that may or may not be identical. When they are broken apart, the released monomers might be used at once for energy, or they might reenter the cellular pools as free molecules.

A VARIETY OF FUNCTIONAL GROUPS

Functional groups, again, are lone atoms or clusters of atoms covalently bonded to carbon atoms of organic compounds. Each has specific chemical and physical properties that are consistent from one molecule to the next. How do such groups differ from hydrocarbon regions? They are more reactive. Important features of carbohydrates, lipids, proteins, and nucleic acids arise from the number, kind, and arrangement of functional groups, such as those shown in Figure 3.4.

Consider: Sugars in your diet belong to a class of organic compounds, the **alcohols**, which have one or more *hydroxyl* groups (—OH). Enzyme action can split

Hydroxyl	—OH		In alcohols (e.g., sugars, amino acids); water soluble
Methyl	H–C(–H)(–H)		In fatty acid chains; insoluble in water
Carbonyl	—CHO (aldehyde)	>CO (ketone)	In sugars, amino acids, nucleotides; water soluble. An *aldehyde* if at end of a carbon backbone; a *ketone* if attached to an interior carbon of backbone
Carboxyl	—COOH (non-ionized)	—COO⁻ (ionized)	In amino acids, fatty acids; water soluble. Highly polar; acts as an acid (releases H^+)
Amino	—NH₂ (non-ionized)	—NH₃⁺ (ionized)	In amino acids and certain nucleotide bases; water soluble; acts as a weak base (accepts H^+)
Phosphate	—O—P(O⁻)(=O)—O⁻	P icon	In nucleotides (e.g., ATP), also in DNA, RNA, many proteins, phospholipids; water soluble, acidic
Sulfhydryl	—SH	—S—S— (disulfide bridge)	In amino acid cysteine; helps stabilize protein structure (at disulfide bridges)

Figure 3.4 *Animated!* Common functional groups in biological molecules, with examples of their occurrences.

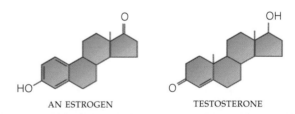

AN ESTROGEN   TESTOSTERONE

Figure 3.5 Observable differences in traits between the male and female wood duck (*Aix sponsa*). Two sex hormones govern the development of feather color and other traits that help males and females recognize each other and so promote reproductive success. Both hormones—testosterone and one of the estrogens—have the same carbon ring structure. They differ in the position of functional groups attached to the ring.

FEMALE WOOD DUCK   MALE WOOD DUCK

molecules or join them at such groups. Also, small alcohols dissolve swiftly because water molecules hydrogen-bond with them. Larger alcohols do not dissolve quickly because they have hydrocarbon chains, which are water insoluble. Such chains are also part of fatty acids, which is why lipids with fatty acid tails resist dissolving in water.

We find *carbonyl* groups—highly reactive and prone to electron transfers—in carbohydrates and fats, and *carboxyl* groups in amino acids and fatty acids. ATP activates other molecules by giving up *phosphate* groups. This group also combines with sugars to form the backbones of DNA and RNA. *Sulfhydryl* groups help stabilize many proteins.

How much can one functional group do? Look at a seemingly minor difference in the functional groups of two structurally similar sex hormones (Figure 3.5). Early on, an embryo of a wood duck, human, or any other vertebrate is neither male nor female. If it starts making testosterone (a hormone), a set of tubes and ducts will develop into male sex organs and later govern male traits. In the *absence* of testosterone, the ducts and tubes will develop into female sex organs. In that case, estrogens will guide the development of female traits.

FIVE CATEGORIES OF REACTIONS

So how do cells actually do the construction work? It will take more than one chapter to sketch out answers (and best guesses) to that question. For now, simply be aware that the reactions by which the cell builds, rearranges, and splits up organic compounds require more than energy inputs. They also require **enzymes**, a class of proteins that cause metabolic reactions to proceed much faster than they would on their own. Different enzymes mediate different reactions. In later chapters, you will come across specific examples of these five categories of reactions:

1. *Functional-group transfer*. One molecule gives up a functional group entirely, and a different molecule immediately accepts it.

2. *Electron transfer*. One or more electrons stripped from one molecule are donated to another molecule.

3. *Rearrangement*. Juggling of internal bonds converts one type of organic compound into another.

4. *Condensation*. Covalent bonds join two molecules into a larger molecule.

5. *Cleavage*. A molecule splits into two smaller ones.

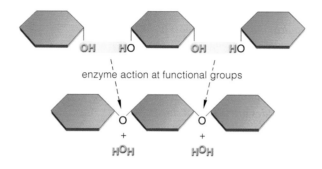

a Two condensation reactions. Enzymes remove an —OH group and an H atom from two molecules, which covalently bond as a larger molecule. Two water molecules form.

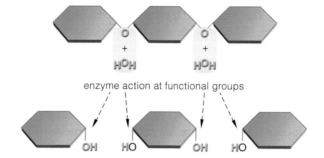

b Hydrolysis, a water-requiring cleavage reaction. Enzyme action splits a molecule into three parts, then attaches an —OH group and an H atom derived from a water molecule to each exposed site.

Figure 3.6 *Animated!* Examples of the metabolic reactions by which most biological molecules are synthesized, rearranged, or broken apart.

To get a sense of these cell activities, think about a **condensation reaction**. Enzymes split an —OH group from one molecule and an H atom from another, and a covalent bond forms at the exposed sites on both of the fragments. The discarded atoms often form water (Figure 3.6*a*). Starch and other large polymers form by way of repeated condensation reactions.

Another example: A type of cleavage reaction called **hydrolysis** is like condensation, but in reverse (Figure 3.6*b*). Enzymes split molecules at specific groups, then attach one —OH group and an H atom derived from a water molecule to the exposed sites. Cells can cleave polymers into smaller molecules when these are required for building blocks or for energy.

Cells build large molecules mainly from four families of small organic compounds called simple sugars, fatty acids, amino acids, and nucleotides.

Functional groups covalently bonded to carbon backbones add enormously to the structural and functional diversity of organic compounds, cells, and multicelled organisms.

Cells continually assemble, rearrange, and degrade organic compounds by enzyme-mediated reactions involving the transfer of functional groups or electrons, rearrangement of internal bonds, and a combining or splitting of molecules.

3.3 The Most Abundant Ones—Carbohydrates

Which biological molecules are most plentiful in nature? Carbohydrates. Most carbohydrates consist of carbon, hydrogen, and oxygen in a 1:2:1 ratio. Cells use them as structural materials and transportable or storable forms of energy. Monosaccharides, oligosaccharides, and polysaccharides are the main classes.

THE SIMPLE SUGARS

"Saccharide" is from a Greek word that means sugar. The *mono*saccharides (one sugar unit) are the simplest carbohydrates. They have at least two —OH groups bonded to their carbon backbone and one aldehyde or ketone group. Most dissolve easily in water. Common types have a backbone of five or six carbon atoms that tends to form a ring structure when dissolved.

Ribose and deoxyribose are the sugar monomers of RNA and DNA, respectively; each has five carbon atoms. Glucose has six (Figure 3.7a). Cells use glucose as an instant energy source, as a building block, and

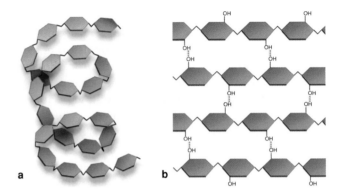

Figure 3.8 Bonding patterns for glucose units in (**a**) starch, and (**b**) cellulose. In amylose, a form of starch, a series of covalently bonded glucose units form a chain that coils. In cellulose, bonds form between glucose chains. The pattern stabilizes the chains, which can become tightly bundled.

as a precursor (parent molecule). For instance, glucose might be remodeled into vitamin C (a sugar acid) or into glycerol, an alcohol with three —OH groups.

SHORT-CHAIN CARBOHYDRATES

Unlike the simple sugars, an *oligo*saccharide is a short chain of covalently bonded sugar monomers. (*Oligo*– means a few.) The *di*saccharides consist of two sugar monomers. Lactose, a disaccharide in milk, consists of one glucose and one galactose unit. Sucrose, the most plentiful sugar in nature, has a glucose and a fructose unit (Figure 3.7c). Table sugar is sucrose extracted from sugarcane and sugar beets. Many proteins and lipids have oligosaccharide side chains. Later in the book, you will learn about oligosaccharide side chains that function in self-recognition, immunity, and other tasks. They are components of diverse molecules that are like docks and flags at the cell surface.

COMPLEX CARBOHYDRATES

The "complex" carbohydrates, or *poly*saccharides, are straight or branched chains of many sugar monomers —often hundreds or thousands. Different kinds have one or more types of monomers. The most common kinds are cellulose, starch, and glycogen. All three consist of glucose but differ in their properties (Figures 3.8 and 3.9). Why? The answer starts with differences in the covalent bonding patterns between their glucose units, which are joined together in chains.

In starch, the pattern of covalent bonding puts each glucose unit at an angle relative to the next unit in line. The chain ends up coiling like a spiral staircase

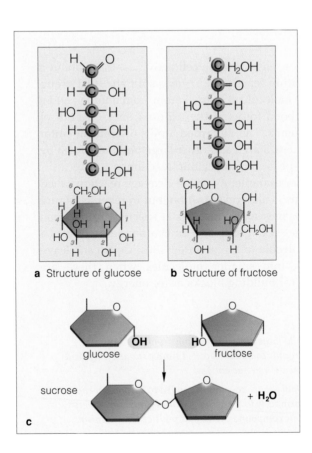

a Structure of glucose   **b** Structure of fructose

glucose   fructose

sucrose   + **H₂O**

c

Figure 3.7 (**a,b**) Straight-chain and ring forms of glucose and fructose. For reference purposes, carbon atoms of these simple sugars are numbered in sequence, starting at the end closest to the molecule's aldehyde or ketone group. (**c**) Condensation of two monosaccharides into a disaccharide.

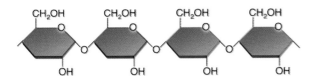

a Structure of amylose, a soluble form of starch. Cells inside tree leaves briefly store excess glucose monomers as starch grains in their chloroplasts, which are tiny, membrane-bound sacs that specialize in photosynthesis.

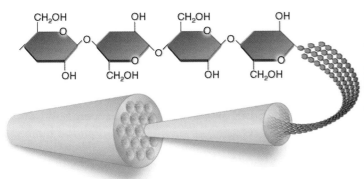

b Structure of cellulose. In cellulose fibers, chains of glucose units stretch side by side and hydrogen-bond at —OH groups. The many hydrogen bonds stabilize the chains in tight bundles that form long fibers. Few organisms produce enzymes that can digest this insoluble material. Cellulose is a structural component of plants and plant products, such as wood and cotton dresses.

c Glycogen. Animal cells build this polysaccharide as a storage form when the body has excess glucose. It is especially abundant in the liver and muscles of highly active animals, including fishes and people.

Figure 3.9 Molecular structure of (**a**) starch, (**b**) cellulose, and (**c**) glycogen, and their typical locations in a few organisms. All three carbohydrates consist only of glucose units.

(Figure 3.8*a*). Many —OH groups project out from the coils, which makes the chains accessible for cleavage reactions. This is important. For example, plants store much of their photosynthetically produced glucose in the form of starch. When free glucose is in short supply, enzymes can quickly hydrolyze the starch.

In cellulose, glucose chains stretch side by side and hydrogen-bond to one another, as in Figure 3.8*b*. The bonding arrangement stabilizes the chains in a tightly bundled pattern, which can resist hydrolysis by most enzymes. Long fibers of cellulose are a structural part of plant cell walls (Figure 3.9*b*). Like the steel rods in reinforced concrete, these fibers are tough, insoluble, and resistant to weight loads and mechanical stress, as when stems are buffeted by strong winds.

In animals, glycogen is the sugar-storage equivalent of starch in plants (Figure 3.9*c*). Muscle and liver cells store a lot of it. When the sugar level in blood falls, liver cells degrade glycogen, and the released glucose enters the blood. Exercise strenuously but briefly, and muscle cells tap glycogen for a burst of energy.

Chitin is a modified polysaccharide, with nitrogen-containing groups attached to its glucose monomers.

Figure 3.10 Molecular structure of chitin, a polysaccharide that occurs in protective body coverings of many animals, including this tick, as well as fungi. You may "hear" chitin when big spider legs clack across a metal oil pan on a garage floor.

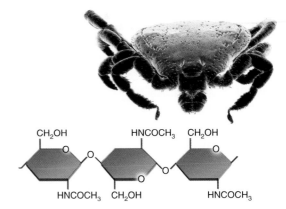

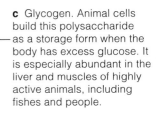

Chitin strengthens the external skeleton and other hard parts of many animals, including crabs, earthworms, insects, spiders, and ticks of the sort shown in Figure 3.10. It also strengthens the cell walls of fungi.

Carbohydrates include simple sugars (such as glucose), oligosaccharides (such as sucrose), and polysaccharides (such as starch). Cells use some carbohydrates as structural materials, others as packets of instant energy, and others as transportable or storable forms of energy.

3.4 Greasy, Oily—Must Be Lipids

Lipids are greasy or oily to the touch. Cells use different lipids as energy reservoirs, structural materials, and signaling molecules. Fats, phospholipids, and waxes have fatty acid tails. Sterols have a backbone of four carbon rings.

FATS AND FATTY ACIDS

Being nonpolar hydrocarbons, **lipids** do not dissolve in water, but they mix with other nonpolar substances—for instance, as butter does in warm cream sauce.

Fats are lipids with one, two, or three fatty acids dangling like tails from a glycerol molecule. A **fatty acid** starts as a carboxyl group attached to a backbone of as many as thirty-six carbon atoms. Each carbon in the backbone has one, two, or three hydrogen atoms covalently bonded to it (Figure 3.11). *Unsaturated* fatty acids contain one or more double covalent bonds. The *saturated* fatty acids have single bonds only.

Weak interactions keep many saturated fatty acids tightly packed in animal fats. These fats are solid at room temperature. Most plant fats stay liquid at room temperature, as "vegetable oils." Their packing is not as stable because of rigid kinks in their fatty acid tails. That is why vegetable oils flow freely.

Neutral fats such as butter, lard, and vegetable oils are mostly **triglycerides**. Each has three fatty acid tails linked to one glycerol (Figure 3.12). Triglycerides are the most abundant lipids in your body and its richest reservoir of energy. Gram for gram, they yield more than twice as much energy as complex carbohydrates such as starches. All vertebrates store triglycerides as droplets in fat cells that make up adipose tissue.

Layers and patches of adipose tissue insulate the body and cushion some of its parts. Like many other

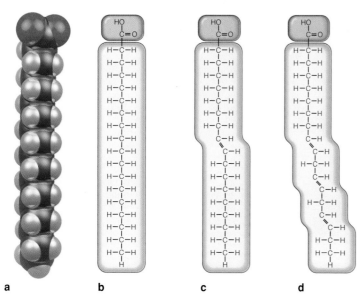

Figure 3.11 Three fatty acids. (**a,b**) Space-filling model and structural formula for stearic acid. The carbon backbone is fully saturated with hydrogen atoms. (**c**) Oleic acid, with a double bond in its backbone, is an unsaturated fatty acid. (**d**) Linolenic acid, also unsaturated, has three double bonds.

Figure 3.12 *Animated!* Condensation of (**a**) three fatty acids and one glycerol molecule into (**b**) a triglyceride. The photograph shows triglyceride-protected emperor penguins during an Antarctic blizzard.

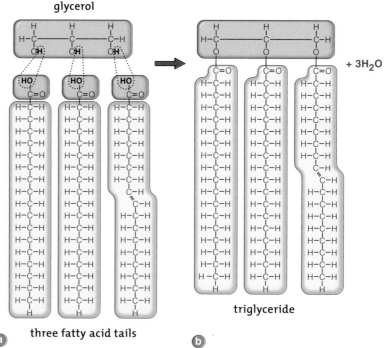

glycerol

$+ 3H_2O$

three fatty acid tails

triglyceride

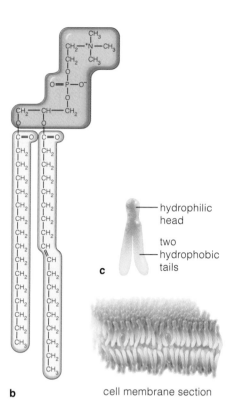

Figure 3.13 (a) Space-filling model, (b) structural formula, and (c) an icon for a phospholipid. This is the most common type in animal and plant cell membranes. Are its two tails saturated or unsaturated?

hydrophilic head

two hydrophobic tails

c

b

cell membrane section

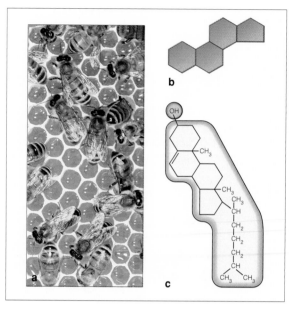

Figure 3.14 (a) Honeycomb—food warehouses and bee nurseries. Bees construct these compartments from their own water-repellent, waxy secretions. (b) Sterol backbone. (c) Structural formula for cholesterol, the main sterol of animal tissues. Your liver makes enough cholesterol for your body. A fat-rich diet may lead to clogged arteries.

animals, penguins of the Antarctic can keep warm in extremely cold winter months thanks to a thick layer of triglycerides beneath their skin (Figure 3.12).

PHOSPHOLIPIDS

Phospholipids have a glycerol backbone, two nonpolar fatty acid tails, and a polar head (Figure 3.13). They are the main component of cell membranes, which consist of two layers of lipids. Phospholipid heads of one layer are dissolved in the cell's fluid interior, and phospholipid heads of the other layer are dissolved in the fluid surroundings. Sandwiched between the two are all of the hydrophobic tails. You will read about membrane structure and function in Chapter 5.

WAXES

Waxes have long-chain fatty acids tightly packed and bonded to long-chain alcohols or carbon rings. All have a firm consistency; all repel water. Surfaces of plants have a cuticle that contains waxes and another lipid, cutin. A plant cuticle restricts water loss and thwarts some parasites. Waxes also protect, lubricate, and lend pliability to skin and to hair. Birds secrete waxes, fats, and fatty acids that waterproof feathers. Bees use beeswax for honeycomb, which houses each new bee generation as well as honey (Figure 3.14a).

CHOLESTEROL AND OTHER STEROLS

Sterols are among the many lipids with no fatty acids. The sterols differ in the number, position, and type of their functional groups, but all have a rigid backbone of four fused-together carbon rings (Figure 3.14b).

Every eukaryotic cell membrane contains sterols. Cholesterol (Figure 3.14c) is the most common type in animal tissues. It is remodeled into compounds as diverse as bile salts, steroids, and vitamin D, which is required for strong bones and teeth. Bile salts play a part in fat digestion inside the small intestine. The steroids called sex hormones are essential for gamete formation and the development of secondary sexual traits. Such traits include the amount and distribution of hair in mammals, and feather color in birds.

Being largely hydrocarbon, lipids can intermingle with other nonpolar substances, but they resist dissolving in water.

Triglycerides, or neutral fats, have a glycerol head and three fatty acid tails. They are the major energy reservoirs. Phospholipids are the main component of cell membranes.

Sterols such as cholesterol are membrane components and precursors of steroid hormones and other compounds. Waxes are firm yet pliable components of water-repelling and lubricating substances.

3.5 Proteins—Diversity in Structure and Function

Of all large biological molecules, proteins are the most diverse. Some kinds speed reactions; others are the stuff of spider webs or feathers, bones, hair, and other body parts. Nutritious types abound in seeds and eggs. Many proteins move substances, help cells communicate, or defend against pathogens. Amazingly, cells assemble thousands of different proteins from only twenty kinds of amino acids.

An **amino acid** is a small organic compound with an amino group ($-NH_3^+$), a carboxyl group ($-COO^-$, the acid), a hydrogen atom, and one or more atoms called its R group. In most cases, these components are attached to the same carbon atom (Figure 3.15a). Appendix V shows all of the biological amino acids.

When a cell constructs a protein, it strings amino acids together, one after the other. Instructions coded

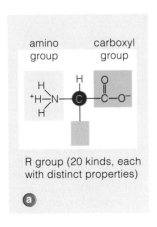

amino group   carboxyl group

R group (20 kinds, each with distinct properties)

a

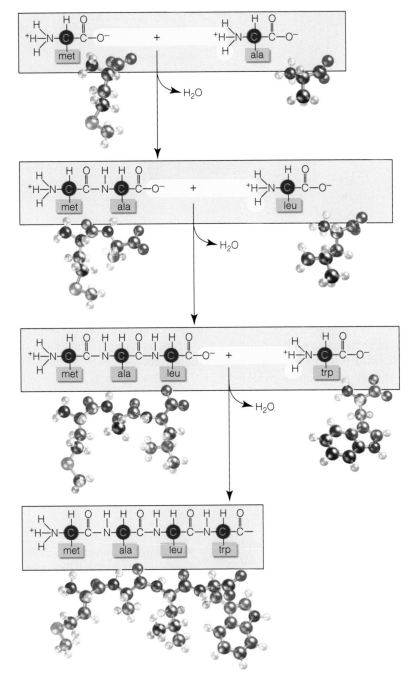

b Instructions encoded in DNA specify the order of amino acids to be joined in a polypeptide chain. The first amino acid is usually methionine (met). Alanine (ala) comes next in this example.

c In a condensation reaction, a peptide bond forms between the methionine and alanine. Leucine (leu) is next in line.

d A peptide bond forms between the alanine and the leucine. Tryptophan (trp) is next.

Figure 3.15 Animated!
(a) Generalized formula for amino acids. The *green* box highlights the R group, one of the side chains that include functional groups. Appendix V shows ball-and-stick models for twenty amino acids.

(b–e) Peptide bond formation during protein synthesis. Section 14.4 offers a closer look at protein synthesis.

e Part of the newly formed polypeptide chain. The sequence of amino acids in this part is met–ala–leu–trp. The reactions may continue until there are hundreds or thousands of amino acids in the chain.

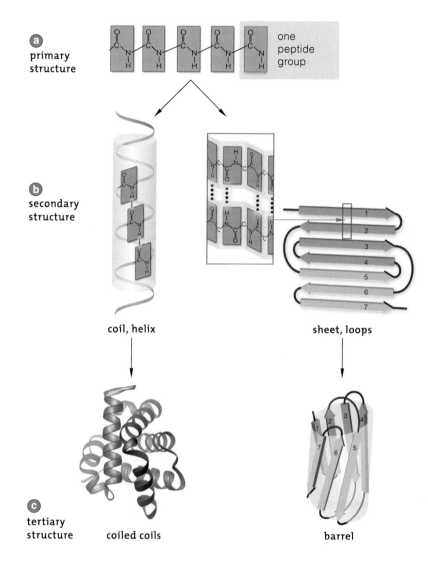

Figure 3.16 The first three of four levels of protein structure. (**a**) Primary structure is a linear sequence of amino acids. (**b**) Many hydrogen bonds (dotted lines) along a polypeptide chain result in a helically coiled or sheetlike secondary structure. (**c**) Coils and sheets packed into stable domains represent a third structural level.

a primary structure

one peptide group

b secondary structure

coil, helix

sheet, loops

c tertiary structure   coiled coils

barrel

in DNA specify the order in which any of the twenty kinds of amino acids will occur. A *peptide* bond forms as a condensation reaction joins the amino group of one amino acid and the carboxyl group of the next in line (Figure 3.15*b–e*). Each **polypeptide chain** consists of three or more amino acids. The carbon backbone of this chain incorporates nitrogen atoms in this regular pattern: —N—C—C—N—C—C—.

A protein's sequence of amino acids is known as its *primary* structure (Figure 3.16*a*). Its *secondary* structure emerges as the chain twists, bends, loops, and folds. Hydrogen bonding between certain R groups makes some stretches of amino acids coil helically, a bit like a spiral staircase, or makes them form sheets or loops as in Figure 3.16*b*. Bear in mind, part of the primary structure for each type of protein is unique in certain respects, but you will come across similar patterns of coils, sheets, and loops among different proteins.

Much as an overly twisted rubber band coils back on itself, the coils, sheets, and loops of a protein fold up even more, into compact domains. A "domain" is a polypeptide chain or a part of it that has become organized as a structurally stable unit. This third level of organization is a protein's *tertiary* structure. The shape of domains and the charge distribution around that shape determines protein function. For instance, the barrel-shaped domains of some proteins function as tunnels through membranes (Figure 3.16*c*).

Many proteins are two or more polypeptide chains bonded together or associating intimately with one another. This is the fourth level of organization, or *quaternary* protein structure. Many enzymes and other proteins are globular, with several polypeptide chains folded into rounded shapes. Hemoglobin, described shortly, is a classic example of such a protein.

Protein structure doesn't stop here. Enzymes often attach short, linear, or branched oligosaccharides to a new polypeptide chain, making a *glyco*protein. Many glycoproteins occur at the cell surface or are secreted from cells. Lipids also get attached to many proteins. The cholesterol, triglycerides, and phospholipids that your body absorbs after a meal are transported about as components of *lipo*proteins.

Many proteins are fibrous, with polypeptide chains organized as strands or sheets. They contribute to cell shape and organization, and help cells and cell parts move about. Other fibrous proteins make up cartilage, hair, skin, and parts of muscles and brain cells.

A protein's primary structure is a sequence of covalently bonded amino acids that make up a polypeptide chain.

Local regions of a polypeptide chain become twisted and folded into helical coils, sheetlike arrays, and loops. These arrangements are the protein's secondary structure.

A polypeptide chain or parts of it become organized as structurally stable, compact, functional domains. Such domains are a protein's tertiary structure.

Many proteins show quaternary structure; they consist of two or more polypeptide chains.

A protein's shape and charge distribution around that shape dictate protein function.

3.6 Why Is Protein Structure So Important?

LINK TO
SECTION
1.4

Cells are good at making proteins that are just what their DNA specifies. But mistakes and mutations happen, and they may alter the protein's primary structure in bad ways. The consequences are sometimes far-reaching.

JUST ONE WRONG AMINO ACID . . .

Four tightly packed polypeptides called globins make up each hemoglobin molecule. Each globin chain is folded into a pocket that cradles a **heme** group, a large organic molecule with an iron atom at its center (Figure 3.17). Heme is an oxygen transporter. During its life span, each of the red blood cells in your body transports billions of oxygen molecules, all bound to the heme in globin molecules.

Globin comes in two slightly different forms, alpha and beta. Two of each form make up one hemoglobin molecule in adult humans. Glutamate is normally the sixth amino acid in the beta globin chain, but a DNA mutation sometimes puts a different amino acid—valine—in the chain's sixth position (Figure 3.18b). Unlike glutamate, which carries an overall negative charge, valine has no net charge. As a result of that one substitution, a tiny patch of the protein changes from polar to nonpolar—which in turn causes globin's behavior to change slightly. Hemoglobin that has this mutation in its beta chain is designated HbS.

. . . AND SICKLE-CELL ANEMIA MAY FOLLOW!

Every human inherits two genes for beta globin, one from each of two parents. (Genes are units of DNA that encode heritable traits.) Cells access both genes when they make beta globin. If one gene is normal and the other has the valine mutation, a person makes enough normal hemoglobin and can lead a relatively normal life. Someone who inherits two mutant genes can only make hemoglobin HbS. The outcome, *sickle-cell anemia*, is a severe genetic disorder.

As blood moves through lungs, hemoglobin in red blood cells binds oxygen and then gives it up in body regions where oxygen levels are low. After the oxygen is released, red blood cells quickly return to the lungs and pick up more. In the few moments when they have no bound oxygen, hemoglobin molecules clump together just a bit. However, HbS molecules do not form such clusters in regions where oxygen levels are low. They form large, stable, rod-shaped aggregates.

Red blood cells containing these aggregates become distorted into sickle shapes (Figure 3.18c). These cells clog tiny blood vessels and disrupt blood circulation. Tissues become oxygen-starved. Figure 3.18d lists the far-reaching effects of sickle-cell anemia.

PROTEINS UNDONE—DENATURATION

Environmental conditions, too, can skew a protein's functioning. Globin cradles heme, an enzyme speeds some reaction, a receptor transduces an energy signal. These proteins and others cannot function unless they stay coiled, folded, and packed in a precise way. Their shape depends on many hydrogen bonds and other interactions that heat, shifts in pH, or detergents can disrupt. At such times, polypeptide chains unwind and change shape in an event called **denaturation**.

Consider albumin, a protein in the white of an egg. When you cook eggs, the heat does not disrupt the covalent bonds of albumin's primary structure. But it destroys albumin's weaker hydrogen bonds, and so the protein unfolds. When the translucent egg white turns opaque, we know albumin has been altered. For a few proteins, denaturation might be reversed if and

Figure 3.17 *Animated!*
(**a**) Globin, a coiled polypeptide chain. The chain cradles heme, a functional group that contains an iron atom. (**b**) Hemoglobin, an oxygen-transport protein in red blood cells. This is one of the proteins with quaternary structure. It consists of four globin molecules held together by hydrogen bonds. To help you distinguish among them, the two alpha globin chains are color-coded yellow and orange, and the two beta globins are color-coded blue and green.

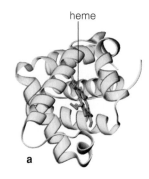

heme

a

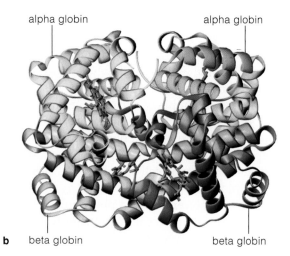

alpha globin   alpha globin

b   beta globin   beta globin

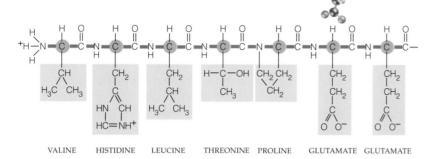

VALINE   HISTIDINE   LEUCINE   THREONINE   PROLINE   GLUTAMATE   GLUTAMATE

a Normal amino acid sequence at the start of a beta chain for hemoglobin.

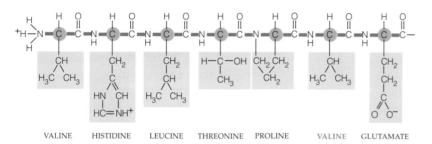

VALINE   HISTIDINE   LEUCINE   THREONINE   PROLINE   VALINE   GLUTAMATE

b One amino acid substitution results in the abnormal beta chain in HbS molecules. Valine was added instead of glutamate at the sixth position of the growing polypeptide chain.

c Glutamate has an overall negative charge; valine has no net charge. This difference gives rise to a water-repellent, sticky patch on HbS molecules. They stick together because of that patch, forming rod-shaped clumps that distort normally rounded red blood cells into sickle shapes. (A sickle is a farm tool that has a crescent-shaped blade.)

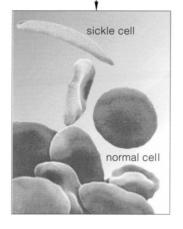

sickle cell

normal cell

Clumping of cells in bloodstream
↓
Circulatory problems, damage to brain, lungs, heart, skeletal muscles, gut, and kidneys
↓
Heart failure, paralysis, pneumonia, rheumatism, gut pain, kidney failure

Spleen concentrates sickle cells
↓
Spleen enlargement
↓
Immune system compromised

Rapid destruction of sickle cells
↓
Anemia, causing weakness, fatigue, impaired development, heart chamber dilation
↓
Impaired brain function, heart failure

d Melba Moore, celebrity spokesperson for sickle-cell anemia organizations. *Right,* range of symptoms for a person with two mutated genes (*Hb^S*) for hemoglobin's beta chain.

Figure 3.18 *Animated!* Sickle-cell anemia's molecular basis and its symptoms. Section 18.6 explores evolutionary and ecological aspects of this genetic disorder.

when normal conditions return, but albumin isn't one of them. There is no way to uncook an egg.

What is the take-home lesson? *A protein's structure dictates its function.* Hemoglobin, hormones, enzymes, transporters—such proteins help us survive. Twists and folds in their polypeptide chains form anchors, or membrane-spanning barrels, or jaws that grip enemy agents in the body. Mutations can alter the chains enough to block or enhance an anchoring, transport, or defensive function. Sometimes the consequences are awful. Yet changes in sequences and functional domains also give rise to variation in traits—the raw material for evolution. *Learn about protein structure and function and you are on your way to comprehending life in its richly normal and abnormal expressions.*

The structure of proteins dictates function. Mutations that alter a protein's structure sometimes have drastic consequences for its function, and for the health of organisms harboring them.

3.7 Nucleotides, DNA, and the RNAs

Certain small organic compounds called nucleotides are energy carriers, enzyme helpers, and messengers. Some are the building blocks for DNA and RNA. They are central to metabolism, survival, and reproduction.

Nucleotides have one sugar, at least one phosphate group, and one nitrogen-containing base. Deoxyribose or ribose is the sugar. Both sugars have a five-carbon ring structure; ribose has an oxygen atom attached to carbon 2 of the ring and deoxyribose does not. The bases have a single or double carbon ring structure.

The nucleotide **ATP** (adenosine triphosphate) has a row of three phosphate groups attached to its sugar (Figure 3.19). ATP can readily transfer the outermost phosphate group to many other molecules and make them reactive. Such transfers are vital for metabolism.

Other nucleotides have different metabolic roles. Some are **coenzymes**, necessary for enzyme function. They move electrons and hydrogen from one reaction site to another. NAD^+ and FAD are major kinds.

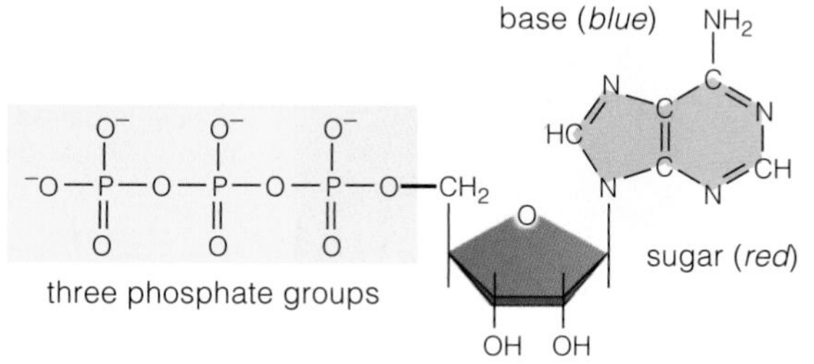

Still other nucleotides act as chemical messengers within and between cells. Later in the book, you will read about one of these messengers, which is known as cAMP (cyclic adenosine monophosphate).

Certain nucleotides also function as monomers for single- and double-stranded molecules called **nucleic acids**. In such strands, a covalent bond forms between the sugar of one nucleotide and the phosphate group of the next (Figure 3.20). The nucleic acids DNA and RNA store and retrieve heritable information.

All cells start life and then maintain themselves with instructions in their double-stranded molecules of deoxyribonucleic acid, or **DNA**. This nucleic acid is made of four kinds of deoxyribonucleotides. Figure 3.20*a* shows their structural formulas. As you can see, the four differ only in their component base, which is adenine, guanine, thymine, or cytosine.

Figure 3.21 shows how hydrogen bonds between bases join the two strands along the length of a DNA molecule. Think of every "base pairing" as one rung of a ladder, and the two sugar–phosphate backbones as the ladder's two posts. The ladder twists and turns in a regular pattern, forming a double helical coil.

The sequence of bases in DNA encodes heritable information about all the proteins that give each new cell the potential to grow, maintain itself, and even to reproduce. Part of that sequence is unique for each species. Some parts are identical, or nearly so, among many species. We return to DNA's structure and its functions in Chapter 13.

Figure 3.19
The structural formula for an ATP molecule.

base (*blue*) NH_2

three phosphate groups

sugar (*red*)

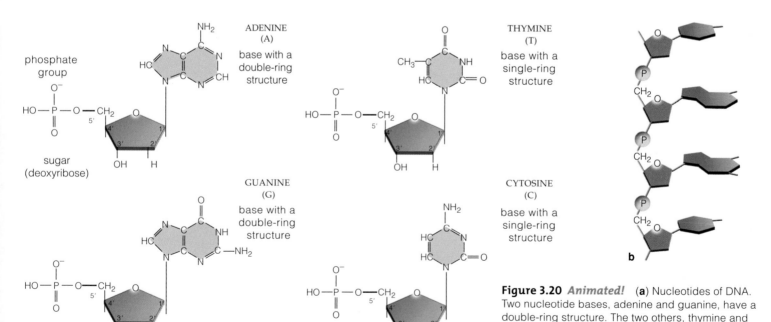

Figure 3.20 *Animated!* (**a**) Nucleotides of DNA. Two nucleotide bases, adenine and guanine, have a double-ring structure. The two others, thymine and cytosine, have a single-ring structure. (**b**) Bonding pattern between successive bases in nucleic acids.

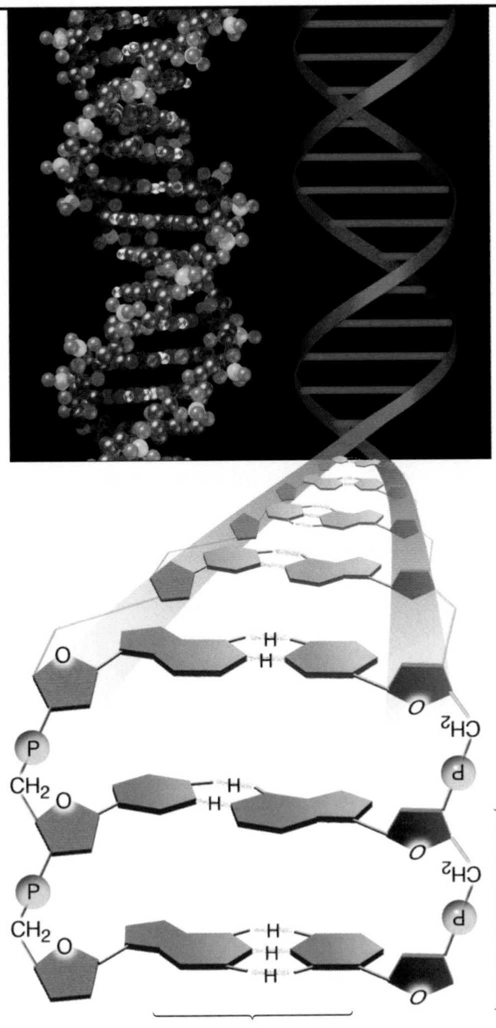

Figure 3.21 Models for the DNA molecule.

The **RNAs** (ribonucleic acids) have four kinds of ribonucleotide monomers. Unlike DNA, most RNAs are single strands, and one base is uracil instead of thymine. One type of RNA is a messenger that carries eukaryotic DNA's protein-building instructions out of the nucleus and into the cytoplasm, where they are translated into proteins by other RNAs. Chapter 14 returns to RNA and its role in protein synthesis.

Different nucleotides function as coenzymes, energy carriers such as ATP, chemical messengers, and building blocks for the nucleic acids DNA and the RNAs.

DNA consists of two nucleotide strands joined by hydrogen bonds and twisted as a double helix. Its nucleotide sequence encodes heritable protein-building information.

RNA usually is a single-stranded nucleic acid. Different RNAs have roles in processes by which a cell retrieves and uses genetic information in DNA to build proteins.

Summary

Section 3.1 Under present-day conditions in nature, only living cells can synthesize complex carbohydrates and lipids, proteins, and nucleic acids—the molecules of life. These molecules differ in their three-dimensional structure and function, starting with the carbon backbone and functional groups. Their structure affords clues to how cells, and multicelled organisms, function.

Biology⊗Now
Read the InfoTrac article "The Form Counts: Proteins, Fats, and Carbohydrates," Beatrice Trum, Consumer's Research Magazine, August 2001.

Section 3.2 All organic compounds have carbon and at least one hydrogen atom. Carbon atoms bond covalently with as many as four other atoms, often in long chains or rings. Functional groups attached to a carbon backbone influence an organic compound's properties. Enzyme-driven reactions synthesize all of the molecules of life from smaller organic molecules. Table 3.1 on the next page summarizes these compounds.

Biology⊗Now
Explore functional groups and view the animation of condensation and hydrolysis on BiologyNow.

Section 3.3 The main carbohydrates are simple sugars, oligosaccharides, and polysaccharides. Cells use carbohydrates as instant energy sources, transportable or storage forms of energy, and structural materials.

Section 3.4 Lipids are greasy or oily compounds that tend not to dissolve in water but mix easily with nonpolar compounds, such as other lipids. Neutral fats (triglycerides), phospholipids, waxes, and sterols are lipids. Cells use lipids as major sources of energy and as structural materials, as in cell membranes.

Biology⊗Now
Watch an animation showing how a triglyceride forms by condensation on BiologyNow.

Section 3.5 Structurally and functionally, proteins are the most diverse molecules of life. Their primary structure is a sequence of amino acids—a polypeptide chain. Such chains twist, coil, and bend into functional domains. Many proteins, including hemoglobin and most enzymes, consist of two or more chains. Certain aggregations of proteins form hair, muscle, connective tissue, and other body parts.

Biology⊗Now
Explore amino acid structure and learn about peptide bond formation with the animation on BiologyNow.
Read the InfoTrac article "Protein Folding and Misfolding," David Gossard, American Scientist, September 2002.

Section 3.6 A protein's overall structure determines its function. Sometimes a mutation in DNA results in an amino acid substitution that alters the protein's structure in ways that cause genetic diseases, including sickle-cell anemia. Weak bonds that hold a protein's

Table 3.1 Summary of the Main Organic Compounds in Living Things

Category	Main Subcategories	Some Examples and Their Functions	
CARBOHYDRATES . . . contain an aldehyde or a ketone group, and one or more hydroxyl groups	**Monosaccharides** (simple sugars) **Oligosaccharides** (short-chain carbohydrates) **Polysaccharides** (complex carbohydrates)	Glucose Sucrose (a disaccharide) Starch, glycogen Cellulose	Energy source Most common form of sugar; the form transported through plants Energy storage Structural roles
LIPIDS . . . are mainly hydrocarbon; generally do not dissolve in water but do dissolve in nonpolar substances, such as other lipids	**Lipids with fatty acids** *Glycerides:* Glycerol backbone with one, two, or three fatty acid tails *Phospholipids:* Glycerol backbone, phosphate group, one other polar group, and (often) two fatty acids *Waxes:* Alcohol with long-chain fatty acid tails **Lipids with no fatty acids** *Sterols:* Four carbon rings; the number, position, and type of functional groups differ among sterols	Fats (e.g., butter), oils (e.g., corn oil) Phosphatidylcholine Waxes in cutin Cholesterol	Energy storage Key component of cell membranes Conservation of water in plants Component of animal cell membranes; precursor of many steroids and vitamin D
PROTEINS . . . are one or more polypeptide chains, each with as many as several thousand covalently linked amino acids	**Fibrous proteins** Long strands or sheets of polypeptide chains; often tough, water-insoluble **Globular proteins** One or more polypeptide chains folded into globular shapes; many roles in cell activities	Keratin Collagen Enzymes Hemoglobin Insulin Antibodies	Structural component of hair, nails Structural component of bone Great increase in rates of reactions Oxygen transport Control of glucose metabolism Tissue defense
NUCLEIC ACIDS (AND NUCLEOTIDES) . . . are chains of units (or individual units) that each consist of a five-carbon sugar, phosphate, and a nitrogen-containing base	**Adenosine phosphates** **Nucleotide coenzymes** **Nucleic acids** Chains of nucleotides	ATP cAMP (Section 36.2) NAD$^+$, NADP$^+$, FAD DNA, RNAs	Energy carrier Messenger in hormone regulation Transfer of electrons, protons (H$^+$) from one reaction site to another Storage, transmission, translation of genetic information

shape can be disrupted by temperature, pH shifts, or exposure to detergent. Usually, the disruptions make the protein unfold permanently.

Biology ⊜ Now

Learn more about hemoglobin structure and sickle-cell mutation by viewing the animation on BiologyNow.

Section 3.7 There are different kinds of nucleotides, but all consist of a sugar, a phosphate group, and a nitrogen-containing base. They have essential roles in metabolism, survival, and reproduction. ATP energizes many kinds of molecules by phosphate-group transfers. Other nucleotides function as coenzymes or chemical messengers. DNA and RNA are nucleic acids, each composed of four kinds of nucleotide subunits.

DNA's nucleotide bases encode information on the primary structure of all of the cell's proteins. Different kinds of RNA molecules interact with DNA and with one another in the translation of that information.

Biology ⊜ Now

Explore DNA with the animation on BiologyNow.

Self-Quiz

Answers in Appendix II

1. Name the molecules of life and the families of small organic compounds from which they are built.

2. Each carbon atom can share pairs of electrons with as many as _____ other atoms.
   a. one   b. two   c. three   d. four

3. Sugars are a class of _____ , which have one or more _____ groups.
   a. proteins; amino   c. alcohols; hydroxyl
   b. acids; phosphate   d. carbohydrates; carboxyl

4. _____ is a simple sugar (a monosaccharide).
   a. Glucose   c. Ribose   e. both a and b
   b. Sucrose   d. Chitin   f. both a and c

5. The fatty acid tails of unsaturated fats incorporate one or more _____ .
   a. single covalent bonds   b. double covalent bonds

6. Sterols are among the many lipids with no _____ .
   a. saturation   c. hydrogens
   b. fatty acids   d. carbons

7. Which of the following is a class of molecules that encompasses all of the other molecules listed?
   a. triglycerides   c. waxes   e. lipids
   b. fatty acids   d. sterols   f. phospholipids

8. _____ are to proteins as _____ are to nucleic acids.
   a. Sugars; lipids   c. Amino acids; hydrogen bonds
   b. Sugars; proteins   d. Amino acids; nucleotides

9. A denatured protein has lost its _____ .
   a. hydrogen bonds   c. function
   b. shape   d. all of the above

10. Nucleotides occur in _____ .
    a. ATP   b. DNA   c. RNA   d. all are correct

11. Which of the following nucleotides is *not* found in DNA?
    a. adenine   b. uracil   c. thymine   d. guanine

12. Match the molecule with the most suitable description.
    ____ long sequence of amino acids   a. carbohydrate
    ____ energy carrier in cells   b. phospholipid
    ____ glycerol, fatty acids, phosphate c. polypeptide
    ____ two strands of nucleotides   d. DNA
    ____ one or more sugar monomers   e. ATP

Additional questions are available on Biology (ⓔ) Now™

Critical Thinking

1. In the following list, identify which is the carbohydrate, the fatty acid, the amino acid, and the polypeptide:

   a. $^+NH_3—CHR—COO^-$ c. (glycine)$_{20}$

   b. $C_6H_{12}O_6$ d. $CH_3(CH_2)_{16}COOH$

2. A clerk in a health-food store tells you that "natural" vitamin C extracts from rose hips are better than synthetic tablets of this vitamin. Given what you know about the structure of organic compounds, how would you respond?

3. It seems there are "good" and "bad" unsaturated fats. The double bonds of both put a bend in their fatty acid tails. The bend in *trans* fatty acid tails keeps them aligned in the same direction along their length. The bend in *cis* fatty acid tails makes them zigzag (Figure 3.22).

   Some *trans* fatty acids occur naturally in beef. But most form by industrial processes that solidify vegetable oils for margarine, shortening, and the like. These substances are widely used in prepared foods (such as cookies) and in french fries and other fast-food products. *Trans* fatty acids are linked to heart attacks. Speculate on why the body handles *cis* fatty acids better than *trans* fatty acids.

4. The shapes of a protein's domains often give us clues to functions. For example, Figure 3.23 is a model for one of the HLAs, a type of recognition protein perched above the surface of all vertebrate body cells. Certain cells of the immune system use HLAs to distinguish self (the body's own cells) from nonself. Each HLA has a jawlike region that can bind bits of an invader or some other threat. It thus alerts the immune defenders that the body has been invaded or otherwise threatened. Speculate on what may happen if a mutation makes the jawlike region misfold.

5. Cholesterol from food or synthesized in the liver is too hydrophobic to circulate in blood; complexes of protein and lipids ferry it around. Low density lipoprotein, or *LDL*, transports cholesterol out of the liver and into cells.

cis fatty acid

trans fatty acid

Figure 3.22 Maybe rethink the french fries?

High density lipoprotein, or *HDL*, ferries the cholesterol that is released from dead cells back to the liver.

High LDL levels are implicated in atherosclerosis, heart problems, and strokes. The main protein in LDL is called ApoA1. A mutant form of ApoA1 has the wrong amino acid (cysteine instead of arginine) at one place in its primary sequence. Carriers of this LDL mutation have very low levels of HDL, which is typically predictive of heart disease. Yet the carriers have no heart problems.

Some heart patients received injections of the mutant LDL, which acted like a drain cleaner. It quickly reduced the size of cholesterol deposits in the patients' arteries.

A few years from now, such a treatment may reverse years of damage. However, many researchers caution that a low-fat, low-cholesterol diet is still the best assurance of long-term health. Would you choose artery-cleansing treatments over a healthy diet?

where the molecule binds and displays "enemies" (*arrow*)

one of the chains spans the plasma membrane and anchors the molecule

Figure 3.23 From structure to function—a protein that helps your body defend itself against bacteria and other foreign agents. HLA-A2 has two polypeptide chains that are like jaws. Another protein anchors it to the plasma membrane.

4 CELL STRUCTURE AND FUNCTION

Animalcules and Cells Fill'd With Juices

Do you ever think of yourself as being close to 1/1,000 of a kilometer tall? Probably not. Yet that is how we refer to cells. We measure them in micrometers—in millionths of a millimeter, which is a thousandth of a meter, which is a thousandth of a kilometer. The bacterial cells in Figure 4.1 are a few micrometers "tall."

Before those cells were fixed on the head of a pin so someone could take their picture, they were the living descendants of a lineage far more ancient than yours. Their line of descent goes back so far they do not even have their DNA housed in a nucleus. They are prokaryotes, which are, at least structurally, the simplest cells of all. Your cells are eukaryotic. Somewhere in time, a nucleus developed in their single-celled ancestors, along with a variety of other internal compartments that help keep metabolic activities organized.

Nearly all cells are invisible to the naked eye. No one knew about them until the seventeenth century, when the first microscopes were being put together in Italy, then in France and England. These were not much to speak of. Galileo Galilei, for instance, simply used two glass lenses inside a cylinder, but the arrangement was good enough to reveal details of an insect's eyes. At midcentury, Robert Hooke focused a microscope on thinly sliced cork from a mature tree and saw tiny compartments (Figure 4.2). He gave them the Latin name *cellulae*, meaning small rooms—hence the origin of the biological term "cell." Actually they were dead plant cell walls, which is what cork is made of, but Hooke did not think of them as being dead because neither he nor anyone else knew cells could be alive. He observed cells "fill'd with juices" in green plant tissues but didn't have a clue to what they were, either.

Given the simplicity of their instruments, it is amazing that the pioneers in microscopy observed as much as they did. Antoni van Leeuwenhoek, a Dutch shopkeeper, had exceptional skill in constructing lenses and possibly the keenest vision. By the late 1600s, he was spying on such wonders as sperm, protists, a bacterium, and "many very small animalcules, the motions of which were very pleasing to behold," in scrapings of tartar from his teeth.

In the 1820s, improved lenses brought cells into sharper focus. Robert Brown, a botanist, was the first to identify a plant cell nucleus. Later, the botanist Matthias Schleiden wondered if a plant cell develops as an independent unit even though it is part of the plant. By 1839, after years of studying animal tissues, the zoologist Theodor Schwann reported that cells and their products make up animals as well as plants. He also reported that cells have an individual

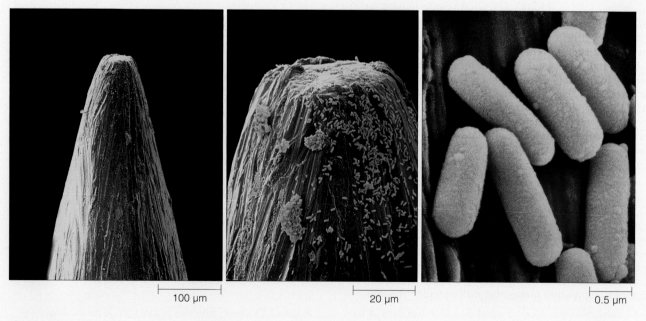

100 µm   20 µm   0.5 µm

Figure 4.1 How small are cells? This example will give you an idea. Shown here at increasingly higher magnifications, a population of rod-shaped bacterial cells peppering the tip of a household pin.

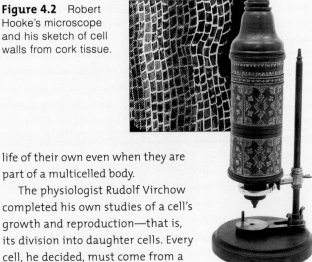

Figure 4.2 Robert Hooke's microscope and his sketch of cell walls from cork tissue.

life of their own even when they are part of a multicelled body.

The physiologist Rudolf Virchow completed his own studies of a cell's growth and reproduction—that is, its division into daughter cells. Every cell, he decided, must come from a cell that already exists.

So microscopic analysis yielded three generalizations, which together constitute the **cell theory**. First, organisms consist of one or more cells. Second, the cell is the smallest unit of organization that still displays the properties of life. Third, the continuity of life arises directly from the growth and division of single cells.

This chapter introduces defining features of prokaryotic and eukaryotic cells. It is not meant for memorization. Read it simply to gain an overview of current understandings of cell structure and function. In later chapters, you might refer back to it as a road map through the details. With its images from microscopy, this chapter and others invite you into otherwise invisible worlds. Why bother to travel there? We are close to creating the simplest form of life in test tubes. That is something worth thinking about.

 How Would You Vote?

Researchers are modifying prokaryotes to identify what it takes to be alive. They are creating "new" organisms by removing genes from living cells, one at a time. What are the potential advantages or bioethical pitfalls of this kind of research? See BiologyNow for details, then vote online.

 Key Concepts

WHAT ALL CELLS HAVE IN COMMON

A plasma membrane is a boundary between the interior of a cell and the surroundings, where the inputs and outputs of substances are controlled. Eukaryotic DNA is enclosed in a nucleus. Prokaryotic DNA is concentrated in a nucleoid. Cytoplasm is everything between the plasma membrane and the region of DNA. Section 4.1

MICROSCOPES

Microscopes employ rays of light or beams of electrons to reveal details at the cellular level of organization. They offer evidence of the cell theory: Each organism consists of one or more cells and their products, a cell has a capacity for independent life, and since the time of life's origin, all cells have arisen from cells that already exist. Section 4.2

PROKARYOTIC CELLS

Structurally, prokaryotic cells are the simplest and most ancient forms of life. Archaeans and bacteria are the only groups. Collectively, they show great metabolic diversity. Different kinds live in or on other organisms and in Earth's waters, soils, sediments, and rocky layers. Section 4.3

EUKARYOTIC CELLS

Organelles—small, membrane-bounded sacs—divide the interior of eukaryotic cells into functional compartments. All cells of protists, plants, fungi, and animals start out life with a nucleus; they are eukaryotic. They differ in the type and number of organelles, in cell structures, and in surface specializations. Sections 4.4–4.9

THE CYTOSKELETON

Arrays of a variety of protein filaments reinforce cell shape and keep its parts organized. Some filaments assemble and disassemble in dynamic ways that can move cells or their inner components to new locations. Sections 4.10–4.11

 Links to Earlier Concepts

Look back on your road map through the levels of organization in nature (Section 1.1). With this chapter you arrive at the level of living cells. You will start to see how lipids are structurally organized as cell membranes (3.4), where DNA and RNA reside in cells (3.7), and where carbohydrates are built and broken apart (3.2–3.3). You will expand your view of how cell structure and function depend on proteins (3.5–3.6).

4.1 So What Is "A Cell"?

LINK TO
SECTION
3.4

*Structurally, bacteria and archaeans are the simplest cells. They are **prokaryotic**, with no nucleus. Cells of all other organisms are **eukaryotic**. They contain a nucleus and other membrane-bound internal compartments.*

THE BASICS OF CELL STRUCTURE

The **cell** is the smallest unit with the properties of life: a capacity for metabolism, controlled responses to the environment, growth, and reproduction. Cells differ in size, shape, and activities, yet are all alike in three respects. They start out life with a plasma membrane, a region of DNA, and cytoplasm (Figure 4.3).

A **plasma membrane** defines the cell as a distinct entity. This thin, outer membrane separates metabolic activities from random events outside, but it does not isolate the cell interior. It is like a house with many doors that do not open for just anyone. Water, carbon dioxide, and oxygen enter and leave freely. Nutrients, ions, and other substances must be escorted.

In eukaryotic cells, the DNA occupies a **nucleus**, a membrane-bound, internal sac. In prokaryotic cells, it occupies a **nucleoid**, a region of the cytoplasm that is not enclosed in a membranous sac.

Cytoplasm is everything in between the plasma membrane and the region of DNA. It has a semifluid matrix and structural components that have roles in protein synthesis, energy conversions, and other vital tasks. For instance, cytoplasm holds many **ribosomes**, the molecular structures on which proteins are built.

PREVIEW OF CELL MEMBRANES

A **lipid bilayer** is a continuous, oily boundary that prevents the free passage of water-soluble substances across it (Figure 4.4). It is the structural basis of the plasma membrane and of various membranes inside

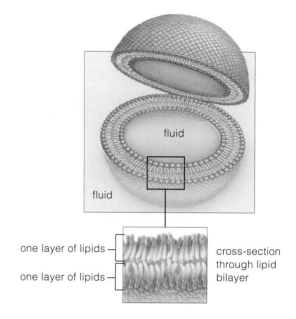

Figure 4.4 Simplified model for the lipid bilayer of all cell membranes. Remember the phospholipids (Section 3.4)? They are the most abundant lipids in cell membranes. Their hydrophobic tails are sandwiched between their hydrophilic heads, which are dissolved in cytoplasm on one side of the bilayer and in extracellular fluid on the other side.

eukaryotic cells. Membranes in the cytoplasm form channels or sacs that compartmentalize the tasks of transporting, synthesizing, modifying, stockpiling, or digesting substances.

Diverse proteins embedded in the lipid bilayer or positioned at one of its surfaces carry out most of the membrane functions (Figure 4.5). For instance, some proteins are channels and others are pumps across the bilayer. Others are receptors; they are like docks for hormones and other signaling molecules that trigger required changes in cell activities. You will read more about membrane proteins in chapters to follow.

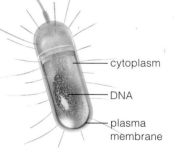

a Bacterial cell (prokaryotic)

Figure 4.3 Overview of the general organization of prokaryotic cells and eukaryotic cells. The three examples are not drawn to the same scale.

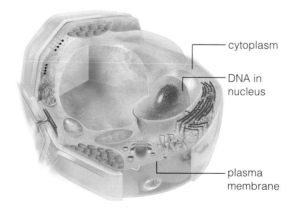

b Plant cell (eukaryotic)

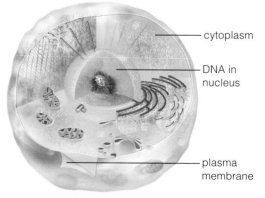

c Animal cell (eukaryotic)

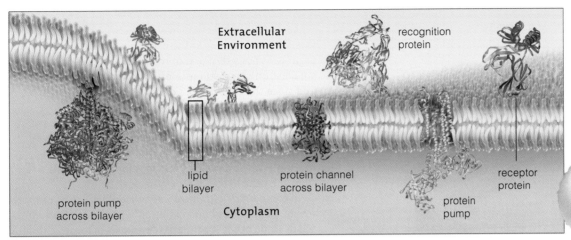

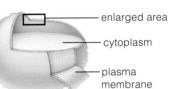

Figure 4.5 Model for a plasma membrane, cutaway view. The lipid bilayer has a variety of proteins spanning or attached to its surfaces. The next chapter focuses on the structure and function of cell membranes.

enlarged area

cytoplasm

plasma membrane

WHY AREN'T CELLS BIGGER?

Are any cells big enough to be seen without the help of a microscope? A few. They include "yolks" of bird eggs, cells in watermelon tissues, and amphibian and fish eggs. These cells can be large because they are not doing too much, metabolically speaking, at maturity. Most of their volume is a nutrient warehouse. If a cell has to perform many tasks, you can expect it to be too tiny to be seen by the unaided eye.

So why aren't all cells big? A physical relationship called the **surface-to-volume ratio** constrains increases in cell size. By this relationship, an object's volume increases with the cube of its diameter, but the surface area increases only with the square.

Apply this constraint to a round cell. As Figure 4.6 shows, *when a cell expands in diameter during growth, its volume increases faster than its surface area.* Suppose you could make a round cell increase four times in diameter. Its volume would increase 64 times (4^3), but its surface area would increase just 16 times (4^2). Each unit of its plasma membrane would now be required to service four times as much cytoplasm as before.

When the girth of any cell becomes too great, the inward flow of nutrients and outward flow of wastes will not be fast enough to keep up with the metabolic activity that keeps the cell alive. The outcome will be a dead cell.

Besides, a big, round cell also would have trouble moving materials through its cytoplasm. The random motion of molecules can distribute substances through tiny cells. When a cell is not tiny, you can expect it to be long or thin, or to have outfoldings or infoldings that increase its surface area relative to its volume. When a cell is smaller, narrower, or frilly surfaced, substances cross its surface and become distributed through the interior with greater efficiency.

Surface-to-volume constraints also shape the body plans of multicelled species. For example, small cells attach end to end in strandlike algae, so each interacts directly with its surroundings. Cells in your muscles are as long as the muscle itself, but each one is thin enough to efficiently exchange substances with fluids in the tissue surrounding them.

diameter (cm):	0.5	1.0	1.5
surface area (cm^2):	0.79	3.14	7.07
volume (cm^3):	0.06	0.52	1.77
surface-to-volume ratio:	13.17:1	6.04:1	3.99:1

Figure 4.6 *Animated!* One example of the surface-to-volume ratio. This physical relationship between increases in volume and surface area puts constraints on the sizes and shapes that are possible in cells.

All living cells have an outermost plasma membrane, an internal region called cytoplasm, and an internal region where DNA is concentrated.

Bacteria and archaeans are two groups of prokaryotic cells. Unlike eukaryotic cells, they do not have an abundance of organelles, particularly a nucleus.

Two layers of lipids are the structural framework for cell membranes. Proteins in the bilayer or positioned at one of its surfaces carry out diverse membrane functions.

A physical relationship called the surface-to-volume ratio constrains increases in cell size. The relationship also influences the shape of individual cells and the body plans of multicelled organisms.

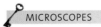

4.2 How Do We "See" Cells?

Like their centuries-old forerunners, modern microscopes are our best windows on the cellular world.

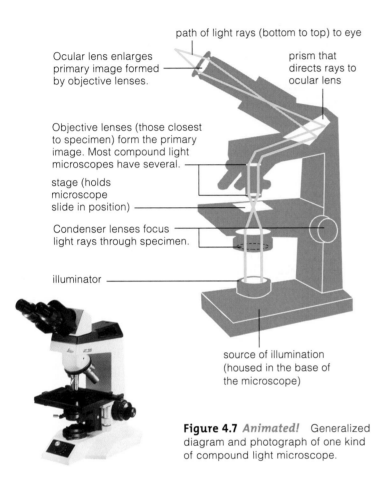

path of light rays (bottom to top) to eye

Ocular lens enlarges primary image formed by objective lenses.

prism that directs rays to ocular lens

Objective lenses (those closest to specimen) form the primary image. Most compound light microscopes have several.

stage (holds microscope slide in position)

Condenser lenses focus light rays through specimen.

illuminator

source of illumination (housed in the base of the microscope)

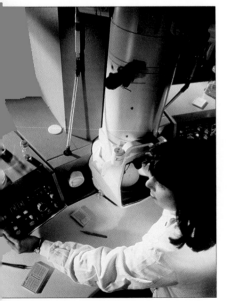

Figure 4.7 *Animated!* Generalized diagram and photograph of one kind of compound light microscope.

Research with modern microscopes still supports the three generalizations of the cell theory. In essence, all organisms consist of one or more cells, the cell is the smallest unit that still retains the characteristics of life, and since the time of life's origin, each new cell is descended from a cell that is already alive.

Like earlier instruments, many microscopes still use waves of light, and the light's wavelengths dictate their capacity to make images. Visualize a series of waves moving across an ocean. Each **wavelength** is the distance from the peak of one wave to the peak behind it.

In *compound light microscopes* (Figure 4.7), two or more sets of glass lenses bend waves of light passing through a cell or some other specimen. The shapes of lenses bend the waves at angles that disperse them in ways that form an enlarged image. *Micrographs* are simply photographs of images that emerge with the help of a microscope.

Cells become visible when they are thin enough for light to pass through them, but most cells are nearly colorless and look uniformly dense. Certain colored dyes can stain cells nonuniformly and make their component parts show up, but stains kill cells. Dead cells break down fast, which is why most cells are preserved before staining.

At present, the best light microscopes can enlarge cells about 2,000 times. Beyond that, cell structures appear larger but they are not clearer. Structures smaller than one-half of a wavelength of light are too small to resolve, so they cannot be distinguished.

Electron microscopes use magnetic lenses to bend and diffract beams of electrons, which cannot be diffracted through a glass lens. Electrons travel in wavelengths about 100,000 times shorter than those of visible light. Hence electron microscopes can resolve details that are 100,000 times smaller than you can see with a light microscope.

In *transmission* electron microscopes, electrons pass through a specimen and are used to make images of its internal details (Figure 4.8). *Scanning* electron microscopes direct a beam of electrons back and forth across a surface of a specimen, which has been given a thin metal coating. The metal responds by emitting electrons and x-rays, which can be converted into an image of the surface.

Figure 4.9 compares the resolving power of microscopes and of the human eye. Figure 4.10 compares the kinds of images that different microscopes offer.

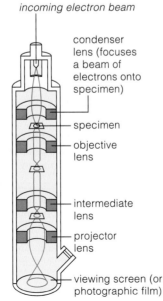

incoming electron beam

condenser lens (focuses a beam of electrons onto specimen)

specimen

objective lens

intermediate lens

projector lens

viewing screen (or photographic film)

Figure 4.8 *Animated!* Generalized diagram of an electron microscope. The photograph gives an idea of the lens diameters for a transmission electron microscope (TEM). When a beam of electrons from an electron gun moves down the microscope column, magnets focus them. With a transmission electron microscope, electrons pass through a thin slice of specimen and illuminate a fluorescent screen on a monitor. Shadows cast by the specimen's internal details appear, as in Figure 4.10c.

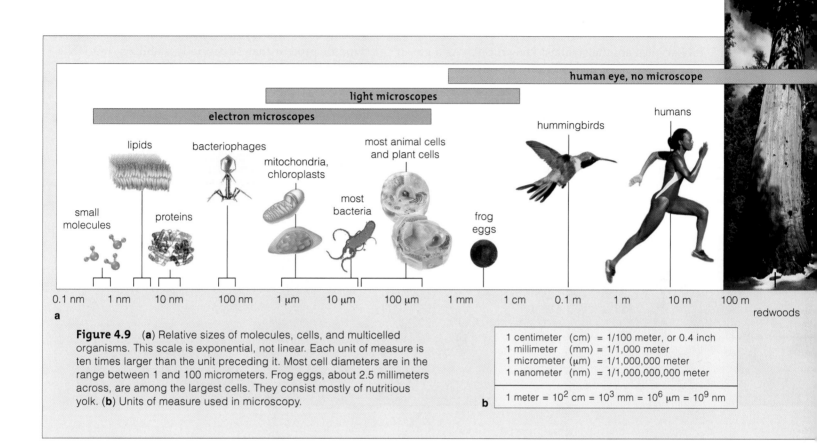

Figure 4.9 (**a**) Relative sizes of molecules, cells, and multicelled organisms. This scale is exponential, not linear. Each unit of measure is ten times larger than the unit preceding it. Most cell diameters are in the range between 1 and 100 micrometers. Frog eggs, about 2.5 millimeters across, are among the largest cells. They consist mostly of nutritious yolk. (**b**) Units of measure used in microscopy.

1 centimeter (cm) = 1/100 meter, or 0.4 inch
1 millimeter (mm) = 1/1,000 meter
1 micrometer (μm) = 1/1,000,000 meter
1 nanometer (nm) = 1/1,000,000,000 meter

1 meter = 10^2 cm = 10^3 mm = 10^6 μm = 10^9 nm

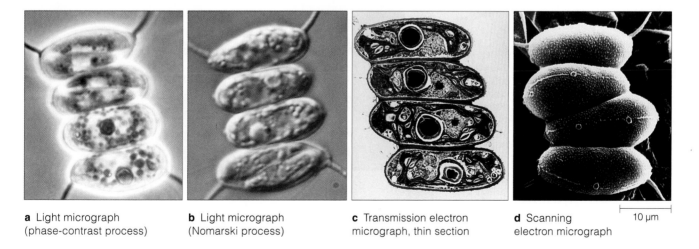

a Light micrograph (phase-contrast process)

b Light micrograph (Nomarski process)

c Transmission electron micrograph, thin section

d Scanning electron micrograph

10 μm

Figure 4.10 How different microscopes reveal different aspects of the same organism: a green alga (*Scenedesmus*). All four images are at the same magnification. (**a**,**b**) Light micrographs. (**c**) Transmission electron micrograph. (**d**) Scanning electron micrograph. A horizontal bar below a micrograph, as in (**d**), provides a visual reference for size. One micrometer (μm) is 1/1,000,000 of 1 meter. Using the scale bar, can you estimate the length and width of a *Scenedesmus* cell?

4.3 Introducing Prokaryotic Cells

LINKS TO
SECTIONS
1.1, 1.3, 3.5

The word prokaryote is taken to mean "before the nucleus." The name reminds us that bacteria and then archaeans originated before cells with a nucleus evolved.

Prokaryotes are the smallest known cells. As a group they are the most metabolically diverse forms of life on Earth. Different kinds exploit energy sources and raw materials in nearly all environments, including dry deserts, deep ocean sediments, and mountain ice.

We recognize two domains of prokaryotic cells—**Bacteria** and **Archaea** (Sections 1.3 and 19.5). Cells of both groups are alike in outward appearance and in size. However, the two groups differ in major ways.

Bacteria start making each new polypeptide chain with formylmethionine, a modified amino acid. Archaeans start a chain with methionine—as eukaryotic cells do (Section 3.5). Eukaryotic cells make many histones, a type of protein that structurally stabilizes the DNA. Archaeans make a few histones. Bacteria make a few histone-like proteins that stabilize the nucleoid.

Most prokaryotic cells are not much wider than one micrometer. The rod-shaped species are no more than a few micrometers long (Figures 4.11 and 4.12). Structurally, these are the simplest cells. A semirigid or rigid wall outside the plasma membrane imparts shape to most species. As Section 4.10 explains, arrays

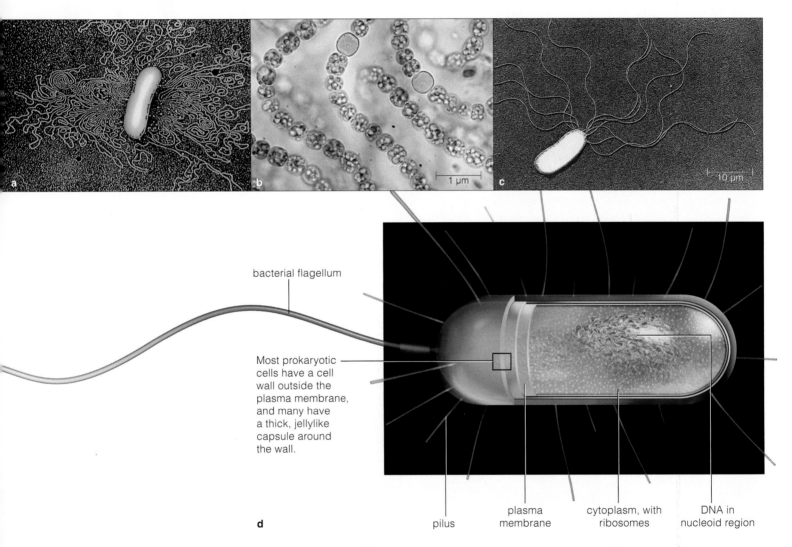

bacterial flagellum

Most prokaryotic cells have a cell wall outside the plasma membrane, and many have a thick, jellylike capsule around the wall.

pilus

plasma membrane

cytoplasm, with ribosomes

DNA in nucleoid region

Figure 4.11 *Animated!* (**a**) Micrograph of *Escherichia coli*. Researchers manipulated this bacterial cell to release its single, circular molecule of DNA. (**b**) Different bacterial species are shaped like balls, rods, or corkscrews. Ball-shaped cells of a photosynthetic bacterium (*Nostoc*) stick together in a thick, jellylike sheath of their own secretions. Chapter 21 offers more examples. (**c**) Like this *Pseudomonas marginalis* cell, many species have one or more bacterial flagella that propel the cell body through fluid environments. (**d**) Generalized sketch of a typical prokaryotic cell.

Figure 4.12 From Bitter Springs, Australia, fossilized bacteria dating to about 850 million years ago, in precambrian times. (**a**) A colonial form, most likely *Myxococcoides minor.* (**b**) Cells of a filamentous species (*Palaeolyngbya*).

(**c**) One of the structural adaptations seen among archaeans. Many of these prokaryotic species live in extremely hostile habitats, such as the ones thought to have prevailed when life originated. Most archaeans and some bacteria have a dense lattice of proteins that are anchored to the outer surface of their plasma membrane. In some species, the unique composition of the lattice may help the cell withstand extreme conditions in the environment. For instance, we find such lattices on archaeans living in near-boiling, mineral-rich water spewing from hydrothermal vents on the ocean floor.

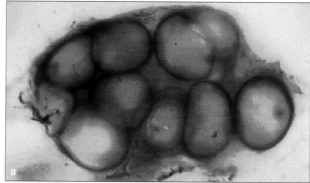

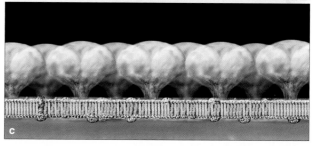

of protein filaments just under the plasma membrane reinforce the cell's shape.

Sticky polysaccharides often envelop bacterial cell walls. They let cells attach to such interesting surfaces as river rocks, teeth, and the vagina. Many disease-causing (pathogenic) bacteria have a thick protective capsule of jellylike polysaccharides around their wall.

Cell walls are permeable to dissolved substances, which cross them on the way to and from the plasma membrane. Some eukaryotic cells also have a wall, but it differs structurally from prokaryotic cell wall.

Many species have one or more bacterial flagella (singular, flagellum). These motile structures do not have a core of microtubules, as eukaryotic cells do (Section 4.11). They start at a tiny rotary motor in the plasma membrane and extend past the cell wall. They help move cells through fluid habitats, such as body fluids of host animals. Many bacteria also have pili (singular, pilus). These protein filaments help the cell cling to surfaces. A "sex" pilus latches on to another cell, then shortens. The attached cell is reeled in, and genetic material is transferred into it (Section 21.3).

As in eukaryotic cells, the plasma membrane of bacteria and archaeans selectively controls the flow of substances into and out of the cytoplasm. Its lipid bilayer bristles with protein channels, transporters, and receptors, and it incorporates built-in machinery for reactions. For example, the plasma membrane in photosynthetic bacterial species has organized arrays of proteins that capture light energy and convert it to bond energy of ATP, which helps build sugars.

The cytoplasm contains many ribosomes on which polypeptide chains are built. DNA is concentrated in an irregularly shaped region of cytoplasm called the nucleoid. Prokaryotic cells inherit one molecule of DNA that is circular, not linear. We call it a bacterial chromosome. The cytoplasm of some species also has plasmids: far smaller circles of DNA that carry just a

few genes. Typically, plasmid genes confer selective advantages, such as antibiotic resistance.

One more intriguing point: In cyanobacteria, part of the plasma membrane projects into the cytoplasm and is repeatedly folded back on itself. As it happens, pigments and other molecules of photosynthesis are embedded in the membrane—as they are in the inner membrane of chloroplasts. Is this a sign that ancient cyanobacteria were the forerunners of chloroplasts? Section 20.4 looks at this possibility. It is one aspect of a remarkable story about how prokaryotes gave rise to all protists, plants, fungi, and animals.

Bacteria and archaeans are two major groups of prokaryotic cells. These cells do not have a nucleus. Most have a cell wall around their plasma membrane. The wall is permeable, and it reinforces and imparts shape to the cell body.

Although structurally simple, prokaryotic cells as a group show the most metabolic diversity. Metabolic activities that are similar to ones occurring in eukaryotic organelles occur at the bacterial plasma membrane or in the cytoplasm.

4.4 Introducing Eukaryotic Cells

All cells synthesize, store, degrade, and transport diverse substances, but eukaryotic cells compartmentalize these operations. Their interior is subdivided into a nucleus and other organelles that have specialized functions.

Like prokaryotes, eukaryotic cells have ribosomes in their cytoplasm. Unlike prokaryotes, they have a well-developed, dynamic "skeleton" of proteins. They all start out life with a nucleus and other membrane-bounded sacs called **organelles**. *Eu–* means true; and *karyon*, meaning kernel, is taken to mean a nucleus. Figures 4.13 and 4.14 show two eukaryotic cells.

What advantages do organelles offer? Their outer membrane encloses and sustains a microenvironment for cell activities. Membrane components control the types and amounts of substances entering or leaving. They concentrate some substances for reactions and isolate or dispose of incompatible or toxic types.

For instance, organelles called mitochondria and chloroplasts concentrate hydrogen ions in a sac, then let them flow out in a way that forms ATP. Enzymes in lysosomes digest large organic compounds. They would even digest the cell if they were to escape.

Also, just as organ systems interact in controlled ways that can keep a whole body running smoothly, specialized organelles interact in ways that help keep a whole cell functioning as it should.

As another example, substances are modified and transported in a specific direction through an entire series of organelles. One series, the *secretory* pathway, moves new polypeptide chains from some ribosomes through ER and Golgi bodies, then on to the plasma membrane for release from the cell. Another series, an *endocytic* pathway, moves ions and molecules into the cytoplasm. In both cases, tiny sacs called **vesicles** act like taxis and move substances from one organelle to the next in line. The sacs form by pinching off from organelle membranes or the plasma membrane.

Figure 4.15 is a visual overview of components that are typical of plant and animal cells.

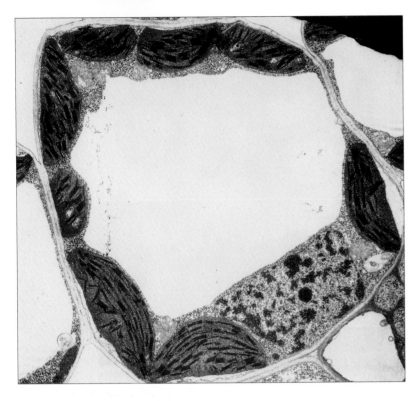

Figure 4.13 Transmission electron micrograph of a plant cell, cross-section. This is a photosynthetic cell from a blade of timothy grass.

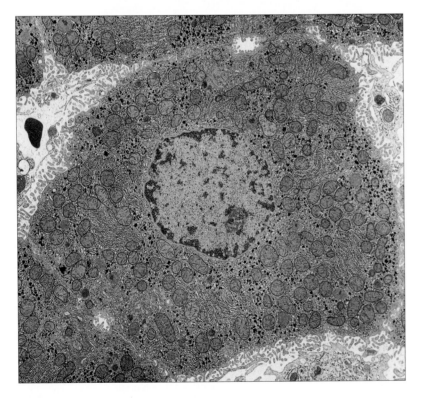

Figure 4.14 Transmission electron micrograph of an animal cell, cross-section. This is one cell from a rat liver.

All eukaryotic cells start out life with a nucleus, ribosomes, and a cytoskeleton. Specialized cells typically incorporate many more kinds of organelles as well as cell structures.

Organelles physically separate chemical reactions, many of which are incompatible.

Organelles organize metabolic events, as when different kinds interact in assembling, storing, or moving substances along pathways to and from the plasma membrane or to specific destinations in the cytoplasm.

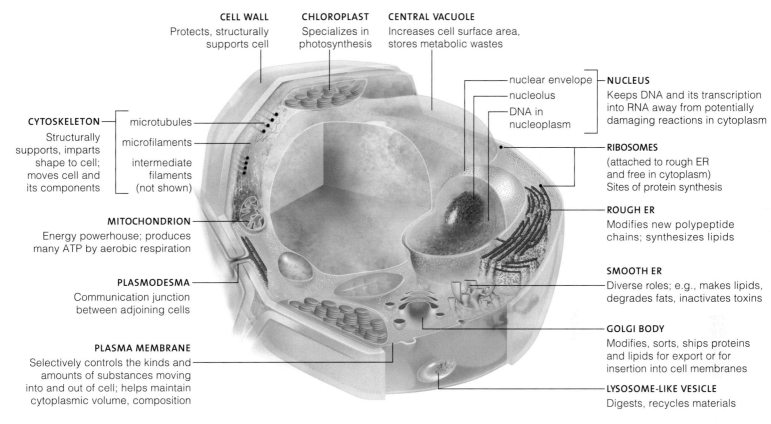

CELL WALL
Protects, structurally supports cell

CHLOROPLAST
Specializes in photosynthesis

CENTRAL VACUOLE
Increases cell surface area, stores metabolic wastes

CYTOSKELETON
Structurally supports, imparts shape to cell; moves cell and its components

- microtubules
- microfilaments
- intermediate filaments (not shown)

MITOCHONDRION
Energy powerhouse; produces many ATP by aerobic respiration

PLASMODESMA
Communication junction between adjoining cells

PLASMA MEMBRANE
Selectively controls the kinds and amounts of substances moving into and out of cell; helps maintain cytoplasmic volume, composition

- nuclear envelope
- nucleolus
- DNA in nucleoplasm

NUCLEUS
Keeps DNA and its transcription into RNA away from potentially damaging reactions in cytoplasm

RIBOSOMES
(attached to rough ER and free in cytoplasm) Sites of protein synthesis

ROUGH ER
Modifies new polypeptide chains; synthesizes lipids

SMOOTH ER
Diverse roles; e.g., makes lipids, degrades fats, inactivates toxins

GOLGI BODY
Modifies, sorts, ships proteins and lipids for export or for insertion into cell membranes

LYSOSOME-LIKE VESICLE
Digests, recycles materials

a Typical plant cell components.

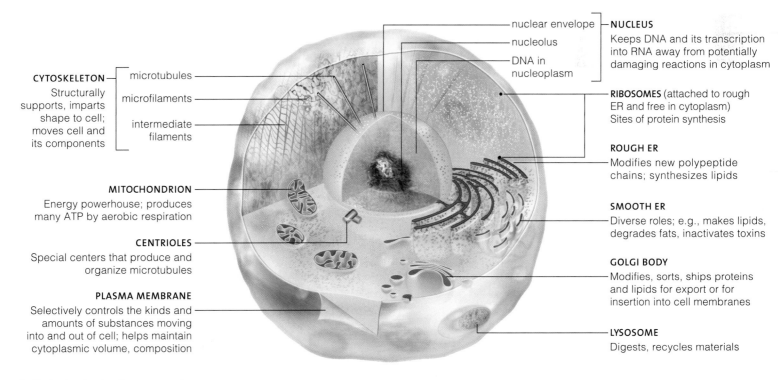

CYTOSKELETON
Structurally supports, imparts shape to cell; moves cell and its components

- microtubules
- microfilaments
- intermediate filaments

MITOCHONDRION
Energy powerhouse; produces many ATP by aerobic respiration

CENTRIOLES
Special centers that produce and organize microtubules

PLASMA MEMBRANE
Selectively controls the kinds and amounts of substances moving into and out of cell; helps maintain cytoplasmic volume, composition

- nuclear envelope
- nucleolus
- DNA in nucleoplasm

NUCLEUS
Keeps DNA and its transcription into RNA away from potentially damaging reactions in cytoplasm

RIBOSOMES (attached to rough ER and free in cytoplasm) Sites of protein synthesis

ROUGH ER
Modifies new polypeptide chains; synthesizes lipids

SMOOTH ER
Diverse roles; e.g., makes lipids, degrades fats, inactivates toxins

GOLGI BODY
Modifies, sorts, ships proteins and lipids for export or for insertion into cell membranes

LYSOSOME
Digests, recycles materials

b Typical animal cell components.

Figure 4.15 *Animated!* Organelles and structures typical of (**a**) plant cells and (**b**) animal cells.

4.5 The Nucleus

Constructing, operating, and reproducing cells cannot be done without carbohydrates, lipids, proteins, and nucleic acids. It takes a class of proteins—enzymes— to build and use these molecules. Instructions for building those proteins are encoded in DNA.

Eukaryotic cells have their genetic material distributed among some number of DNA molecules of different lengths. For instance, the nucleus of a human body cell normally holds forty-six DNA molecules. If you could stretch them out end to end, the line would be about 2 meters (6–1/2 feet) long. That is a lot of DNA —a lot more than the one circular molecule found in prokaryotic cells, which are smaller and less complex.

The nucleus has two main functions. First, it keeps the DNA from getting tangled with the cytoplasmic machinery and isolates it from potentially damaging reactions. Second, the outer membranes of a nucleus are a boundary where cells control the movement of substances to and from the cytoplasm. This structural and functional separation makes it far easier to keep DNA molecules organized and to copy them before a cell divides.

Figure 4.16 shows the components of the nucleus. Table 4.1 lists their functions.

NUCLEAR ENVELOPE

The **nuclear envelope** is a double-membrane system in which two lipid bilayers are pressed against each other (Figure 4.17). Its outer membrane merges with the membrane of ER, an organelle in the cytoplasm. Like rough ER membranes, this outer membrane has a profusion of ribosomes bound to it.

Table 4.1 Components of the Nucleus	
Nuclear envelope	Pore-riddled double-membrane system that selectively controls which substances enter and leave the nucleus
Nucleoplasm	Semifluid interior portion of the nucleus
Nucleolus	Rounded mass of proteins and copies of genes for ribosomal RNA used to construct ribosomal subunits
Chromosome	One DNA molecule and the many proteins that are intimately associated with it
Chromatin	Total collection of all DNA molecules and their associated proteins in the nucleus

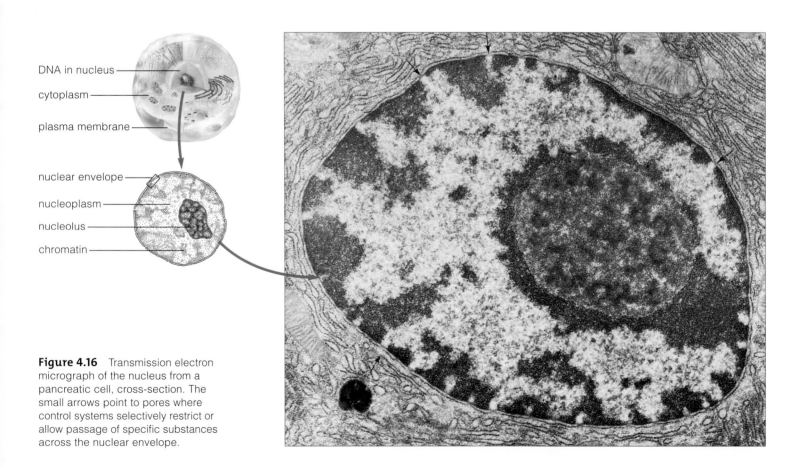

DNA in nucleus
cytoplasm
plasma membrane

nuclear envelope
nucleoplasm
nucleolus
chromatin

Figure 4.16 Transmission electron micrograph of the nucleus from a pancreatic cell, cross-section. The small arrows point to pores where control systems selectively restrict or allow passage of specific substances across the nuclear envelope.

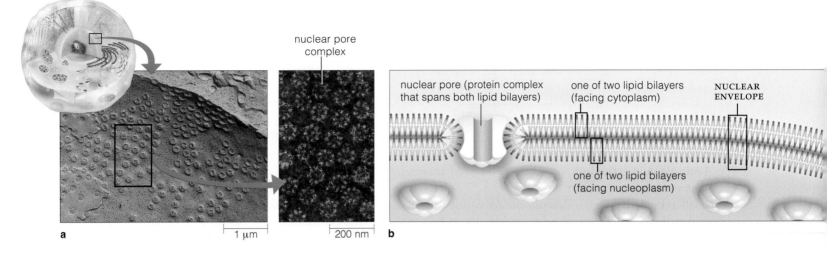

nuclear pore
complex

a   1 µm   200 nm   b

nuclear pore (protein complex that spans both lipid bilayers)

one of two lipid bilayers (facing cytoplasm)

NUCLEAR ENVELOPE

one of two lipid bilayers (facing nucleoplasm)

Figure 4.17 *Animated!* Proteins of the nuclear envelope. (**a**) *Left*, the outer surface of a nuclear envelope was frozen and then fractured. The envelope split apart, revealing the pores that span the two lipid bilayers. *Right*, each pore across the envelope is an organized cluster of membrane proteins. It permits the selective transport of substances into and out of the nucleus. (**b**) Sketch of the nuclear envelope's structure.

The inner surface of the nuclear envelope is bathed in a semifluid matrix called nucleoplasm. The surface has attachment sites for fibrous proteins that anchor the DNA molecules and help keep them organized.

Membrane proteins that span both bilayers have diverse functions. Many are receptors or transporters; many others form pore complexes (Figure 4.17). Ions and small, water-soluble molecules cross the nuclear envelope only at the pores, which span both bilayers.

NUCLEOLUS

The ribosomes mentioned earlier consist of subunits that are constructed in the nucleus from ribosomal RNA and proteins. The construction site is a **nucleolus** (plural, nucleoli). At least one occurs in all nuclei. It is a dense mass of proteins and multiple copies of genes coding for ribosomal RNA. The subunits do not join together until after they move out through pores and enter the cytoplasm. One large and one small subunit join as an intact ribosome during protein synthesis.

GRAINY, THREADLIKE, RODLIKE—NUCLEAR DNA'S CHANGING APPEARANCE

When a eukaryotic cell is not dividing, you cannot see individual DNA molecules, nor can you see that each consists of two strands twisted together. The nucleus just looks grainy, as in Figure 4.16. When the cell is preparing to divide, however, it copies all of its DNA. Soon, the duplicated molecules become visible as long threads that condense further into compact structures.

Early microscopists named that seemingly grainy substance chromatin and called the condensed forms chromosomes. Today we define **chromatin** as the cell's collection of DNA and all proteins associated with it. A **chromosome** is a double-stranded DNA molecule and its associated proteins, regardless of whether it is in dispersed or condensed form:

one chromosome (one dispersed DNA molecule + proteins; not duplicated)

one chromosome (threadlike and now duplicated; two DNA molecules + proteins)

one chromosome (duplicated and also condensed tightly)

In other words, a chromosome's appearance changes over the life of a eukaryotic cell. In chapters to come, you will be looking at different aspects of eukaryotic chromosomes, so you may find it useful to remember that chromosome structure is dynamic, not fixed.

The nucleus has an outer envelope of two lipid bilayers. The envelope keeps DNA molecules separated from the cytoplasmic machinery and controls access to a cell's hereditary information.

With this separation, DNA is easier to keep organized and to copy before a parent cell divides into daughter cells.

Pores across the nuclear envelope help control the passage of many substances between the nucleus and cytoplasm.

At different stages in the life of a eukaryotic cell, nuclear DNA looks different; its structural organization changes.

4.6 The Endomembrane System

*New polypeptide chains fold and twist into proteins. Some proteins are used at once or stockpiled in the cytoplasm. Others enter the **endomembrane system**—ER, Golgi bodies, and vesicles. All proteins to be exported or inserted into cell membranes pass through this system (Figure 4.18).*

ENDOPLASMIC RETICULUM

Endoplasmic reticulum, or **ER**, is a flattened channel that starts at the nuclear envelope and folds back on itself repeatedly in the cytoplasm. At various points inside the channel, many polypeptide chains become modified into final proteins, and lipids are assembled.

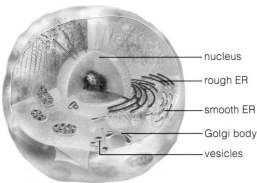

- nucleus
- rough ER
- smooth ER
- Golgi body
- vesicles

Vesicles that pinch off from ER membranes deliver many of the proteins and lipids to Golgi bodies.

Rough ER has many ribosomes attached to its outer surface (Figure 4.18c). Many of the polypeptide chains being translated on ribosomes have an entry code—a sequence of fifteen to twenty amino acids—that lets them cross the ER membrane and enter the channel, where enzymes may modify them. Other chains do not get all the way across; they become inserted into the ER membrane, as diverse membrane proteins.

Cells that make, store, and secrete proteins have a lot of rough ER. For instance, your pancreas has ER-rich cells that make and secrete digestive enzymes.

Smooth ER is ribosome-free (Figure 4.18d). It makes lipid molecules that become part of cell membranes. The ER also takes part in fatty acid breakdown and degrades some toxins. Sarcoplasmic reticulum, a type of smooth ER, functions in muscle contraction.

GOLGI BODIES

Patches of ER membrane bulge and break away as vesicles, each with proteins inside or incorporated in

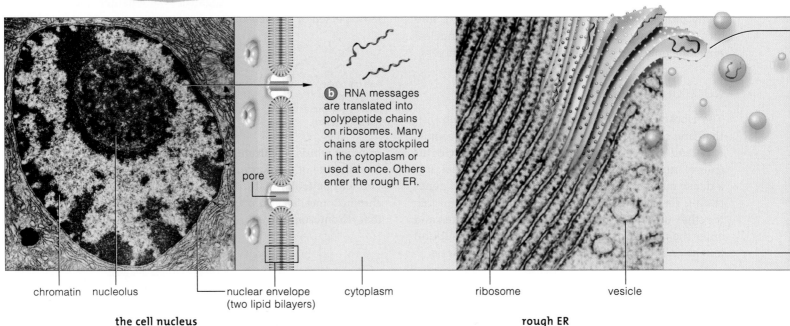

b RNA messages are translated into polypeptide chains on ribosomes. Many chains are stockpiled in the cytoplasm or used at once. Others enter the rough ER.

pore

chromatin   nucleolus   nuclear envelope   cytoplasm   ribosome   vesicle
(two lipid bilayers)

the cell nucleus

a DNA instructions for making proteins are transcribed in the nucleus and moved to the cytoplasm. RNAs are the messengers and protein builders.

rough ER

c Flattened sacs of rough ER form one continuous channel between the nucleus and smooth ER. Polypeptide chains that enter the channel undergo modification. They will be inserted into organelle membranes or will be secreted from the cell.

Figure 4.18 *Animated!* Endomembrane system. With this system's components, many proteins are processed, lipids are assembled, and both products are sorted and shipped to cellular destinations or to the plasma membrane for export.

its membrane. Many vesicles fuse with **Golgi bodies**, organelles in which the membrane channel folds back on itself like a stack of pancakes (Figure 4.18e). Golgi bodies attach sugar side chains to proteins and lipids that they received from the ER. They also cleave some proteins. Finished products are packaged in vesicles.

DIVERSE MEMBRANOUS SACS

Vesicles, again, are tiny sacs that form as buds from the ER, Golgi bodies, and plasma membrane. Many act as storage sacs in the cytoplasm. Others transport substances to or from another organelle or the plasma membrane. As Section 5.6 explains, *exocytic* vesicles release substances to the outside. *Endocytic* ones form at the plasma membrane. They enclose substances at the surface, then sink through the cytoplasm.

Lysosomes are vesicles that bud from Golgi bodies and take part in intracellular digestion. They store inactive forms of diverse hydrolytic enzymes that can digest almost all biological molecules. The enzymes are activated when another vesicle moves through the cytoplasm and fuses with the lysosome. The contents of the vesicle are digested into bits. As you will see in Section 5.6, that is what happens after macrophages have engulfed cells, particles, and assorted debris.

Peroxisomes hold enzymes that digest fatty acids, amino acids, and hydrogen peroxide (H_2O_2), a toxic metabolic product. Enzymes convert H_2O_2 to water and oxygen or use it in reactions that degrade alcohol and other toxins. Drink alcohol, and peroxisomes in liver and kidney cells usually degrade nearly half of it.

Vesicles also fuse and form larger membranous sacs, or vacuoles. The central vacuole of mature plant cells, described in Section 4.8, is one of them.

Inside the highly folded channel formed by ER membranes, many new polypeptide chains are modified and lipids are assembled. In the channels formed by Golgi membranes, many of the proteins and lipids are further modified and packaged for export or shipment to locations in the cell.

Vesicles are small sacs that help integrate cell activities. Many kinds store or transport substances. Lysosomes and peroxisomes start out as vesicles and become organelles of digestion in the cytoplasm.

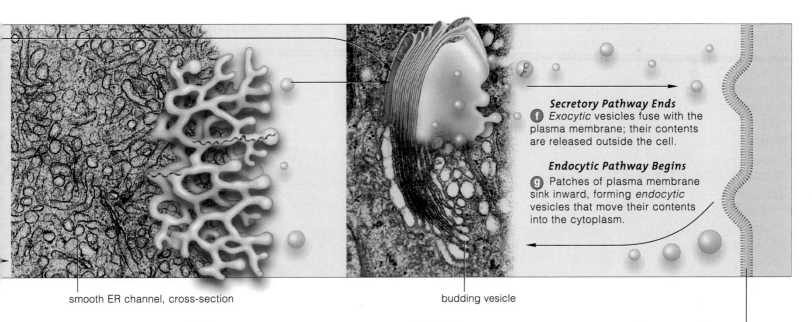

smooth ER channel, cross-section

budding vesicle

Secretory Pathway Ends
f *Exocytic* vesicles fuse with the plasma membrane; their contents are released outside the cell.

Endocytic Pathway Begins
g Patches of plasma membrane sink inward, forming *endocytic* vesicles that move their contents into the cytoplasm.

smooth ER

d Some proteins in the channel continue on to smooth ER, becoming membrane proteins or smooth ER enzymes. Some of these enzymes make lipids and inactive toxins.

Golgi body

e A Golgi body receives, processes, and repackages substances that arrive in vesicles from the ER. Different vesicles transport the substances to other parts of the cell.

plasma membrane

h Exocytic vesicles release cell products and wastes to the outside. Endocytic vesicles move nutrients, water, and other substances into the cytoplasm from outside (Section 5.6).

4.7 Mitochondria

LINK TO
SECTION
3.7

Recall, from Section 3.7, that ATP is an energy carrier. It delivers energy, in the form of phosphate-group transfers, that drives reactions at sites throughout the cell. Without energy deliveries from ATP, cells could not grow, survive, or reproduce. ATP formation is essential for life.

The **mitochondrion** (plural, mitochondria) is a type of organelle that specializes in ATP formation. Reactions in this organelle help cells extract more energy from organic compounds than they can get by any other means. The reactions, called *aerobic* respiration, require free oxygen. With each breath, you take in oxygen mainly for mitochondria in your trillions of cells.

Typical mitochondria are 1 to 4 micrometers long; a few are 10 micrometers long. Some are branching. These organelles change shape, split in two, and fuse together. Figure 4.19 shows an example.

Each mitochondrion has an outer membrane and another one inside that most often is highly folded (Figure 4.19). The membrane arrangement creates two compartments. Hydrogen ions become stockpiled in the outer compartment. The ions then flow to the inner compartment in a controlled way. The energy inherent in the flow drives ATP formation.

No prokaryotic cells have mitochondria. Nearly all eukaryotic cells do. A single-celled yeast might have only one. Cells that have huge demands for energy may have a thousand or more. Skeletal muscle cells are one example. Liver cells, too, have a profusion of

mitochondria. Take a closer look at Figure 4.14. That micrograph alone should tell you that the liver is an energy-demanding organ.

In size and biochemistry, mitochondria resemble bacteria. Like bacteria, they have their own DNA, and they divide on their own. They have some ribosomes. Did mitochondria evolve by way of endosymbiosis in ancient prokaryotic cells? By this theory, one cell was engulfed by another cell, or entered it as an internal parasite, but it escaped digestion. That cell and its descendants kept their plasma membrane intact and reproduced in the host. In time, they became protected, permanent residents. Structures and functions once required for independent life were no longer essential and were lost. Later descendants evolved into double-membraned mitochondria. We return to their possible endosymbiotic origins in Section 20.4.

All organisms require ATP, which carries energy (in the form of phosphate bonds) from one reaction site to another. ATP drives nearly all cell activities.

The organelles called mitochondria are the ATP-producing powerhouses of all eukaryotic cells.

Energy-releasing reactions proceed at the compartmented, internal membrane system of mitochondria. The reactions, which require oxygen, produce far more ATP than can be produced by any other cellular reaction.

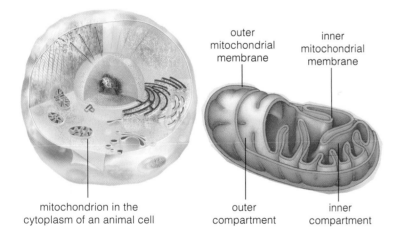

outer mitochondrial membrane

inner mitochondrial membrane

mitochondrion in the cytoplasm of an animal cell

outer compartment

inner compartment

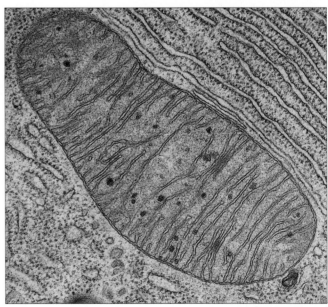

0.5 μm

Figure 4.19 Sketch and transmission electron micrograph, thin section, of a typical mitochondrion. This organelle specializes in forming large quantities of ATP, the main carrier of energy between reaction sites in cells. ATP formation in mitochondria cannot occur without free oxygen.

4.8 Specialized Plant Organelles

Two kinds of organelles are prominent in many plant cells. They are plastids, such as chloroplasts, and central vacuoles.

CHLOROPLASTS AND OTHER PLASTIDS

Plastids are organelles that function in photosynthesis or storage in plants. Three types are common in plant tissues: chloroplasts, chromoplasts, and amyloplasts.

Chloroplasts are organelles that are specialized for photosynthesis. Most have oval or disk shapes. Two outer membranes enclose their semifluid interior, the stroma (Figure 4.20). In the stroma, a third membrane forms a single compartment that is commonly folded in intricate ways. Often the folds resemble a stack of flattened disks, called a granum (plural, grana).

Photosynthesis proceeds at the innermost, *thylakoid* membrane, which incorporates light-trapping pigments and other proteins. The most abundant photosynthetic pigments are chlorophylls, which reflect green light. Many kinds of accessory pigments assist chlorophylls in capturing light energy. The energy drives reactions in which ATP and an enzyme helper, NADPH, form. The ATP and NADPH are then used at sites in the stroma where sugars, starch, and other compounds are assembled. The new starch molecules may briefly accumulate in the stroma, as starch grains.

In many ways, chloroplasts are like photosynthetic bacteria. Like mitochondria, they may have evolved by endosymbiosis (Section 20.4).

Chromoplasts have no chlorophylls. They have an abundance of carotenoids, the source of red-to-yellow colors of many flowers, autumn leaves, ripe fruits, and carrots and other roots. The colors attract animals that pollinate the plants or disperse seeds.

Amyloplasts are pigment-free. They typically store starch grains. They are notably abundant in cells of stems, potato tubers (underground stems), and seeds.

CENTRAL VACUOLES

Many mature, living plant cells contain a fluid-filled **central vacuole**. This organelle stores amino acids, sugars, ions, and toxic wastes. Also, it expands during growth and increases fluid pressure on the pliable cell wall. The cell surface area is forced to increase, which favors absorption of water and other substances. Often the central vacuole takes up 50 to 90 percent of the cell's interior, with cytoplasm confined to a narrow zone in between this large organelle and the plasma membrane (Figures 4.15*a* and 4.20).

> Photosynthetic cells of plants and many protists contain chloroplasts and other plastids that function in food production and storage.
>
> Many plant cells have a central vacuole. When this storage vacuole enlarges during growth, cells are forced to enlarge, which increases the surface area available for absorption.

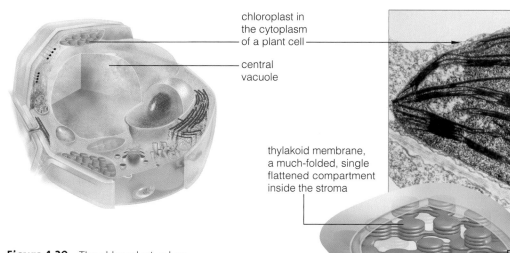

chloroplast in the cytoplasm of a plant cell

central vacuole

thylakoid membrane, a much-folded, single flattened compartment inside the stroma

stroma (semifluid interior)

two outer membranes

Figure 4.20 The chloroplast, a key defining character of all photosynthetic eukaryotic cells. *Right*, transmission electron micrograph of a chloroplast from corn (*Zea mays*), thin section.

4.9 Cell Surface Specializations

Turn now to a brief glimpse into some specialized surface structures of eukaryotic cells. Many of these architectural marvels are made primarily of cell secretions. Others are clusters of membrane proteins that connect neighboring cells, structurally and functionally.

EUKARYOTIC CELL WALLS

Single-celled eukaryotic species are directly exposed to the environment. Many have a **cell wall** around the plasma membrane. A cell wall protects and physically supports a cell, and imparts shape to it. The wall is porous, so water and solutes easily move to and from the plasma membrane. Any cell would die without these exchanges. Different plant cells form one or two walls around their plasma membrane. Cells of many protists and fungi have a wall. Animal cells do not.

Consider the growing parts of multicelled plants. New cells are secreting molecules of pectin and other gluelike polysaccharides, which form a matrix around them. They also secrete ropelike strands of cellulose

molecules into the matrix. These materials make up the plant cell's **primary wall** (Figure 4.21). The sticky primary wall cements abutting cells together. Being thin and pliable, it allows the cell to enlarge under the pressure of incoming water.

Cells that have only a thin primary wall retain the capacity to divide or change shape as they grow and develop. Many types stop enlarging when they are mature. Such cells secrete material on the primary wall's inner surface. These deposits form a lignified, rigid **secondary wall** that reinforces cell shape (Figure 4.21*d*). The secondary wall deposits are extensive and contribute more to structural support.

In woody plants, up to 25 percent of the secondary wall is made of lignin. This organic compound makes plant parts more waterproof, less susceptible to plant-attacking organisms, and stronger.

At plant surfaces exposed to air, waxes and other cell secretions build up as a protective cuticle. This semitransparent surface covering limits water losses on hot, dry days (Figure 4.22*a*).

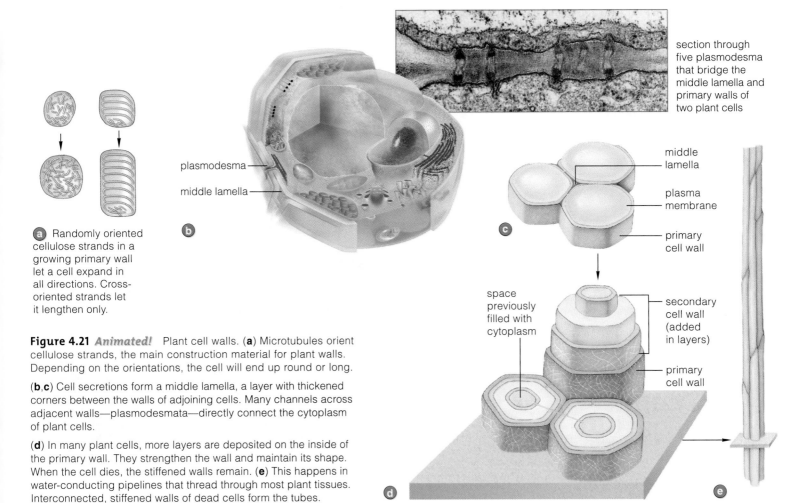

section through five plasmodesma that bridge the middle lamella and primary walls of two plant cells

plasmodesma

middle lamella

middle lamella

plasma membrane

primary cell wall

space previously filled with cytoplasm

secondary cell wall (added in layers)

primary cell wall

(a) Randomly oriented cellulose strands in a growing primary wall let a cell expand in all directions. Cross-oriented strands let it lengthen only.

Figure 4.21 *Animated!* Plant cell walls. (**a**) Microtubules orient cellulose strands, the main construction material for plant walls. Depending on the orientations, the cell will end up round or long.

(**b**,**c**) Cell secretions form a middle lamella, a layer with thickened corners between the walls of adjoining cells. Many channels across adjacent walls—plasmodesmata—directly connect the cytoplasm of plant cells.

(**d**) In many plant cells, more layers are deposited on the inside of the primary wall. They strengthen the wall and maintain its shape. When the cell dies, the stiffened walls remain. (**e**) This happens in water-conducting pipelines that thread through most plant tissues. Interconnected, stiffened walls of dead cells form the tubes.

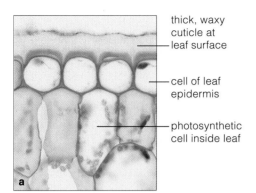

thick, waxy cuticle at leaf surface

cell of leaf epidermis

photosynthetic cell inside leaf

a

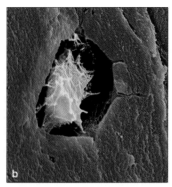

b

Figure 4.22 (**a**) Section through a plant cuticle, a surface covering made of cell secretions.

(**b**) A living cell imprisoned in hardened bone tissue, the stuff of vertebrate skeletons.

MATRIXES BETWEEN ANIMAL CELLS

Animal cells have no cell walls. Intervening between many of them are matrixes made of cell secretions and of materials absorbed from the surroundings. For example, the cartilage at the knobby ends of leg bones contains scattered cells and protein fibers embedded in a ground substance of firm polysaccharides. Living cells also secrete the extensive, hardened matrix that we call bone tissue (Figure 4.22*b*).

CELL JUNCTIONS

Even when a wall or some other structure imprisons a cell in its own secretions, the cell interacts with the outside world at its plasma membrane. In multicelled species, structures extend into neighboring cells or into a matrix. **Cell junctions** are molecular structures where a cell sends or receives signals or materials, or recognizes and glues itself to cells of the same type.

In plants, for instance, channels extend across the primary wall of adjacent living cells and interconnect the cytoplasm of both (Figure 4.21*b*). Each channel is a plasmodesma (plural, plasmodesmata). Substances flow quickly from cell to cell across these junctions.

In most tissues of animals, three types of cell-to-cell junctions are common (Figure 4.23). *Tight* junctions link the cells of most body tissues, including epithelia that line outer surfaces, internal cavities, and organs. These junctions seal abutting cells together so water-soluble substances cannot leak between them. That is why gastric fluid does not leak across the stomach lining and damage internal tissues. *Adhering* junctions occur in skin, the heart, and other organs subjected to continual stretching. At *gap* junctions, the cytoplasm of certain kinds of adjacent cells connect directly. Gap junctions function as open channels for a rapid flow of substances, most notably in heart muscle.

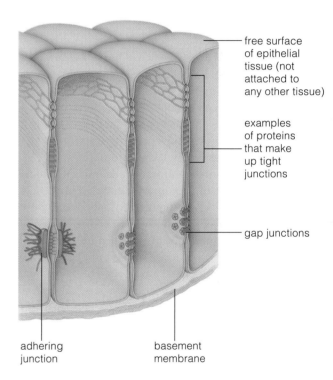

free surface of epithelial tissue (not attached to any other tissue)

examples of proteins that make up tight junctions

gap junctions

adhering junction

basement membrane

Figure 4.23 *Animated!* The three most common types of cell junctions in animal tissues.

A variety of protist, plant, and fungal cells have a porous wall that surrounds the plasma membrane.

Young plant cells have a thin primary wall pliable enough to permit expansion. Some mature cells also form a lignin-reinforced secondary wall that provides structural support.

Animal cells have no walls, but they and many other cells often secrete substances that form matrixes between cells. In multicelled species, different junctions commonly serve as structural and functional connections between cells.

4.10 Even Cells Have a Skeleton

What does the word "skeleton" mean to you? A collection of bones, such as a rib cage? That is one kind of skeleton in nature. However, anything that forms a structural framework is a skeleton—and cells, too, have one.

COMPONENTS OF THE CYTOSKELETON

The **cytoskeleton** of eukaryotic cells is an organized system of protein filaments that extends between the nucleus and plasma membrane. Different portions of it reinforce, organize, and move internal cell parts or the cell body. Many parts are permanent, and others form only at certain times in the life of a cell.

Microtubules and **microfilaments** are two classes of cytoskeletal elements in nearly all eukaryotic cells. Some cells also have ropelike **intermediate filaments** (Figures 4.24 and 4.25).

Microtubules Microtubules, the largest cytoskeletal elements, help keep organelles and cell structures in place or move them to new locations. For instance, before a cell divides, microtubules form a spindle that harnesses chromosomes and moves them about. Other microtubules move chloroplasts, vesicles, and other organelles through the cytoplasm.

In plant and animal cells, microtubules are hollow cylinders of tubulin monomers (Figure 4.24a). Tubulin consists of two chemically distinct polypeptide chains, each folded into a rounded shape. As a microtubule is being assembled, all of its monomers are oriented in the same direction. The assembly pattern puts slightly different chemical properties at opposite ends of the cylinder. Like bricks being stacked into a new wall, monomers join the cylinder's *plus* (fast-growing) end, which at first grows freely through the cytoplasm.

Microtubules of animal cells normally grow in all directions from small patches of dense material called centrosomes. Their *minus* (slow-growing) ends remain anchored in it. Microtubules can abruptly fall apart in controlled ways; they are not permanently stable.

In any given interval, some of the microtubules are being allowed to disassemble. Others get capped with proteins that stabilize them. For instance, microtubules in the advancing end of an amoeba stay intact when the cell is hot on a chemical trail to a potential meal. Nothing is stimulating microtubules to grow in that cell's trailing end, so they are allowed to fall apart.

Some plants make poisons that act on microtubules of plant-eating animals. By binding to tubulins, the poisons make microtubules fall apart and stop new ones from forming. For instance, the autumn crocus (*Colchicum autumnale*) makes colchicine (Figure 4.26a). The plant has an evolved insensitivity to colchicine, which can't bind well to its own tubulins. The western yew (*Taxus brevifolia*) makes the microtubule poison taxol (Figure 4.26b). A synthetic taxol stops the growth of certain tumors. It keeps microtubule spindles from forming, and thereby suppresses the uncontrolled cell divisions that form abnormal tissue masses. Taxol has the same effect on normal cells; it is just that tumor cells divide far more often than most of them.

Microfilaments Microfilaments are the thinnest of all cytoskeletal elements inside eukaryotic cells (Figure 4.24b). Two helically coiled polypeptide chains of actin monomers make up each filament.

Microfilaments are assembled and disassembled in controlled ways. Often they are organized in bundles or networks. As an example, just beneath the plasma membrane is a **cell cortex**: bundled up, crosslinked, and gel-like meshes of microfilaments. Some parts of the cortex reinforce a cell's shape. Others reconfigure the surface. For instance, during animal cell division, a ring of microfilaments around the cell's midsection contracts and pinches the cell in two. Microfilaments also anchor membrane proteins and are components of muscle contraction. In some large cells, *cytoplasmic streaming* gets under way as microfilament meshes loosen up. Local gel-like regions become more fluid and flow strongly, thereby redistributing substances and cell components through the cell interior.

Myosin and Other Accessory Proteins Genes for tubulin and actin were not drastically modified over time; they have been highly conserved in the DNA. All eukaryotic cells make similar forms of these monomers for microtubules and microfilaments. In spite of the structural uniformity, microtubules and microfilaments have different functions, thanks to other proteins that associate with them.

tubulin subunits

a 25 nm

actin subunit

b 5–7 nm

one polypeptide chain

8–12 nm

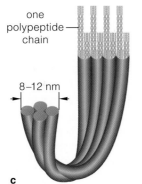

c

Figure 4.24 Structural arrangement of subunits in (**a**) microtubules, (**b**) microfilaments, and (**c**) one of the intermediate filaments.

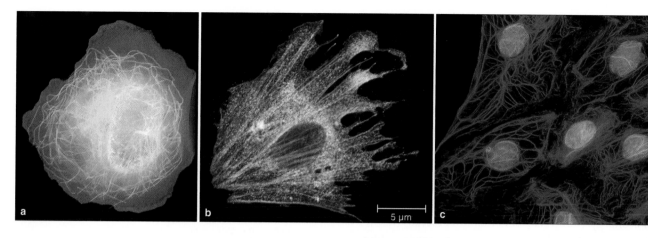

Figure 4.25 Cells stained to reveal the cytoskeleton. (**a**) Distribution of microtubules (*gold*) and actin filaments (*red*) in a pancreatic epithelial cell. This cell secretes bicarbonate, which functions in digestion by neutralizing stomach acid. The *blue* region is DNA. (**b**) Actin filaments (*red*) and accessory proteins (*green, blue*) help new epithelial cells such as this one migrate to proper locations in tissues as human embryos are developing. (**c**) Intermediate filaments of keratin (*red*) of cultured kangaroo rat cells. Each *blue*-stained organelle is a nucleus.

For example, as you will read in the next section, kinesin and myosin are two kinds of *motor* proteins. Inputs of ATP energy make them move along tracks of cytoskeletal elements and put cell components in new locations. As another example, the *crosslinking* proteins interconnect microfilaments at the cell cortex.

Intermediate Filaments Intermediate filaments are between microtubules and microfilaments in size. They are eight to twelve nanometers wide and are the most stable elements of some cytoskeletons (Figure 4.25c). Six known groups strengthen and maintain the shape of cells or cell parts. For example, lamins help form a basketlike mesh that reinforces the nucleus. They anchor adjoining actin and myosin filaments as units of contraction inside muscle cells. Desmins and vimentins help hold these contractile units in position. Different cytokeratins reinforce cells that make nails, claws, horns, and hairs. Intermediate filaments may reinforce the nuclear envelope of all eukaryotic cells.

Different animal cells have different intermediate filaments in their cytoplasm. Because each type of cell contains one or at most two kinds, researchers can use these intermediate filaments to identify which type of cell it is. This typing is a useful tool in diagnosing the tissue origin of different forms of cancerous cells.

WHAT ABOUT PROKARYOTIC CELLS?

Unlike eukaryotic cells, bacteria and archaeans do not have a well-developed cytoskeleton. Until recently, we thought they did not have any cytoskeletal elements. However, reinforcing filaments have been identified in certain bacteria. What is more, the filaments are made of protein subunits that are a lot like tubulin and actin. Also, their repetitive pattern of assembly is similar to the assembly patterns for microtubules and microfilaments in eukaryotic cells.

For instance, just beneath the plasma membrane of rod-shaped bacterial cells, protein monomers (MreB) form filaments that help determine cell shape. These filaments, which are structurally like one of the actins, suggest that the eukaryotic cytoskeleton had its origin in ancestral prokaryotes.

Figure 4.26
Two sources of microtubule poisons: (**a**) Autumn crocus, *Colchicum autumnale*. (**b**) Western yew, *Taxus brevifolia*.

> A cytoskeleton is the basis of cell shape, internal structure, and movement. In eukaryotic cells, its components are microtubules, microfilaments, and in some cell types, intermediate filaments. Accessory proteins associate with these filaments and extend their range of functions.
>
> Microtubules, cylinders of tubulin monomers, organize the cell interior and have roles in moving cell components.
>
> Microfilaments consist of two helically coiled polypeptide chains of actin monomers. They form flexible, linear bundles and networks that reinforce or restructure the cell surface.
>
> Intermediate filaments strengthen and maintain cell shapes. Some are present only in certain animal cells. Others may help reinforce the nuclear envelope of all eukaryotic cells.
>
> Cytoskeletal elements similar to the microtubules and microfilaments have been identified in prokaryotic cells.

4.11 How Do Cells Move?

The skeleton of eukaryotic cells differs notably from your skeleton in a key respect. The cytoskeleton has elements that are not permanently rigid. At prescribed times, they assemble and disassemble.

MOVING ALONG WITH MOTOR PROTEINS

Think of a train station at the busiest holiday season, and you get an idea of what goes on in cells. Many of the cell's microtubules and microfilaments are like train tracks. The kinesins, dyneins, myosins, and other motor proteins function as the freight engines (Figure 4.27). Energy from ATP fuels the movement.

Some motor proteins move chromosomes. Others slide one microtubule over another; still others inch along tracks inside nerve cells that extend from your spine to your toes. Many engines are organized one after another, and each moves a vesicle partway along the track before giving it up to the next engine in line. From dawn to dusk, kinesins inside plant cells drag chloroplasts to new positions, the better to intercept light as the angle of the sun changes overhead.

Different kinds of myosins can move structures along microfilaments or slide one microfilament over another. As Sections 37.5 and 37.6 show, muscle cells form long fibers that are functionally divided into many contractile units along their length. Each unit has many parallel rows of microfilaments and myosin filaments. Myosin activated by ATP shortens the unit by sliding the microfilaments toward the unit's center. When all units shorten, the cell shortens; it contracts.

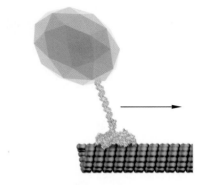

Figure 4.27 *Animated!* Kinesin (*brown*). This motor protein inches along a microtubule and drags a vesicle (*pink*) or some other cellular freight with it.

CILIA, FLAGELLA, AND FALSE FEET

Besides moving internal parts, many cells move their body or extend parts of it. Consider **flagella** (singular, flagellum) and **cilia** (singular, cilium). Both are motile structures that project from the surface of many types of cells. Both are completely sheathed by a membrane that is an extension of the plasma membrane.

Ciliated protists swim by beating their many cilia in synchrony. Cilia in certain airways to your lungs beat nonstop. Their coordinated movement sweeps out the airborne bacteria and particles that otherwise might cause disease (Figure 4.28*a*). Eukaryotic flagella usually are longer and not as profuse as cilia. Many single eukaryotic cells, such as sperm, swim with the help of whiplike flagella (Figure 4.28*b*).

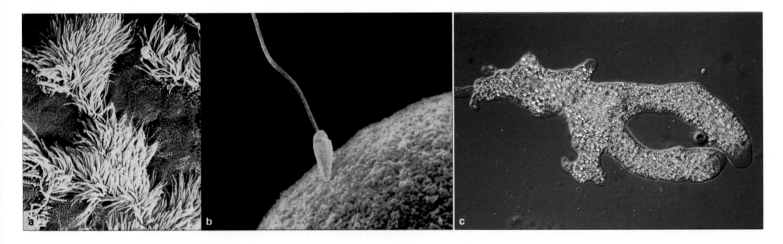

Figure 4.28 Cilia, a flagellum, and false feet. (**a**) Light micrograph of cilia (*gold*) projecting from the surface of some of the cells that line an airway to human lungs. (**b**) Scanning electron micrograph of a human sperm about to penetrate an egg. (**c**) Light micrograph of a predatory amoeba (*Chaos carolinense*) extending two pseudopods around a single-celled green alga (*Pandorina*).

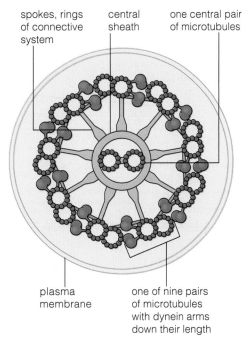

spokes, rings of connective system

central sheath

one central pair of microtubules

plasma membrane

one of nine pairs of microtubules with dynein arms down their length

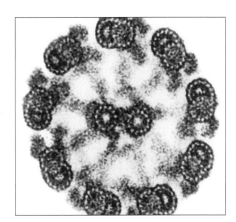

cross-section through one cilium

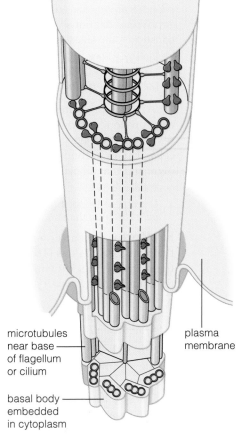

microtubules near base of flagellum or cilium

plasma membrane

basal body embedded in cytoplasm

Figure 4.29 Internal organization of cilia and flagella. Both motile structures have a 9+2 array, an internal ring of nine pairs of microtubules around one pair at the core. Spokes and linking elements connected to the array stabilize it and keep it from slipping sideways, out of alignment.

Dynein arms projecting from microtubules in the ring incorporate ATPases. Repeated phosphate-group transfers from ATP cause them to bind briefly and reversibly to the microtubule pair in front of them. Each time, the arms force the pair to slide down a bit. The short, sliding strokes occur all around the ring, down the length of the microtubule, and it makes the motile structure bend.

Extending down the length of a cilium or flagellum is a *9+2 array*. Nine pairs of microtubules form a ring around a central pair, all stabilized by protein spokes and links. First a centrosome gives rise to a **centriole**. This barrel-shaped structure produces and organizes microtubules into the 9+2 array, then it remains below the finished array as a **basal body** (Figure 4.29).

Flagella and cilia move by a sliding mechanism. All pairs of microtubules extend the same distance into the motile structure's tip. Stubby dynein arms project from each pair in the outer ring. When ATP energizes them, the arms grab the microtubule pair in front of them, tilt in a short, downward stroke, then let go. As the bound pair slides down, its arms bind the pair in front of it, forcing it to slide down also— and so on around the ring. The microtubules cannot slide too far, but each *bends* a bit. Their sliding motion is converted to a bending motion.

As one more example, macrophages and amoebas form **pseudopods**, or "false feet." These temporary, irregular lobes bulge out from the cell. They move the cell and also engulf prey or some other target (Figure 4.28c). The pseudopods advance in a steady direction as microfilaments inside them are elongating. Motor proteins that are attached to the microfilaments are dragging the plasma membrane along with them in the direction of interest.

Cell contractions and migrations, chromosome movements, and other forms of cell movements arise by interactions among organized arrays of microtubules, microfilaments, and accessory proteins.

When energized by ATP, motor proteins move in specific directions, along tracks of microtubules and microfilaments. They deliver cell components to new locations.

Interactions among cytoskeletal elements bring about the movement of the motile structures called cilia and flagella, as well as the dynamic movements of pseudopods.

Summary

Section 4.1 Cells start life with a plasma membrane, cytoplasm that contains ribosomes and other structures, and DNA in a nucleus or nucleoid. Their membranes are a lipid bilayer with diverse kinds and numbers of proteins embedded in it or positioned at its surfaces. A physical relationship, the surface-to-volume ratio, constrains the sizes and shapes of cells.

Biology ⊛ Now
Use the interaction on BiologyNow to investigate the physical limits on cell size.

Section 4.2 Different microscopes use light or electrons to reveal cell shapes and structures. Microscopy reinforces the theory that all organisms are made of cells, that an individual cell has a capacity to live on its own, and that all cells now arise from preexisting cells.

Biology ⊛ Now
Learn how different types of microscopes function with the animation on BiologyNow.

Section 4.3 Bacteria and archaeans are prokaryotic; they have no nucleus (Table 4.2). They are structurally the simplest cells known, but collectively they show great metabolic diversity.

Biology ⊛ Now
View the animation about prokaryotic cell structure on BiologyNow.

Section 4.4 Organelles are membranous sacs that divide the interior of eukaryotic cells into functional compartments. Table 4.2 lists the major types.

Biology ⊛ Now
Introduce yourself to the major types of eukaryotic organelles with the interaction on BiologyNow.

Section 4.5 The nucleus keeps DNA molecules separated from metabolic reactions in the cytoplasm and controls access to a cell's hereditary information. It helps keep the DNA organized and easier to copy before a cell divides into daughter cells.

Biology ⊛ Now
Take a close-up look at the nuclear membrane with the animation on BiologyNow.

Section 4.6 In the endomembrane system's ER and Golgi bodies, new polypeptide chains are modified and lipids are assembled. Many proteins and lipids become part of membranes or packaged inside vesicles that function in transport, storage, and other cell activities.

Biology ⊛ Now
Follow a path through the endomembrane system with the animation on BiologyNow.

Section 4.7 The mitochondrion is the organelle that specializes in forming many ATP by aerobic respiration.

Section 4.8 The chloroplast is an organelle that specializes in photosynthesis.

Biology ⊛ Now
Look inside a chloroplast with the animation on BiologyNow.

Section 4.9 Most prokaryotic cells, cells of many protists and fungi, and all plant cells have a porous wall around their plasma membrane. In multicelled species, structural and functional connections link cells.

Biology ⊛ Now
Study the structure of cell walls and junctions with the animation on BiologyNow.

Sections 4.10, 4.11 Microtubules, microfilaments, and intermediate filaments make up a cytoskeleton in eukaryotic cells. They reinforce cell shapes, organize parts, and often move cells or structures, as by flagella.

Biology ⊛ Now
Learn more about cytoskeletal elements and their actions with the animation on BiologyNow.

Self-Quiz
Answers in Appendix II

1. Cell membranes consist mainly of a _____ .
   a. carbohydrate bilayer and proteins
   b. protein bilayer and phospholipids
   c. lipid bilayer and proteins

2. Identify the components of the cells shown in the two sketches at the bottom of the next page.

3. Organelles _____ .
   a. are membrane-bound compartments
   b. are typical of eukaryotic cells, not prokaryotic cells
   c. separate chemical reactions in time and space
   d. All of the above are features of organelles.

4. You will not observe _____ in animal cells.
   a. mitochondria   c. ribosomes
   b. a plasma membrane   d. a cell wall

5. Is this statement false: The plasma membrane is the outermost component of all cells. Explain your answer.

6. Unlike eukaryotic cells, prokaryotic cells _____ .
   a. lack a plasma membrane   c. have no nucleus
   b. have RNA, not DNA   d. all of the above

7. Match each cell component with its function.
   ____ mitochondrion   a. protein synthesis
   ____ chloroplast   b. initial modification of new
   ____ ribosome   polypeptide chains
   ____ rough ER   c. modification of new proteins;
   ____ Golgi body   sorting, shipping tasks
                         d. photosynthesis
                         e. formation of many ATP

Additional questions are available on **Biology ⊛ Now™**

Critical Thinking

1. Why is it likely that you will never meet a two-ton amoeba on a sidewalk?

2. Your professor shows you an electron micrograph of a cell with many mitochondria, Golgi bodies, and a lot of rough ER. What kinds of cellular activities would require such an abundance of the three kinds of organelles?

3. *Kartagener syndrome* is a genetic disorder caused by a mutated form of the protein dynein. Affected people have chronically irritated sinuses, and mucus builds up in the

Table 4.2 Summary of Typical Components of Prokaryotic and Eukaryotic Cells

Cell Component	Function	Prokaryotic Bacteria, Archaea	Eukaryotic Protists	Fungi	Plants	Animals
Cell wall	Protection, structural support	✔*	✔*	✔	✔	None
Plasma membrane	Control of substances moving into and out of cell	✔	✔	✔	✔	✔
Nucleus	Physical separation and organization of DNA	None	✔	✔	✔	✔
DNA	Encoding of hereditary information	✔	✔	✔	✔	✔
RNA	Transcription, translation of DNA messages into polypeptide chains of specific proteins	✔	✔	✔	✔	✔
Nucleolus	Assembly of subunits of ribosomes	None	✔	✔	✔	✔
Ribosome	Protein synthesis	✔	✔	✔	✔	✔
Endoplasmic reticulum (ER)	Initial modification of many of the newly forming polypeptide chains of proteins; lipid synthesis	None	✔	✔	✔	✔
Golgi body	Final modification of proteins, lipids; sorting and packaging them for use inside cell or for export	None	✔	✔	✔	✔
Lysosome	Intracellular digestion	None	✔	✔*	✔*	✔
Mitochondrion	ATP formation	**	✔	✔	✔	✔
Photosynthetic pigments	Light–energy conversion	✔*	✔*	None	✔	None
Chloroplast	Photosynthesis; some starch storage	None	✔*	None	✔	None
Central vacuole	Increasing cell surface area; storage	None	None	✔*	✔	None
Bacterial flagellum	Locomotion through fluid surroundings	✔*	None	None	None	None
Flagellum or cilium with 9+2 microtubular array	Locomotion through or motion within fluid surroundings	None	✔*	✔*	✔*	✔
Complex cytoskeleton	Cell shape; internal organization; basis of cell movement and, in many cells, locomotion	Rudimentary***	✔*	✔*	✔*	✔

* Known to be present in cells of at least some groups.
** Many groups use oxygen-requiring (aerobic) pathways of ATP formation, but mitochondria are not involved.
*** Protein filaments form a simple scaffold that helps support the cell wall in at least some species.

airways to their lungs. Bacteria form huge populations in the thick mucus. Their metabolic by-products and the inflammation they trigger combine to damage tissues. Males affected by the syndrome can produce sperm, but they are infertile (Figure 4.30). Some have still become fathers with the help of a procedure that injects sperm cells directly into eggs. Explain how an abnormal dynein molecule could cause the observed effects.

4. As they grow and develop, many kinds of plant cells form a secondary wall on the *inner* surface of the primary wall that formed earlier. Speculate on the reason why the secondary wall does not form on the outside.

5. Reflect on Table 4.2. Notice how most prokaryotes, all plant cells, and many protist and fungal cells have walls, and that animal cells have none. Why do you suppose animal cells alone do not form walls?

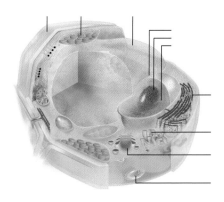

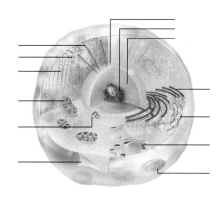

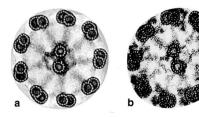

Figure 4.30 Cross-section through the flagellum of a sperm cell from (**a**) a male affected by Kartagener syndrome and (**b**) an unaffected male. Check the dynein arms projecting from the microtubule pairs.

5 A CLOSER LOOK AT CELL MEMBRANES

One Bad Transporter and Cystic Fibrosis

Each living cell is engaged in risky business. Think of how it has to move something as ordinary as water in one direction or the other across its plasma membrane. If all goes well, it takes in or sends out water in just the right amounts—not too little, not too much. But who is to say life always goes well?

CFTR is one of the protein channels across the plasma membrane of epithelial cells. Sheets of these cells line sweat glands, airways and sinuses, and ducts in the digestive and reproductive systems. Chloride ions move through them, and water follows to form a thin film on the free surface of the linings. Mucus, which lubricates tissues and helps prevent infection, slides freely on the watery film.

Sometimes mutation changes how CFTR works. Not enough chloride and water reach the lining's free surface, so the film does not form. Mucus dries out and thickens. Among other things, it clogs ducts from the pancreas, so digestive enzymes cannot get to the small intestine where most food is digested and absorbed. Weight loss follows. Sweat glands secrete too much salt and alter the water–salt balance for the internal environment, which affects the heart and other organs. Males become sterile.

Problems also develop in airways to the lungs, where ciliated cells are supposed to sweep away bacteria and other particles stuck in mucus. Now the mucus makes cilia too sticky, and **biofilms** form. Biofilms are microbial populations anchored to one epithelial lining or another by stiff, sticky polysaccharides of their own making. They resist the body's defenses and antibiotics. *Pseudomonas aeruginosa*, the most efficient of the colonizers, cause low-grade infections that may last for years. Most patients can expect to live no longer than thirty years, at which time their lungs usually fail. At present there is no cure.

These symptoms—outcomes of mutation in the CFTR protein—characterize *cystic fibrosis* (CF), the most common fatal genetic disorder in the United States. More than 10 million people carry a mutant form of the gene. CF develops when they inherited a mutated gene from both parents. This happens in about 1 of every 3,300 live births (Figure 5.1).

CFTR is one of the ABC transporters in all prokaryotic and eukaryotic cells (Figure 5.2). Some of these proteins, including CFTR, are channels that let hydrophobic substances cross a membrane. Others pump substances across. By their action, some types affect what other membrane proteins are doing.

In all but 10 percent of CF patients, loss of a single amino acid during protein synthesis causes the disorder. Before a new CFTR protein is shipped to the plasma membrane, it is supposed to be modified in that endomembrane system you read about in Chapter 4. Copies of the mutant protein do enter the ER, but enzymes destroy 99 percent of them before they reach Golgi bodies. Thus few chloride channels reach their normal destinations.

Mutant CFTR may also contribute to the sinus problems of an estimated 30 million people in the United States

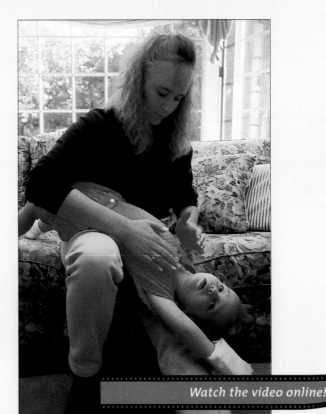

Watch the video online!

Figure 5.1 Child affected by cystic fibrosis, or CF, who each day endures chest thumps, back thumps, and repositionings to dislodge thick mucus that collects in airways to the lungs. Symptoms vary from one affected individual to the next, partly because the abnormal protein that causes CF has mutated in more than 500 ways. Environmental factors and a person's genetic makeup also affect the outcome.

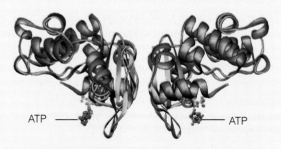

ATP ———— ———— ATP

Figure 5.2 Model for part of an ABC transporter, a category of membrane proteins that includes CFTR. The parts shown here are ATP-driven motors that can widen an ion channel across the plasma membrane.

alone. In *sinusitis*, the linings of cavities inside the skull (around the nose) are chronically inflamed. In one study at Johns Hopkins University, researchers found a single copy of a mutant CFTR gene in 10 of 147 sinusitis patients. And they were only looking for 16 of more than 500 known mutant forms of the CFTR gene!

Think about it. A startling percentage of the human population can develop problems when the copies of even one kind of membrane protein don't work.

Your life depends on the functions of thousands of kinds of proteins and other molecules. Breathing, eating, moving, sleeping, crying, thinking—whatever you might be doing starts at the level of individual cells. And each cell functions properly only if it can be responsive to conditions in the microenvironments on both sides of its plasma membrane. Each eukaryotic cell also has to be responsive to conditions on both sides of its organelle membranes. *Cell membranes—* these thin boundary layers make the difference between organization and chaos.

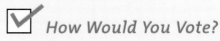

How Would You Vote?

The ability to detect mutant genes that cause severe disorders raises bioethical questions. Should we encourage the mass screening of prospective parents for mutant genes that cause cystic fibrosis? Should society encourage women to give birth only if their child will not develop severe medical problems? See BiologyNow for details, then vote online.

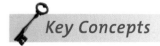

Key Concepts

MEMBRANE STRUCTURE AND FUNCTION

Cell membranes have a thin, oily, water-insoluble lipid bilayer that functions as a boundary between the outside environment and the cell interior.

The lipid bilayer consists primarily of phospholipids. Many diverse proteins are embedded in the bilayer or are positioned at one of its surfaces. The proteins carry out most membrane functions, such as transport across the bilayer and cell-to-cell recognition. Sections 5.1, 5.2

DIFFUSION ACROSS MEMBRANES

Metabolism requires concentration gradients that drive the directional movements of substances. Cells have built-in mechanisms for increasing or decreasing water and solute concentrations across the plasma membrane and internal cell membranes. Section 5.3

TRANSPORT ACROSS MEMBRANES

In passive transport, a solute crosses a membrane by diffusing through a channel inside a transport protein. In active transport, a different kind of transport protein pumps the solute across a membrane, against its concentration gradient. An input of energy, typically from ATP, jump-starts active transport. Section 5.4

OSMOSIS

By a molecular behavior called osmosis, water diffuses across any selectively permeable membrane to a region where its concentration is lower. Section 5.5

MEMBRANE TRAFFIC

Larger packets of substances and, in some cases, engulfed cells move across the plasma membrane by processes of endocytosis and exocytosis. Membrane cycling pathways extend from the plasma membrane to organelles of the endomembrane system. Section 5.6

Links to Earlier Concepts

Reflect again on the road map in Section 1.1. Here you will see how complex lipids and proteins become organized in cell membranes (3.4, 4.1). Remember the different levels of protein organization? You will consider some examples of how protein structure translates into specific functions (3.6). You will be applying your knowledge of the properties of water molecules to the movement of water across membranes (2.5). You will see how the endomembrane system (4.6) helps cycle membranes.

5.1 Organization of Cell Membranes

LINKS TO
SECTIONS
3.4, 4.1

Cell membranes consist of a lipid bilayer in which many different kinds of proteins are embedded. The membrane is a continuous boundary layer that selectively controls the flow of substances across it.

REVISITING THE LIPID BILAYER

Think back on the phospholipids, the most abundant components of cell membranes (Section 3.4 and Figure 5.3a). Each has a phosphate-containing head and two fatty acid tails attached to one glycerol backbone. The head is hydrophilic, meaning it dissolves fast in water. The tails are hydrophobic; water repels them.

Immerse a lot of phospholipids in water, and they interact with water molecules and with one another until they spontaneously cluster into a sheet or film at the water's surface. Some line up as two layers, with all fatty acid tails sandwiched between the outward-facing hydrophilic heads. This is a **lipid bilayer**, the basic framework for cell membranes (Figure 5.3c).

THE FLUID MOSAIC MODEL

By the **fluid mosaic model**, every cell membrane has a mixed composition—or a *mosaic*—of phospholipids, glycolipids, sterols, and proteins. The lipids form an oily bilayer that serves as a barrier to water-soluble substances. Diverse proteins are either embedded in the bilayer or attached to one of its surfaces. They carry out most membrane functions.

The membrane is *fluid* because of interactions and motions of its components. The phospholipids differ in their heads and the length of their fatty acid tails. At least one of the tails is usually kinked, or unsaturated. Remember, an unsaturated fatty acid has one or more double covalent bonds in its carbon backbone; a fully saturated type has none. Also, most phospholipids drift sideways, spin around their long axis, and flex their tails, so they do not bunch up as a solid layer.

Figure 5.4 shows the fluid mosaic model. Section 5.2 is an overview of the membrane proteins that you will be reading about in many chapters to come.

DO MEMBRANE PROTEINS STAY PUT?

Some time ago, researchers figured out how to split a frozen plasma membrane down the middle of its bilayer. They found that proteins were not spread like a coat on the bilayer, as some had thought, but rather that many were embedded in it (Figure 5.5a). Were those proteins rigidly positioned in the membrane? No one knew until researchers designed an ingenious

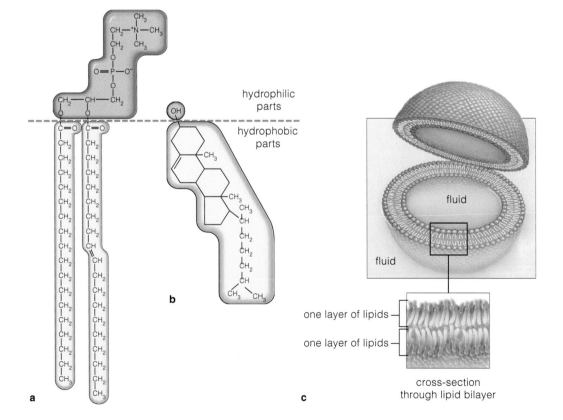

Figure 5.3 (**a**) Structural formula for phosphatidylcholine. This phospholipid is one of the most common molecules of animal cell membranes. *Orange* signifies its hydrophilic head, and *yellow*, its hydrophobic tails.

(**b**) Structural formula for cholesterol, the main sterol in animal tissues. Phytosterols are its equivalent in plant tissues.

(**c**) Spontaneous organization of lipid molecules into two layers (a bilayer structure). When immersed in liquid water, their hydrophobic tails become sandwiched between their hydrophilic heads, which dissolve in the water.

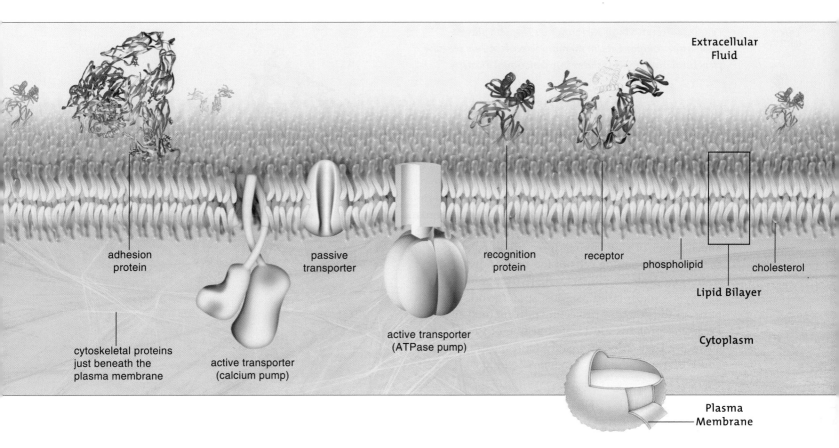

Extracellular Fluid

adhesion protein

passive transporter

recognition protein

receptor

phospholipid

cholesterol

Lipid Bilayer

cytoskeletal proteins just beneath the plasma membrane

active transporter (calcium pump)

active transporter (ATPase pump)

Cytoplasm

Plasma Membrane

Figure 5.4 *Animated!* Fluid mosaic model for the plasma membrane of an animal cell.

experiment. They induced an isolated human cell and an isolated mouse cell to fuse. The plasma membranes from the two species merged to form one continuous membrane in a new, hybrid cell. Most of the proteins mixed together in less than an hour (Figure 5.5b).

As we now know, many proteins are free to move laterally through the lipid bilayer, but others stay put. Some unite in complexes and do not move relative to one another. Receptors for acetylcholine, a signaling molecule, are like this. Cytoskeletal elements tether other proteins and restrict their lateral movements. For instance, a mesh of cross-lined spectrin proteins anchor glycophorin, a type of recognition protein, to the surface of all red blood cells. A transport protein that moves chloride one way and bicarbonate the other across the plasma membrane is similarly anchored.

All cell membranes consist of two layers of lipids—mainly phospholipids—and diverse proteins. Hydrophobic parts of the lipids are sandwiched between hydrophilic parts, which are dissolved in cytoplasmic fluid or in extracellular fluid.

All cell membranes have protein receptors, transporters, and enzymes. The plasma membrane also incorporates adhesion, communication, and recognition proteins.

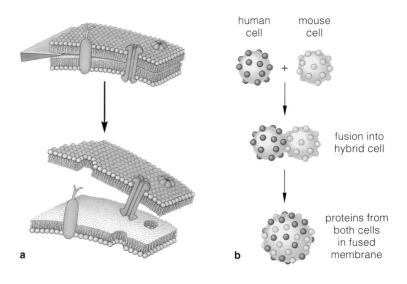

human cell   mouse cell

fusion into hybrid cell

proteins from both cells in fused membrane

a   b

Figure 5.5 *Animated!* Studying membranes. (**a**) Researchers split the two layers of a cell membrane's lipid bilayer apart, which revealed that proteins are embedded in the bilayer. (**b**) Result of an experiment in which plasma membranes from cells of two species were induced to fuse. Membrane proteins from both drifted laterally and became mixed.

5.2 Overview of the Membrane Proteins

LINK TO
SECTION
3.6

Cells interact with their surroundings through plasma membrane components. In membrane proteins, we see how structural diversity translates into functional diversity.

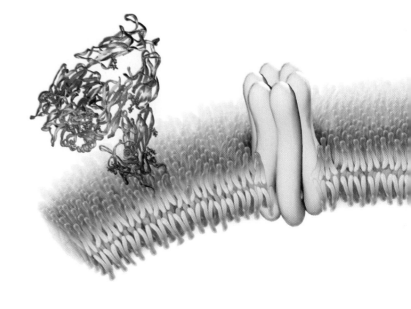

HOW ARE THE PROTEINS ORIENTED?

The fluid mosaic model is a good starting point for exploring membranes. But membranes differ in their composition and organization. Even the two surfaces of the same bilayer differ. For instance, many proteins (and lipids) of a plasma membrane have side chains of oligosaccharides and other carbohydrates, but only on the outward-facing surface (Figure 5.6). The kinds and number of side chains differ from one species to the next, even among cells of the same individual.

Integral proteins interact with hydrophobic parts of a bilayer's phospholipids. Most span the bilayer, with hydrophilic domains projecting beyond both surfaces. *Peripheral* proteins are located at one of the bilayer's surfaces. They interact weakly with integral proteins and with polar regions of membrane lipids.

WHAT ARE THEIR FUNCTIONS?

Figure 5.6 shows the main membrane proteins, lists their defining features, and gives some examples. The **transport proteins** either passively let specific solutes diffuse through a membrane-spanning channel in their interior or actively pump them through. Transporters are incorporated into all cell membranes.

The other proteins shown are typical of the plasma membrane. The **receptor proteins** bind extracellular substances, such as hormones, that can trigger change in cell activities. For example, certain enzymes control cell growth and division. They are switched on when somatotropin binds with receptors for it. Cells differ in their combinations of receptors.

Multicelled organisms have **recognition proteins** that are unique identity tags for each species; they are like molecular fingerprints. **Adhesion proteins** help cells of the same type locate each other and remain in the proper tissues. The **communication proteins** form channels that match up across the plasma membranes of two cells. They let signals and substances rapidly flow from the cytoplasm of one into the other.

All cell membranes have transporters that passively and actively assist water-soluble substances across the lipid bilayer. The plasma membrane, especially of multicelled species, has diverse receptors and proteins that function in self-recognition, adhesion, and communication.

**Adhesion
Proteins**

These proteins are embedded in the plasma membrane. They help one cell adhere to another or to a protein, such as collagen, that is part of an extracellular matrix.

Integrins, including this one, relay signals across the cell membrane. Cadherins of one cell bind with identical cadherins in adjoining cells. Selectins, which hold cells together, are abundant in endothelium, the special lining of blood vessels and the heart.

**Communication
Proteins**

Communication proteins of one cell match up with identical proteins in the plasma membrane of an adjoining cell. Fingerlike projections of both intertwine in the space between the two cells. The result is a channel that directly connects the cytoplasm of both. Chemical and electrical signals flow fast through the channel.

This protein is one-half of a cardiac gap junction in heart muscle. The other half is in the lipid bilayer of another heart muscle cell (not shown) positioned above it. Signals flow so fast across such channels that heart muscle cells contract as a single functional unit.

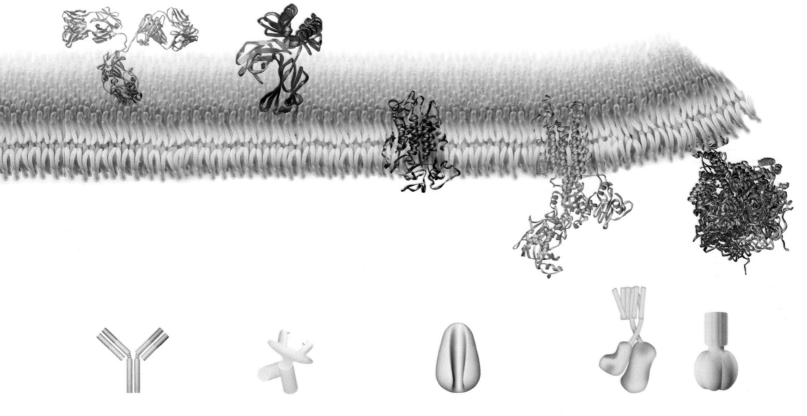

Receptor Proteins

Receptors embedded in a membrane are docks for hormones and other signaling molecules that may cause target cells to change their activities.

A signal might make a cell synthesize a certain protein, block or speed a reaction, secrete a substance, or get ready to divide.

Shown above, an antibody, a type of receptor made only by the type of white blood cell known as the B lymphocytes. These receptors are vital for all immune responses (Chapter 39).

Recognition Proteins

Certain glycoproteins (and glycolipids) project above the plasma membrane and identify a cell as *nonself* (foreign) or *self* (belonging to one's own body or a tissue).

Some, such as the HLAs (page 49), function in tissue defense. Foreign fragments bound to HLA sound the alarm for cells that defend the body. Other recognition proteins help cells stick to one another in tissues.

Passive Transporters

Passive transporters have a channel through their interior. Different kinds assist solutes or water simply by letting them diffuse through the channel, down concentration or electric gradients (Section 5.4). They do not require activation by energy inputs.

Shown here, GluT1; when its channel changes shape, glucose can cross a membrane. Aquaporins are open channels for water (page 89).

One cotransporter helps chloride and bicarbonate ions across a membrane at the same time, in opposite directions.

Ion-selective channels have molecular gates. Some gates open or close fast if a small molecule binds to them or if the charge distribution across the membrane shifts. Nerve and muscle cells have gated channels for sodium, calcium, potassium, and chloride ions.

Active Transporters

Active transport proteins pump a solute across the membrane to the side where it is more concentrated and less likely to move on its own. They require energy inputs to do this. Some are cotransporters that let one kind of solute flow passively "downhill" even as they pump a different kind "uphill."

Left, a calcium pump. Like the sodium–potassium pump, it is one of the ATPases.

Right, a type of ATPase that pumps H^+ through its interior channel, against gradients. It also can let H^+ diffuse back through the channel in a way that drives ATP synthesis. Hence its more precise name, ATP synthase (Chapters 7 and 8).

Figure 5.6 *Animated!* Major categories of membrane proteins. Included are simple icons and descriptions for membrane proteins that you will encounter in later chapters. The transporters span the lipid bilayer of all cell membranes. The other proteins shown are components of plasma membranes. Bear in mind, cell membranes also incorporate additional kinds of proteins, including some enzymes.

5.3 Diffusion, Membranes, and Metabolism

LINKS TO
SECTIONS
2.3, 2.5, 3.4

What determines whether a substance will move one way or another to and from a cell, across that cell's membranes, or through the cell itself? Diffusion down concentration gradients is part of the answer.

WHAT IS A CONCENTRATION GRADIENT?

A **concentration gradient** is a difference in the number per unit volume of molecules (or ions) of a substance between two adjoining regions. In the absence of other forces, the molecules move from a region where they are more concentrated to a region where they are not as concentrated. Why? Their inherent thermal energy keeps them in constant motion, so that they collide at random and bounce off one another millions of times each second. This happens more in regions where the molecules are most concentrated, and when you add it all up, the *net* movement is toward the region where they are not colliding and bouncing around as much. The molecules flow down their concentration gradient.

Diffusion is the name for the net movement of like molecules or ions down a concentration gradient. It is a factor in how substances move into, through, and out of cells. In multicelled species, it moves substances between body regions and between the body and its environment. For instance, when photosynthesis is going on in leaf cells, oxygen builds up and diffuses out of the cells and into air spaces in the leaf, where its concentration is lower. It then diffuses into the air outside the leaf, where its concentration is lower still.

Like other substances, oxygen tends to diffuse in a direction set by its *own* concentration gradient, not by gradients of other solutes. You can see the outcome by squeezing a drop of dye into water. The dye molecules diffuse slowly into the region where they are not as concentrated, and the water molecules move into the region where *they* are not as concentrated. Figure 5.7 shows simple examples of diffusion.

WHAT DETERMINES DIFFUSION RATES?

How fast a particular solute diffuses depends on the steepness of its concentration gradient, its size, the temperature, and electric or pressure gradients that may be present.

First, rates are high with steep gradients, because more molecules are moving out of a region of greater concentration compared with the number moving into it. Second, more heat energy makes molecules move faster and collide more often in warmer regions. Third, smaller molecules diffuse faster than large ones do.

Fourth, an electric gradient may alter the rate and direction of diffusion. An **electric gradient** is simply a difference in electric charge between adjoining regions. For example, each ion dissolved in fluids bathing a cell membrane contributes to a local electric charge. Opposite charges attract. Therefore, the fluid having more negative charge overall exerts the greatest pull on positively charged substances, such as sodium ions. Later chapters explain how many cell activities, such as ATP formation and the sending and receiving of signals in nervous systems, require the driving force of electric and concentration gradients.

Fifth, diffusion also may be affected by a **pressure gradient**. This is a difference in pressure exerted per unit volume (or area) between two adjoining regions.

MEMBRANE CROSSING MECHANISMS

Now think about the water bathing the surfaces of a cell membrane. Plenty of substances are dissolved in it, but the kinds and amounts close to its two surfaces

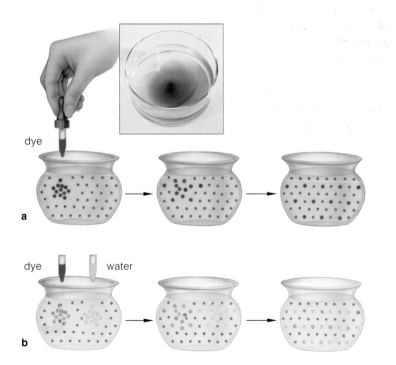

dye

a

dye   water

b

Figure 5.7 *Animated!* Two examples of diffusion. (**a**) A drop of dye enters a bowl of water. Gradually, the dye molecules become evenly dispersed through the molecules of water. (**b**) The same thing happens with the water molecules. Here, dye (*red*) and water (*yellow*) are added to the same bowl. Each substance will show a net movement down its own concentration gradient.

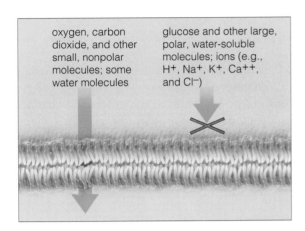

oxygen, carbon dioxide, and other small, nonpolar molecules; some water molecules

glucose and other large, polar, water-soluble molecules; ions (e.g., H+, Na+, K+, Ca++, and Cl−)

Figure 5.8 *Animated!* Selective permeability of cell membranes. Small, nonpolar molecules and some water molecules cross the lipid bilayer. Ions and large, polar, water-soluble molecules and the water dissolving them cross with the help of transport proteins. Also, proteins called aquaporins specifically enhance the diffusion of water across the plasma membrane of certain cells.

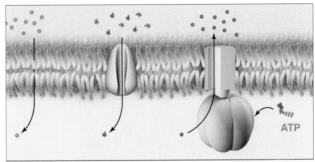

High

Concentration gradient across cell membrane

Low

| Diffusion of lipid-soluble substances across bilayer | Passive transport of water-soluble substances through channel protein; no energy input needed | Active transport through ATPase; requires energy input from ATP |

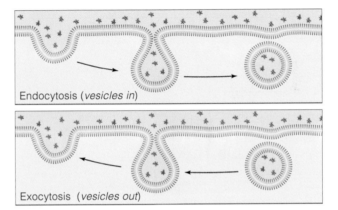

Endocytosis (*vesicles in*)

Exocytosis (*vesicles out*)

Figure 5.9 Overview of membrane crossing mechanisms.

differ. The membrane itself helps set up and maintain these differences. How? Its diverse lipid and protein components show **selective permeability**. They allow some substances but not others to enter and leave a cell. They also control when each substance can cross and how much crosses at a given time (Figure 5.8).

Membrane barriers and crossings are vital, because metabolism depends on the cell's capacity to increase, decrease, and maintain concentrations of substances required for reactions. That capacity also supplies the cell or organelles with raw materials, removes wastes, and maintains the cell volume and pH within ranges that favor reactions.

Lipids of a membrane's bilayer are mostly nonpolar, so they let small, nonpolar molecules such as O_2 and CO_2 slip across. Water molecules are polar, but some can slip through gaps that form when hydrophobic tails of lipids flex and bend (Section 5.1).

The lipid bilayer is impermeable to ions and large, polar molecules, including glucose. These substances cross a membrane by diffusing through the interior of transport proteins that span the bilayer. In many cells, proteins called aquaporins allow molecules of water to quickly cross the plasma membrane.

The passive transporters help specific solutes move down their concentration gradients but do not expend energy doing so. The mechanism, described shortly, is called *passive transport* or "facilitated" diffusion.

The active transporters help specific solutes diffuse across membranes, but they are not passive about it.

They move solutes against concentration and electric gradients, and they require an input of energy to do so. We call this mechanism *active transport*.

Other mechanisms move large particles into or out of cells. In *endocytosis* a vesicle forms around particles when a patch of plasma membrane sinks inward and seals back on itself. In *exocytosis*, a vesicle that formed in the cytoplasm fuses with the plasma membrane, so that its contents are released to the outside.

Before getting into these diverse mechanisms, you may wish to study the overview in Figure 5.9.

Diffusion is the net movement of molecules or ions of a substance into an adjoining region where they are not as concentrated. The steepness of such a concentration gradient as well as temperature, molecular size, and electric and pressure gradients affect diffusion rates.

Cellular mechanisms increase and decrease concentration gradients across cell membranes.

5.4 Working With and Against Gradients

LINK TO
SECTION
4.6

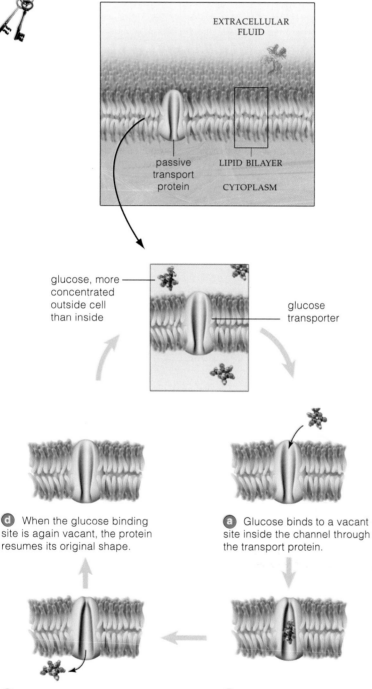

EXTRACELLULAR FLUID

passive transport protein

LIPID BILAYER

CYTOPLASM

glucose, more concentrated outside cell than inside

glucose transporter

d When the glucose binding site is again vacant, the protein resumes its original shape.

a Glucose binds to a vacant site inside the channel through the transport protein.

c Glucose becomes exposed to fluid on other side of the membrane. It detaches from the binding site and diffuses out of the channel.

b Bound glucose makes the protein change shape. Part of the channel closes behind the solute. Another part opens in front of it.

Figure 5.10 *Animated!* Passive transport. This model shows one of the glucose transporters that span the plasma membrane. Glucose crosses in both directions. The *net* movement of this solute is down its concentration gradient.

Large, polar molecules and ions cannot diffuse across a lipid bilayer. They require the help of transport proteins.

Many kinds of solutes cross a membrane by diffusing through a channel or tunnel inside transport proteins. When one solute molecule or ion enters the channel and weakly binds to the protein, the protein's shape changes. The channel closes behind the solute and opens in front of it, which exposes the solute to fluid on the other side of the membrane. Now the solute is released; the binding site reverts to its original shape.

PASSIVE TRANSPORT

In **passive transport**, a concentration gradient, electric gradient, or both drive diffusion of a substance across a cell membrane, through the interior of a transport protein. The protein does not require an energy input to assist the directional movement. That is why this mechanism is also known as facilitated diffusion.

Some passive transporters are open channels; others open or close as conditions change. Figure 5.10 shows how a glucose transporter works. When one end of its channel is shut, the other is open and invites glucose in. The channel closes behind the glucose and opens in front of it, on the other side of the membrane.

The *net* direction of a solute's movement depends on how many of its molecules or ions are randomly colliding with the transporters. Encounters simply are more frequent on the side of the membrane where its concentration is greatest. The solute's *net* movement tends to be toward the side of the membrane where it is less concentrated.

If nothing else were going on, passive transport would continue until concentrations on both sides of a membrane were equal. However, other events affect the outcome. For example, the bloodstream moves glucose to all tissues. There, glucose transporters help molecules of glucose get into cells. But as fast as some glucose molecules are diffusing into the cells, others are being used as building blocks and energy sources. By *using* glucose, then, cells help maintain a gradient that favors the uptake of *more* glucose molecules.

ACTIVE TRANSPORT

Solute concentrations continually shift across the cell membrane. Living cells never stop expending energy to pump solutes into and out of their interior. With **active transport**, energy-driven protein motors help a particular kind of solute cross a cell membrane *against* its concentration gradient.

Only specific solutes can bind to functional groups that line the interior channel of an active transporter, which is activated by a phosphate group from an ATP molecule. The phosphate-group transfer changes the transporter's shape in a way that releases the solute on the other side of the membrane.

Figure 5.11 focuses on a **calcium pump**. This active transporter helps keep the concentration of calcium in a cell at least a thousand times lower than outside. What is so great about that? You will find out later, but for now think of how one of your muscles moves. The nervous system commands calcium ions to flood out from a specialized ER compartment that threads around muscle fibers inside the muscle. Calcium ions clear the way for trillions of motor proteins (myosins) to interact with actin filaments in ways that bring about contraction (Section 37.7). That muscle will go on contracting until staggering numbers of calcium pumps move those ions back inside the compartment, against their concentation gradient.

The **sodium–potassium pump** is a cotransporter that moves two kinds of ions in opposite directions. Sodium ions (Na^+) from the cytoplasm diffuse into the pump's channel and bind to functional groups. A phosphate-group transfer by ATP activates the pump, which changes shape. The change opens the channel on other side of the membrane, where Na^+ is released and potassium (K^+) diffuses in—down *its* gradient. The phosphate group is released. The channel closes behind the K^+, which is released to the other side of the membrane. As you will see, the nervous, digestive, and urinary systems of vertebrates cannot function without cellular pumps that respond to signals and to chemical changes (Sections 34.3, 41.5, and 42.3).

All cells incorporate membrane pumps. In Section 32.3, you will read about an H^+ pump that controls the transport of a hormone in growing plant parts.

Many membrane transport proteins act as open or gated channels across cell membranes. They undergo reversible changes in shape that assist solutes across the membrane.

In passive transport, a transporter allows a solute to cross a cell membrane simply by diffusing through its interior.

In active transport, the net diffusion of a specific solute is against its gradient. The transporter must be activated, usually by an energy input from ATP, which counters the force inherent in the gradient.

Passive and active transport continually help lower or raise gradients across a membrane, which helps the cell respond to signals and to chemical changes.

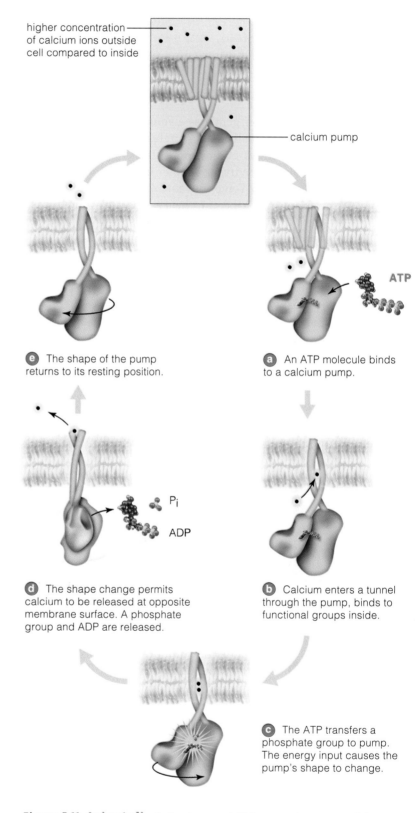

higher concentration of calcium ions outside cell compared to inside

calcium pump

e The shape of the pump returns to its resting position.

a An ATP molecule binds to a calcium pump.

Pi

ADP

d The shape change permits calcium to be released at opposite membrane surface. A phosphate group and ADP are released.

b Calcium enters a tunnel through the pump, binds to functional groups inside.

c The ATP transfers a phosphate group to pump. The energy input causes the pump's shape to change.

Figure 5.11 *Animated!* Active transport. This example uses a calcium pump that spans the plasma membrane. This sketch shows its channel for calcium ions. After two calcium ions bind to the pump, ATP transfers a phosphate group to it, thus providing energy that drives the movement of calcium *against* a concentration gradient across the cell membrane.

5.5 Which Way Will Water Move?

LINKS TO
SECTIONS
2.5, 4.8, 4.9

By far, more water diffuses across cell membranes than any other substance, so the main factors that influence its directional movement deserve special attention.

MOVEMENT OF WATER

Something as gentle as a running faucet or as mighty as Niagara Falls demonstrates **bulk flow**, or the mass movement of one or more substances in response to pressure, gravity, or another external force. Bulk flow accounts for some movement of water in multicelled organisms. A beating heart generates fluid pressure that pumps blood, which is mostly water. Sap flows inside tubes in trees, and this, too, is bulk flow.

What about the movement of water into and out of cells and organelles? If the concentration of water is not equal on both sides of a membrane, osmosis will probably occur. **Osmosis** is the diffusion of water across a selectively permeable membrane, to a region where the water concentration is lower.

You might be wondering: How can water —a liquid—be more or less concentrated? For the answer, you have to think of water in terms of its concentration relative to the amounts of solutes that may be dissolved in it. The greater the solute concentration, the lower the water concentration.

Visualize yourself pouring some glucose or another solute to a glass of water, so that you increase the volume of liquid. The glass has the same number of water molecules but in a greater volume of liquid.

Now visualize yourself using a membrane to divide the inside of another glass of water into two compartments. The membrane lets water but not glucose diffuse across it. Next, add glucose on one side of the membrane. Water follows its concentration gradient into the glucose solution until its concentration is the same on both sides of the membrane (Figure 5.12).

In cases of osmosis, "solute concentration" refers to the total number of molecules or ions in a volume of a solution. It does not matter whether the dissolved substance is glucose, urea, or anything else. The type of solute does not dictate water concentration.

EFFECTS OF TONICITY

Suppose you decide to test the statement that water tends to move into a region where solutes are more concentrated. You make three sacs from a membrane that water but not sucrose can cross, and fill each one with a solution that is 2 percent sucrose. You immerse the first sac in a liter of distilled water, the second sac in a solution that is 10 percent sucrose, and the third sac in a solution that is 2 percent sucrose.

In each experiment, tonicity dictates the extent and direction of water movement across the membrane, as Figure 5.13 shows. *Tonicity* refers to the relative solute concentrations of two fluids. When two fluids that are on opposing sides of a membrane differ in their solute concentrations, the **hypotonic solution** is the one with fewer solutes. The one having more solutes is a **hypertonic solution**. Water tends to diffuse from a hypotonic fluid into a hypertonic fluid. **Isotonic solutions** show no net osmotic movement.

Most cells have built-in mechanisms that counter shifts in tonicity. Red blood cells do not. Figure 5.13 shows what would happen to them if tonicity were to change. Normally, fluid in red blood cells is isotonic with tissue fluid. If the tissue fluid became hypotonic, too much water would diffuse into the cells, which would burst. If that tissue fluid became hypertonic, water would diffuse out, and the cells would shrivel.

EFFECTS OF FLUID PRESSURE

Most cells do not swell and burst from an influx of water by osmosis. For one thing, they can selectively transport solutes out. For another thing, the cells of plants and many protists, fungi, and bacteria have a

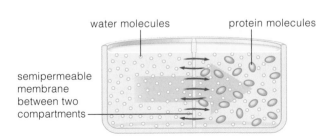

water molecules   protein molecules

semipermeable
membrane
between two
compartments

Figure 5.12 Solute concentration gradients and osmosis. A membrane divides this container into two compartments. Water but not proteins can cross it. Pour 1 liter of water in the left compartment and 1 liter of a protein-rich solution in the right compartment. The proteins occupy some of the space available, and the net diffusion of water in this case is from left to right (large *gray* arrow).

wall that helps keep them from rupturing when they become turgid, or swollen with fluid.

In later chapters, you will see how osmosis affects the water and solutes inside plants and animals. For now, just think about the hypotonic and hypertonic solutions in Figure 5.14. Water molecules move back and forth until the water concentration is equal on both sides of a membrane that separates them. But the volume of the formerly hypertonic solution has now increased, because its solutes cannot diffuse out.

The same thing happens in plant cells, which tend to be hypertonic relative to soil water. When a young plant cell grows, water moves into it by osmosis and exerts fluid pressure on its primary wall (Section 4.9). Up to a point, this pliable wall expands under fluid pressure, and the cell increases in volume. Continued expansion ends when the wall shows enough resistance to stop the further inward movement of water.

Any volume of fluid exerts **hydrostatic pressure**, or *turgor* pressure, against the wall or membrane that contains it. The **osmotic pressure** of any fluid is one measure of the tendency of water to follow its water concentration gradient and move into that fluid. When hydrostatic pressure and osmotic pressure are equal in magnitude, osmosis stops completely.

Plant cells also are vulnerable to the loss of water, which can occur when soil dries or becomes too salty. Water stops diffusing in and starts diffusing out, so hydrostatic pressure falls and the cytoplasm shrinks.

Osmosis is a net diffusion of water between two solutions that differ in solute concentration and are separated by a selectively permeable membrane. The greater the number of molecules and ions dissolved in a given amount of water, the lower the water concentration will be.

Water tends to move osmotically to regions of greater solute concentration (from hypotonic to hypertonic solutions). There is no net diffusion between isotonic solutions.

Fluid pressure that a solution exerts against a membrane or wall influences the osmotic movement of water.

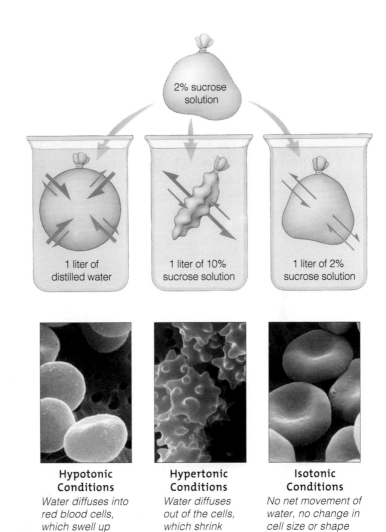

Hypotonic Conditions	Hypertonic Conditions	Isotonic Conditions
Water diffuses into red blood cells, which swell up	*Water diffuses out of the cells, which shrink*	*No net movement of water, no change in cell size or shape*

Figure 5.13 *Animated!* Tonicity and the direction of water movement between two adjoining regions. In each of three containers, arrow widths signify the direction and the relative amounts of flow. The micrographs below each sketch show the shape of a human red blood cell that is immersed in fluids of higher, lower, or equal concentrations of solutes. The solutions inside and outside red blood cells are normally balanced. This type of cell has no way to adjust to drastic change in solute levels in its fluid surroundings.

Figure 5.14 *Animated!* Experiment showing an increase in fluid volume as an outcome of osmosis. A selectively permeable membrane separates two compartments. Over time, the net diffusion will be the same in both directions across the membrane, but the fluid volume in the second compartment will be greater because there are more solute molecules in it.

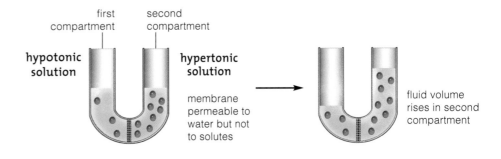

5.6 Membrane Traffic To and From the Cell Surface

LINKS TO
SECTIONS
4.6, 4.11

We leave this chapter with another look at exocytosis and endocytosis. By these mechanisms, vesicles move substances to and from the plasma membrane. Vesicles help the cell take in and expel materials in larger packets than transport proteins would be able to handle.

ENDOCYTOSIS AND EXOCYTOSIS

Think back on the lipid bilayer and how it minimizes the number of hydrophobic groups exposed to water. When the arrangement is disrupted—as when part of the plasma membrane or an organelle pinches off as a

vesicle—the bilayer becomes self-sealing. Why? The disruption exposes too many of hydrophobic groups to the surroundings. When a patch of membrane is budding off, its phospholipids are being repelled by water on both sides of it. The water molecules "push" the phospholipids together, which rounds off the bud as a vesicle and also seals the rupture.

The lipid bilayer's self-sealing behavior is the basis of membrane traffic to and from a cell surface (Figure 5.15). That traffic moves in two directions.

By **endocytosis**, a small patch of plasma membrane balloons inward and pinches off inside the cytoplasm. It forms an endocytic vesicle that moves its contents to some organelle or stores them in a cytoplasmic region. By **exocytosis**, a vesicle moves to the cell surface, and then the protein-studded lipid bilayer of its membrane fuses with the plasma membrane. While this exocytic vesicle is losing its identity, its contents are released to the outside (Figure 5.15).

There are three endocytic pathways. With *receptor-mediated* endocytosis, a hormone, vitamin, mineral, or another substance binds to receptors at the plasma membrane. A slight depression, or pit, forms in the plasma membrane beneath the receptors. The pit sinks into the cytoplasm as hydrophobic interactions cause a vesicle to form (Figure 5.16).

Phagocytosis ("cell eating") is a common endocytic pathway. Phagocytes such as amoebas engulf microbes, food particles, or cellular debris. In multicelled species, macrophages and some other white blood cells engulf pathogenic viruses and bacteria, cancerous body cells, and other threats. Receptors play a different role in phagocytosis. When they bind to a specific substance, they cause microfilaments to become rearranged into a mesh just beneath the phagocyte's plasma membrane. The microfilaments contract and a bulging volume of cytoplasm is squeezed toward the cell periphery. The bulge, still enclosed in the plasma membrane, extends outward as a pseudopod (Section 4.11 and Figure 5.17).

endocytosis   exocytosis

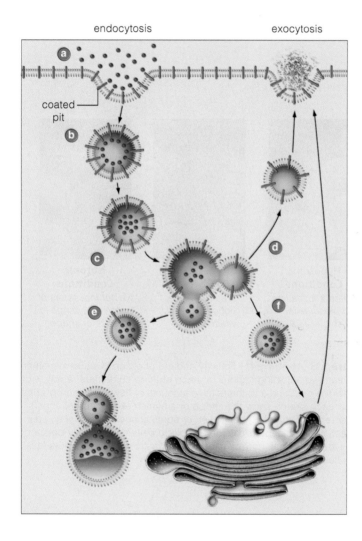

coated pit

Figure 5.15 *Animated!* Endocytosis and exocytosis. This sketch starts with receptor-mediated endocytosis. (**a**) Molecules get concentrated inside coated pits at the plasma membrane. (**b**) The pits sink inward and become endocytic vesicles. (**c**) The vesicle contents are sorted and often released from receptors. (**d**) Many sorted molecules are cycled back to the plasma membrane. (**e,f**) Many others are delivered to lysosomes and stay there or are degraded. Still others are routed to spaces in the nuclear envelope and inside ER membranes, and others to Golgi bodies.

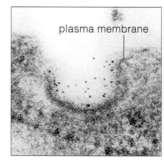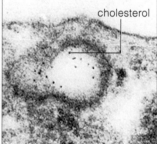

plasma membrane   cholesterol

Figure 5.16 Endocytosis of cholesterol molecules.

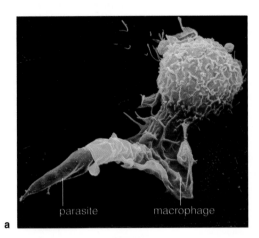

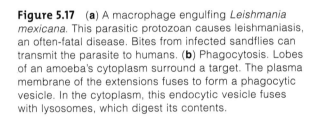

b   bacterium   phagocytic vesicle

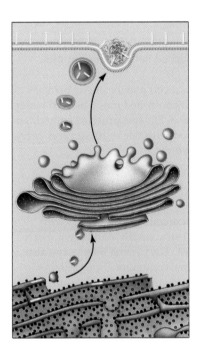

Figure 5.18 Example of how the asymmetric distribution of proteins, carbohydrates, and lipids in cell membranes originates. Proteins of the plasma membrane start out as new polypeptide chains, which become modified inside the channels of the ER and Golgi bodies. Many depart in vesicles that bud off, move to the plasma membrane, and fuse with it. The proteins inside automatically become oriented in the proper direction in the plasma membrane.

Figure 5.17 (**a**) A macrophage engulfing *Leishmania mexicana*. This parasitic protozoan causes leishmaniasis, an often-fatal disease. Bites from infected sandflies can transmit the parasite to humans. (**b**) Phagocytosis. Lobes of an amoeba's cytoplasm surround a target. The plasma membrane of the extensions fuses to form a phagocytic vesicle. In the cytoplasm, this endocytic vesicle fuses with lysosomes, which digest its contents.

Pseudopods flow completely around their target and then form a cytoplasmic vesicle. The vesicle sinks into the cytoplasm and fuses with lysosomes (Section 4.6). Lysosomal enzymes digest the vesicle's contents into fragments and smaller, reusable molecules.

Bulk-phase endocytosis is not as selective. A vesicle forms around a small volume of the extracellular fluid regardless of the kinds of substances dissolved in it.

MEMBRANE CYCLING

As long as a cell is alive, exocytosis and endocytosis are continually replacing and withdrawing patches of its plasma membrane, as in Figure 5.15. Apparently they do so at rates that maintain the total surface area of the plasma membrane. Steady losses in the form of endocytic membranes are balanced by replacements in the form of exocytic membranes.

For example, neurons release neurotransmitters in bursts of exocytosis. Each neurotransmitter is a type of signaling molecule that acts on neighboring cells.

An intense burst of endocytosis counterbalances each major burst of exocytosis.

The membranes are not shipped any which way. As an example, the composition and organization of the plasma membrane start inside the ER membranes, where many polypeptide chains become modified before being packaged and moved on to their final destinations (Section 4.6). Proteins that will become part of the plasma membrane are shipped in vesicles that fuse with a Golgi body. There, they are further modified, then sent off in other vesicles that fuse with the plasma membrane. As Figure 5.18 shows, fusion releases the proteins to the membrane surface that faces outside. There they will perform their functions.

Whereas transport proteins in a plasma membrane deal with ions and small molecules, exocytosis and endocytosis move large packets of materials in bulk across a plasma membrane.

By exocytosis, a cytoplasmic vesicle fuses with the plasma membrane, and its contents are released outside the cell.

By endocytosis, a small patch of plasma membrane sinks into the cytoplasm and pinches off as a vesicle. Membrane receptors often activate cytoskeletal elements that take part in endocytosis.

Phagocytosis is a form of endocytosis by which predatory amoebas engulf prey and certain white blood cells actively engulf tissue invaders, tissue debris, and cancer cells.

Summary

Section 5.1 Animal cell membranes consist mainly of phospholipids, along with glycolipids and sterols. The lipids are organized as a double layer, with all of their hydrophobic tails sandwiched between hydrophilic heads at both surfaces.

The lipid bilayer gives a cell membrane its primary structure and prevents uncontrolled movement of water-soluble substances across it. Diverse proteins embedded in the bilayer or associated with one of its surfaces carry out most membrane functions.

Biology⊗Now
Learn about membrane structure and the experiments that elucidated it with the animation on BiologyNow.

Section 5.2 Each cell membrane associates with cytoplasmic proteins that structurally reinforce it. Each has receptors at its surface. The plasma membrane also contains adhesion proteins, communication proteins, recognition proteins, and diverse receptors (Figure 5.19). Differences in the number and types of proteins affect responsiveness to substances at the membrane, as well as cell metabolism, pH, and volume.

Water-soluble substances cross cell membranes by passing through the interior of transport proteins, which open to both sides of the membrane.

Receptor proteins bind extracellular substances, and binding triggers alterations in cell activities.

Recognition proteins are molecular fingerprints; they identify cells as being of a given type. Adhesion proteins help cells of tissues adhere to one another and to proteins of the extracellular matrix.

Communication junctions extend across the plasma membranes of adjoining cells; they let substances and signals travel swiftly from one into the other.

Biology⊗Now
Use the animation on BiologyNow to familiarize yourself with the functions of receptor proteins.

Section 5.3 A concentration gradient is a difference in the number per unit volume of molecules (or ions) of a substance between two regions. The molecules tend to show a net movement down such a gradient, to the region where they are less concentrated. This behavior is called diffusion. The steepness of a concentration gradient, temperature, molecular size, and gradients in electrical charge and pressure influence diffusion rates.

Built-in cellular mechanisms work with and against gradients to move solutes across membranes.

Molecular oxygen, carbon dioxide, and other small, nonpolar molecules easily diffuse across a membrane's lipid bilayer. Ions and large, polar molecules such as glucose cross it through the interior of transport proteins that span the bilayer. Water molecules slip through gaps that briefly open in the bilayer. Aquaporins selectively assist water molecules across certain cell membranes.

Biology⊗Now
Investigate diffusion across membranes with the interaction on BiologyNow.

Section 5.4 Many solutes cross membranes through transport proteins that act as open or gated channels or that reversibly change shape. Passive transport does not require energy input; a solute is free to follow its own concentration gradient across the membrane. Active transport requires an energy input from ATP to move a specific solute against its concentration gradient.

Biology⊗Now
Compare the processes of passive and active transport, using the animation on BiologyNow.

Section 5.5 Osmosis is the diffusion of water across a selectively permeable membrane. The water molecules move down a water concentration gradient, which is influenced by solute concentrations and pressure.

Biology⊗Now
Explore the effects of osmosis with the interaction and animation on BiologyNow.

Section 5.6 By exocytosis, a cytoplasmic vesicle fuses with the plasma membrane, and its contents are released outside. By endocytosis, a patch of plasma membrane forms a vesicle that sinks into the cytoplasm.

Biology⊗Now
Use the animation on BiologyNow to discover how membrane components are cycled.

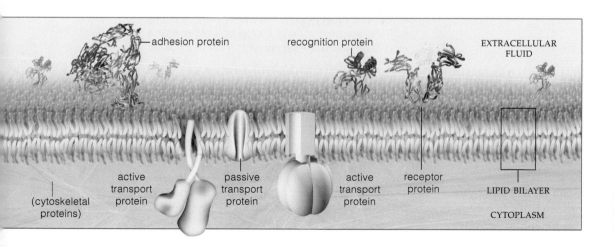

adhesion protein   recognition protein   EXTRACELLULAR FLUID

active transport protein   passive transport protein   active transport protein   receptor protein   LIPID BILAYER

(cytoskeletal proteins)   CYTOPLASM

Figure 5.19 Summary of major types of membrane proteins.

 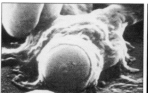 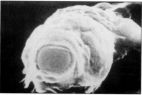

Figure 5.20 Go ahead, name the mystery membrane mechanism.

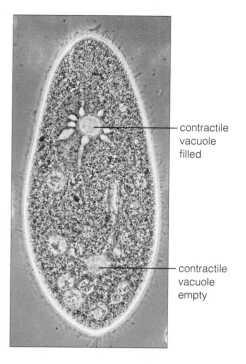

Figure 5.21 Light micrograph of one of the ciliated protozoans (*Paramecium*). This tiny single-celled body is crammed with diverse organelles, including contractile vacuoles.

— contractile vacuole filled

— contractile vacuole empty

Self-Quiz

Answers in Appendix II

1. Cell membranes consist mainly of a _____ .
   a. carbohydrate bilayer and proteins
   b. protein bilayer and phospholipids
   c. lipid bilayer and proteins

2. In a lipid bilayer, _____ of all of the lipid molecules are sandwiched between all of the _____ .
   a. hydrophilic tails; hydrophobic heads
   b. hydrophilic heads; hydrophilic tails
   c. hydrophobic tails; hydrophilic heads
   d. hydrophobic heads; hydrophilic tails

3. Most membrane functions are carried out by _____ .
   a. proteins   c. nucleic acids
   b. phospholipids   d. hormones

4. Plasma membranes incorporate _____ .
   a. transport proteins   c. recognition proteins
   b. adhesion proteins   d. all of the above

5. Diffusion is the movement of ions or molecules from one region to another where they are less concentrated. The rate of diffusion is affected by _____ .
   a. temperature   c. molecular size
   b. electrical gradients d. all of the above

6. _____ can readily diffuse across a lipid bilayer.
   a. Glucose   c. Carbon dioxide
   b. Oxygen   d. b and c

7. Some sodium ions cross a cell membrane through transport proteins that first must be activated by an energy boost. This is an example of _____ .
   a. passive transport   c. facilitated diffusion
   b. active transport   d. a and c

8. Immerse a living cell in a hypotonic solution, and water will tend to _____ .
   a. move into the cell   c. show no net movement
   b. move out of the cell d. move in by endocytosis

9. Vesicles form by way of _____ .
   a. membrane cycling   d. halitosis
   b. exocytosis   e. a through c
   c. phagocytosis   f. all of the above

10. Match the term with its most suitable description.
   ____ phagocytosis   a. molecular fingerprint
   ____ passive transport   b. basis of diffusion
   ____ recognition protein   c. big in membranes
   ____ active   d. one cell engulfs another
        transport   e. requires energy boost
   ____ phospholipid   f. docks for signals and
   ____ concentration   substances at cell surface
        gradient   g. no energy boost required
   ____ receptors   to move solutes

Additional questions are available on **Biology ⊛ Now™**

Critical Thinking

1. Is the white blood cell shown in Figure 5.20 disposing of a worn-out red blood cell by endocytosis, phagocytosis, or both?

2. Water moves osmotically into *Paramecium*, a single-celled aquatic protist. If unchecked, the influx would bloat the cell and rupture its plasma membrane, and the cell would die. An energy-requiring mechanism that involves contractile vacuoles expels excess water (Figure 5.21). Water enters the vacuole's tubelike extensions and collects inside. A full vacuole contracts and squirts water out of the cell through a pore. Are *Paramecium*'s surroundings hypotonic, hypertonic, or isotonic?

3. Water crosses cell membranes by diffusing past lipids that are jostling apart from one another in the bilayer. In many tissues, it also crosses faster through the interior channels of *aquaporins* (*white* arrow in Figure 5.22). As many as 3 billion water molecules per second flow through an aquaporin. Researchers already have found similar aquaporins in bacteria, plants, and insects.

Different aquaporins help different tissues respond to shifting conditions in the internal environment. They have roles in how the kidneys conserve or get rid of excess water. They play a part in producing and maintaining the fluid that bathes the spinal cord and brain, in producing saliva and tears, in keeping the lining of the lungs moist, and in keeping red blood cells from bursting or shriveling as the body's water–solute balance shifts.

If the gene for one of these water channels mutates, the outcome may be serious. Mutation in *aquaporin–0* results in cataracts, and mutation in *aquaporin–2* leads to a form of diabetes insipidus. Yet *aquaporin–1* seems less essential. In its absence, affected adults tend to produce unusually dilute urine but remain in good health as long as they drink plenty of water. Even so, affected individuals are rare. Speculate on the reasons why.

extracellular fluid

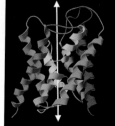

cytoplasm

Figure 5.22 Model for one of the four aquaporin subunits.

Alcohol, Enzymes, and Your Liver

The next time someone asks you to have a drink or two, or three, stop for a moment and think about the challenge confronting the cells that are supposed to keep a drinker alive—especially heavy drinkers. It makes little difference whether someone gulps down 12 ounces of beer, 5 ounces of wine, or 1–1/2 ounces of eighty-proof vodka. Each of these drinks has the same amount of "alcohol" or, more precisely, ethanol.

Ethanol molecules—CH_3CH_2OH—have water-soluble and fat-soluble components. Once they reach the stomach and the small intestine, they are quickly absorbed into the internal environment. The bloodstream transports more than 90 percent of ethanol's components to liver cells. There, enzymes speed their breakdown to a nontoxic form called acetate, or acetic acid. The liver has great numbers of alcohol-metabolizing enzymes, but they can detoxify only so much in a given hour.

One of the enzymes you will read about in this chapter is catalase, a foot soldier against toxins that can attack the body (Figure 6.1). Catalase assists another enzyme, alcohol dehydrogenase. When alcohol circulates through the liver, these enzymes convert it to acetaldehyde. The reactions cannot end there, because acetaldehyde becomes toxic when it accumulates in high concentrations. In healthy people at least, still another kind of enzyme speeds its breakdown to nontoxic forms.

Given the liver's central role in alcohol metabolism, habitually heavy drinkers gamble with alcohol-induced liver diseases. Over time, the capacity to tolerate alcohol diminishes because there are fewer and fewer liver cells —hence fewer enzymes—for detoxification.

What are some of the possible outcomes? A big one is *alcoholic hepatitis*, an all-too-common disease characterized by inflammation and destruction of liver tissue. Another disease, *alcoholic cirrhosis*, permanently scars the liver. In time, the liver stops working, with devastating effects.

The liver is the largest gland in the human body, and its activity impacts everything else. You would have a really hard time digesting and absorbing food without it. Your cells would have a hard time synthesizing and taking up carbohydrates, lipids, and proteins, and staying alive.

There is more to think about. The liver gets rid of a lot more toxic compounds than just acetaldehyde. It also makes certain plasma proteins that circulate freely in blood. In their absence, your body would not be able to defend itself well from attacks or to stop bleeding even from small cuts.

Figure 6.1 Something to think about—ribbon model for catalase, an enzyme that helps detoxify many substances that can damage the body, such as the alcohol in beer, martinis, and other drinks.

It would not be able to maintain the fluid volume of the internal environment that all of your cells depend upon.

Now think about a self-destructive behavior known as *binge drinking*. The idea is to consume large amounts of alcohol in a brief period. Binge drinking is now the most serious drug problem on campuses throughout the United States. Consider this finding from one study: Of nearly 17,600 students surveyed at 140 colleges and universities, 44 percent said they are caught in the culture of drinking. They report having five alcoholic drinks a day, on average.

Binge drinking can do far more than damage the liver. Put aside the related 500,000 injuries from accidents, the 70,000 cases of date rape, the 400,000 cases of (whoops) unprotected sex among students in an average year. Binge drinking can kill before you know what hit you. Drink too much, too fast, and you can abruptly end the beating of your heart.

With this example, we turn to **metabolism**, the cell's capacity to acquire energy and use it to build, degrade, store, and release substances in controlled ways. At times, the activities of your cells may be the last thing you want to think about. But they help define who you are and what you will become, liver and all.

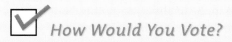
Watch the video online!

How Would You Vote?

Some people have damaged their liver because they drank too much alcohol. Others have a diseased liver. There are not enough liver donors for all the people waiting for liver transplants. Should life-style be a factor in deciding who gets a transplant? See BiologyNow for details, then vote online.

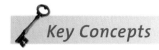
Key Concepts

THE NATURE OF ENERGY FLOW

Energy cannot be created or destroyed. It can only be converted from one form to another. Concentrated forms of energy tend to spread out, or disperse, spontaneously to less concentrated and less usable forms, such as dispersed heat. Collectively, chemical bonds resist this tendency and help organisms maintain their complex organization. Section 6.1

ENERGY CHANGES AND ATP

Metabolic reactions require an energy input before they can run spontaneously to completion. Some end with products that have more energy than the reactants. Others release more usable energy than the amount invested to start the reactions. ATP couples reactions that can release usable energy with reactions that require it. Section 6.2

ENERGY CHANGES AND ENZYMES

On their own, chemical reactions proceed too slowly to sustain life. Enzymes increase reaction rates enormously. They lower the amount of energy it takes to align reactive groups, destabilize electric charges, and break bonds so that products can form from reactants.

Temperature, pH, salinity, and other environmental factors influence enzyme activity. So does the availability of cofactors, or enzyme helpers. Sections 6.3, 6.4

THE NATURE OF METABOLISM

Metabolic pathways are enzyme-mediated sequences of reactions. By controlling enzymes that govern key steps in these pathways, cells build up, maintain, and decrease amounts of thousands of substances. ATP-forming pathways require redox reactions, or electron transfers. Section 6.5

FROM CONCEPT TO APPLICATION

We can simply absorb information about nature, including how enzymes work. We also can interpret information in novel ways that may have practical application. Section 6.6

Links to Earlier Concepts

Reflect again on the road map for life's organization (Section 1.1). Here you will gain insight into how organisms tap into a grand, one-way flow of energy to maintain that organization (1.2). You will start thinking about how cells use the chemical behavior of electrons and protons in ways that help make ATP (2.3). Remember what you learned about acids and bases (2.6)? Here you will see how pH influences enzyme activity.

6.1 Energy and Time's Arrow

LINK TO
SECTION
1.2

You know, almost without thinking about it, that your life does not stand still and will not start all over again. You have a sense of time's arrow—that everything we have observed and experienced, and all we expect will happen, goes forward. But <u>why</u> does it go forward?

WHAT IS ENERGY?

A dictionary definition of **energy** is "the capacity to do work," which doesn't say much. It is no more than a clue to two of the most sweeping laws of nature we humans have ever tried to wrap our minds around. In itself, the **first law of thermodynamics** seems simple enough: *Energy cannot be created or destroyed.*

Basically, this law deals with the *quantity* of energy in the universe. There is a finite amount distributed in different forms. However, one form may be converted to another.

One form, *potential* energy, is a capacity to do work because of something's location and the arrangement of its parts. While the skydivers shown in Figure 6.2 were still inside the plane, each had a store of potential energy because of their position above the ground. ATP and other molecules in their body had potential energy because of how their atoms were held together in particular arrangements by chemical bonds.

When the skydivers jumped, potential energy was transformed into *kinetic* energy, or energy of motion. ATP in muscle cells gave up some potential energy to molecules of contractile units and set them in motion. The motion of many thousands of muscle cells made whole muscles move. With each energy transfer from ATP, a bit of energy slipped off into the surroundings as *thermal* energy, or heat.

Figure 6.2 Forms of energy. How many can you identify?

The potential energy of molecules has its own name, *chemical* energy. It is measurable, as in kilocalories. A **kilocalorie** is the same as 1,000 calories, or the amount of energy it takes to heat 1,000 grams of water from 14.5°C to 15.5°C at standardized pressure.

THE ONE-WAY FLOW OF ENERGY

Now imagine all the energy changes going on inside the skydivers, between their hot bodies and molecules of air they are falling through, between the sun and those plants in the fields below them. Zoom out past the sun, past our solar system, and all the way to the boundary of the known universe, and you will pass staggeringly diverse energy conversions.

Amazingly, another law of thermodynamics helps explain every conversion by dealing with the *quality* of energy. It tells us why the sun will eventually burn out, why animals eat plants and one another, why nitroglycerin spontaneously explodes but you don't, why it rains, why skydivers fall down instead of up and splat if their parachutes don't open, and why your life can't start over again.

The **second law of thermodynamics** sounds simple enough: *Energy tends to flow from concentrated to less concentrated forms—so the cost of concentrating it in one area comes at a greater cost of energy dispersal or dilution somewhere else.* This tendency gives us our sense of "time's arrow." Skydivers in free fall do not have a concentrated form of energy that can get them back to the plane. A hot pan gives off heat as it cools, and that heat cannot diffuse back from the air to the pan. You can't rewind the winds flowing around the eye of a hurricane. You can't take back a scream.

The second law says only that concentrated energy tends to disperse spontaneously. It does not say when or how slowly or fast that might happen. In our own world, *the collective strength of chemical bonds resists the spontaneous direction of energy flow.* Think of the energy in uncountable numbers of chemical bonds between all the atoms making up the Rocky Mountains. Millions of years will pass before attacks by winds, rain, ice, and other assaults will break enough bonds to return the mountains to the seas. Or think of all the chemical bonds in your skin, heart, liver, and other body parts. The concentration of energy definable as "you" stays together as long as they do.

Entropy is the measure of how much and how far a concentrated form of energy has been dispersed after an energy change. It is not a measure of order versus disorder. In 2004, at one of the most active subduction zones on the seafloor, a major earthquake generated a

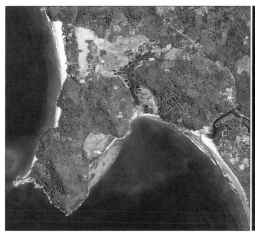

Figure 6.3 Aftermath of an appalling change in entropy. Gleebruk village, Indonesia, shown before and after a tsunami struck in December 2004. A major earthquake on the seafloor generated giant ocean waves—a form of concentrated mechanical energy that spread out after the waves hit land.

tsunami that spread across the Indian Ocean. Few of us missed the images of monstrous waves lashing at coasts all the way from Indonesia to Africa (Figure 6.3). Orderly rows of houses, hotels, and crops were ripped apart and washed away. But this horrific mess was not an increase in entropy; *it was its aftermath.* The earthquake had generated waves of pressure—a form of mechanical energy—in ocean water. There was a huge concentration of pressure that dissipated as the waves slammed against shorelines. The before/after difference in mechanical energy—from the original concentration on the seafloor to the last fingers of surf that fizzled away on land—was the entropy change.

The point is, *energy tends to flow in one direction, from concentrated to less concentrated forms.* The world of life is responsive to the flow. Photosynthetic cells of plants and other producers tap into a concentrated store of energy: light from the sun. They convert it to chemical bond energy in sugars and other organic compounds. Consumers as well as producers access that stored energy by breaking and rearranging chemical bonds. With each conversion, though, some energy is lost as heat, a dispersed form of energy that cells can't gather up again. The inevitable losses mean that the world of life must maintain its complex organization through ongoing replenishments of energy—which is being lost from someplace else (Section 1.2 and Figure 6.4).

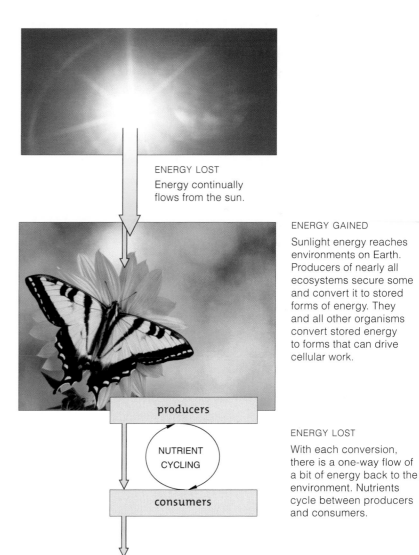

ENERGY LOST
Energy continually flows from the sun.

ENERGY GAINED

Sunlight energy reaches environments on Earth. Producers of nearly all ecosystems secure some and convert it to stored forms of energy. They and all other organisms convert stored energy to forms that can drive cellular work.

producers

NUTRIENT CYCLING

consumers

ENERGY LOST

With each conversion, there is a one-way flow of a bit of energy back to the environment. Nutrients cycle between producers and consumers.

Energy is the capacity to do work, and it cannot be created or destroyed. It can be converted from one form to another.

Energy concentrated in one place tends to spread out, or disperse, on its own. The collective strength of chemical bonds resists this spontaneous direction of energy flow.

Life continues as long as organisms tap into concentrated energy sources and use them to build complex molecules even as they continually lose energy in the form of heat.

Figure 6.4 A one-way flow of energy into an ecosystem compensates for the one-way flow of energy out of it.

6.2 Time's Arrow and the World of Life

LINKS TO
SECTIONS
1.5, 3.7, 5.4

Once the two laws of thermodynamics were just hypotheses. They became accepted as theories so powerful and wide-ranging that they became known as laws. Remember, the best theories—and laws—are supported by predictions based on them (Section 1.5). With respect to metabolism, the key prediction is this: It takes net inputs of energy to force small molecules to combine into larger ones, such as glucose, that are more concentrated forms of energy.

The molecules of life do not form spontaneously, and they are not that stable in the presence of free oxygen, which reacts with them and disrupts their structure. Cellular mechanisms safeguard these molecules. They control when and where energy changes will occur; that is, when energy will flow from one substance to another during chemical reactions.

You already know about some of the participants in metabolic reactions. Starting substances are called *reactants*. Substances formed before a reaction ends are *intermediates*, and those remaining are *products*. ATP

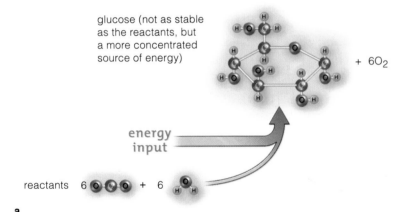

glucose (not as stable as the reactants, but a more concentrated source of energy)

+ $6O_2$

energy input

reactants   6   +   6

a

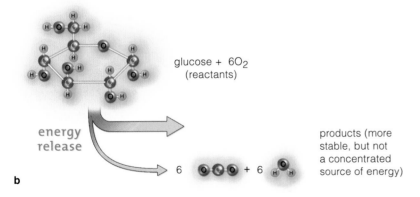

glucose + $6O_2$ (reactants)

energy release

products (more stable, but not a concentrated source of energy)

6   +   6

b

Figure 6.5 *Animated!* Two categories of energy changes in chemical work. (**a**) Endergonic reactions require a net input of energy. (**b**) Exergonic reactions end with a net release of usable energy.

and other *energy carriers* activate enzymes and other molecules by phosphate-group transfers. *Enzymes* are catalysts; they speed specific reactions enormously. *Cofactors* are metal ions or coenzymes. They assist the enzymes by accepting and donating electrons, atoms, and functional groups. *Transport proteins* help solutes across membranes. Their action affects concentrations of substances, which in turn affects how, when, and whether a reaction can proceed.

ACTIVATION ENERGY—WHY THE WORLD DOESN'T GO UP IN SMOKE

When substances react, some of the chemical energy required to break bonds is conserved as new bonds form. Some energy is lost as heat, light, or both. Think of what happens after a spark from a campfire ignites tinder-dry plants. Plants are mostly cellulose—which has three reactive hydroxyl groups *in each one of many repeating units*: $C_6H_7O(OH)_3$. Once this reaction starts, it proceeds swiftly on its own. Most of the cellulose breaks down fully to carbon dioxide (CO_2) and water (H_2O), with the release of light and heat. Remember Figure 1.1? A single match can start a firestorm.

Why doesn't the world go up in flames on its own? It takes a boost of energy to overcome the strength of chemical bonds in reactants. The boost has to be big for some reactions but not much for others. Like other explosives, nitroglycerin has a lot of oxygen atoms. A hard shake or jarring is all it takes for nitroglycerin to start falling apart—explosively fast—into hot gases.

Each kind of reaction has a characteristic **activation energy**, the minimum amount of energy that can get the reaction to the point that it will run on its own. By controlling the energy inputs into reactions, cells are able to control when and how fast the reactions occur.

UP AND DOWN THE ENERGY HILLS

Let's see how this works when photosynthetic cells build glucose and other carbohydrates. An input of energy from the sun triggers two stages of reactions. Ultimately, six CO_2 and six H_2O molecules are used in the formation of one glucose molecule ($C_6H_{12}O_6$) and six oxygen molecules (O_2). Figure 6.5*a* is a simple way to think about these reactants and products.

The synthesis of glucose is an example of a series of reactions that just will not happen on their own. Why not? Carbon dioxide and water do have energy stored in their covalent bonds, but the bonds are so stable that it is as if they are at the base of an "energy hill." It takes inputs of energy to break those bonds

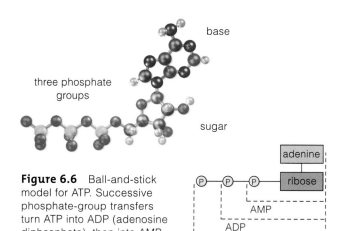

Figure 6.6 Ball-and-stick model for ATP. Successive phosphate-group transfers turn ATP into ADP (adenosine diphosphate), then into AMP (adenosine monophosphate).

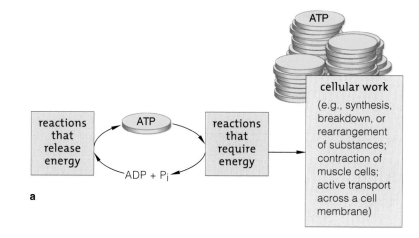

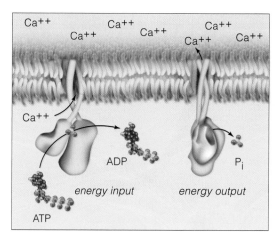

a

b A case of cellular work. ATP transfers a phosphate group to a transport protein spanning a plasma membrane. As you read in Section 5.4, the energy boost causes this active transport protein to change its shape in ways that pump calcium ions out of the cell, against their concentration gradient.

Figure 6.7 *Animated!* (**a**) ATP function. Recurring phosphate-group transfers convert ATP into ADP, and back to ATP. (**b**) Example of ATP-driven cellular work.

and convert them to something higher up on the hill. Any reactions that require a net input of energy are said to be *endergonic* (meaning energy in).

Glucose is a concentrated source of energy. When it is broken apart, energy is released. Cells capture some of the energy to do cellular work. Energy also ends up in covalent bonds of smaller, more stable products. With aerobic respiration, those products are six CO_2 and six H_2O molecules (Figure 6.5b).

Aerobic respiration releases energy bit by bit, with many conversion steps, so cells can capture some of it efficiently. This metabolic process is like a downhill run, from a concentrated form of energy (glucose) to less concentrated forms. Any reactions that end with a net release of energy are *exergonic* (meaning energy out). They, too, require an energy boost to get past the activation energy barrier. However, the net amount of energy released is more than the amount invested.

ATP—THE CELL'S ENERGY CURRENCY

It doesn't take a huge leap of the imagination to sense that cells stay alive by *coupling* reactions that require energy with reactions that release it. For nearly all metabolic reactions, adenosine triphosphate, or **ATP**, is the energy carrier, or coupling agent. All cells make this nucleotide, which consists of a five-carbon sugar (ribose), the base adenine, and three phosphate groups (Figure 6.6). ATP easily gives up phosphate groups to other molecules and thus primes them to react. Any phosphate-group transfer is called **phosphorylation**.

ATP is the currency in a cell's economy. Cells spend it in energy-requiring reactions and also invest it in energy-releasing reactions that help keep them alive. We use a cartoon coin to symbolize ATP (Figure 6.7a). Because ATP is the main energy carrier for so many reactions, you might infer—correctly—that cells have

ways of renewing it. When ATP gives up a phosphate group, ADP (adenosine diphosphate) forms. ATP can re-form when ADP binds to inorganic phosphate (P_i) or to a phosphate group that was split from a different molecule. Regenerating ATP by this **ATP/ADP cycle** helps drive most metabolic reactions (Figure 6.7b).

Activation energy is the minimum amount of energy required to get any reaction to the point where it will run spontaneously, with no further energy input. The amount differs for different reactions.

Reactions that build large organic compounds from smaller ones cannot run without a net input of energy. Reactions that degrade large molecules to smaller ones end with a net output of usable energy.

ATP, the main energy carrier in all cells, couples energy-releasing and energy-requiring reactions. It primes other molecules to react by phosphate-group transfers.

6.3 How Enzymes Make Substances React

If you left a cupful of glucose out in the open, years would pass before you would see its conversion to carbon dioxide and water. Yet that same conversion takes just a few seconds in your body. Enzymes make the difference.

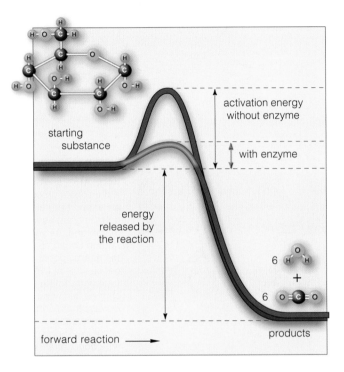

Enzymes are catalysts that can make reactions occur hundreds to millions of times faster than they would on their own. Enzyme molecules can work again and again; a reaction does not use them up or irreversibly alter them. Each type of enzyme chemically recognizes, binds, and alters specific reactants only. For instance, thrombin recognizes and cleaves only a peptide bond between arginine and glycine in a particular protein, thereby converting the protein into a factor that helps clot blood. Finally, nearly all enzymes are proteins. (A few kinds of RNAs also show enzymatic activity.)

Reactions cannot proceed until the reactants have a minimum amount of internal energy—the activation energy. Visualize activation energy as a barrier—a hill or brick wall (Figures 6.8 and 6.9). Enzymes lower it. How? *Compared with the surrounding environment, they offer a stable microenvironment, more favorable for reaction.*

Consider: Enzymes are far larger than **substrates**, another name for the reactants that bind to a specific enzyme. Their polypeptide chains are folded in ways that afford structural stability. Certain folds also form one or more chemically stable **active sites**: pockets or crevices where the substrates bind and where specific reactions can proceed rapidly and repeatedly.

Part of a substrate is complementary in shape, size, solubility, and charge to the active site. Because of the

Figure 6.8 *Animated!*
Activation energy: the minimum amount of internal energy that reactants must have before a reaction will run to products. An enzyme enhances the reaction rate by lowering the required amount; it lowers the energy hill.

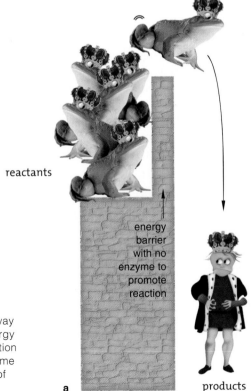

Figure 6.9 Simple way to think about the energy required to get a reaction going without an enzyme (**a**) and with the help of an enzyme (**b**).

one of four heme groups cradled
in one of four polypeptide chains

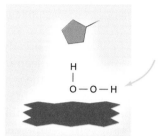 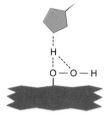 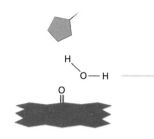

a Hydrogen peroxide (H_2O_2) enters a cavity in catalase. It is the substrate for a reaction aided by an iron molecule in a heme group (*red*).

b A hydrogen of the peroxide is attracted to histidine, an amino acid projecting into the cavity. One oxygen binds the iron.

c This binding destabilizes the peroxide bond, which breaks. Water (H_2O) forms. In a later reaction, another H_2O_2 will pull the oxygen from iron, which will then be free to act again.

Figure 6.10 *Animated!* How catalase works. This enzyme has four polypeptide chains and four heme groups (coded *red*).

fit, each enzyme chemically recognizes and binds its substrate among thousands of substances in cells.

Think back on the main types of enzyme-mediated reactions (Section 3.2). With *functional group transfers*, one molecule gives up a functional group to another. With *electron transfers*, one or more electrons stripped away from a molecule are donated elsewhere. With *rearrangements*, a juggling of internal bonds converts one kind of molecule to another. With *condensation*, two or more molecules become covalently bound into a larger molecule. Finally, with *cleavage* reactions, a larger molecule splits into smaller ones.

When we talk about activation energy, *we really are talking about the energy it takes to align reactive chemical groups, destabilize electric charges, and break bonds.* These events put a substrate at its **transition state**. Then, its bonds are at the breaking point, and the reaction can run easily to product (Figure 6.10).

The binding between an enzyme and its substrate is weak and temporary (that is why the reaction does not change the enzyme). However, energy is released when these weak bonds form. This "binding energy" stabilizes the transition state long enough to keep the enzyme and its substrate together for the reaction to be completed.

With enzymes, four mechanisms work alone or in combination to lower the activation energy and move substrates to the transition state:

Helping substrates get together. When they are at low concentrations, molecules of substrates rarely react. Binding at an active site is as effective as a localized boost in concentration, by as much as ten millionfold.

Orienting substrates in positions favoring reaction. On their own, substrates collide from random directions. By contrast, the weak but extensive bonds at an active site put reactive groups close together.

Shutting out water molecules. Because of its capacity to form hydrogen bonds so easily, water can interfere with the breaking and formation of chemical bonds in reactions. Certain active sites have an abundance of nonpolar amino acids with hydrophobic groups, which repel water and keep it away from the reactions.

Inducing a fit between enzyme and substrate. By the **induced-fit model**, a substrate is almost but not quite complementary to an active site. An enzyme restrains the substrate and stretches or squeezes it into a certain shape, often next to another molecule or to a reactive group. By optimizing the fit between them, it moves the substrate to the transition state.

On their own, chemical reactions occur too slowly to sustain life. Enzymes greatly increase reaction rates by lowering the activation energy. That is the minimum amount of energy required to align reactive groups, destabilize electric charges, and break bonds so that products can form from reactants.

In an enzyme's active site, substrates move to a transition state, when their bonds are at the breaking point and the reaction can run spontaneously to completion.

The transition state is reached by various mechanisms that concentrate and orient substrates, exclude water from the active site, and induce an optimal fit between the active site and substrate.

6.4 Enzymes Don't Work In a Vacuum

LINK TO SECTION 2.6

Many factors influence what an enzyme molecule does at any given time or whether it is built in the first place. Here we highlight a few of the major factors.

CONTROLS OVER ENZYMES

What happens when one or another of the thousands of substances in cells becomes too abundant or scarce? Many controls over enzymes help cells respond fast by adjusting specific reactions. Feedback mechanisms can activate or inhibit enzymes in ways that conserve energy and resources. Cells produce what conditions require—no more, no less.

Controls maintain, lower, or raise concentrations of substances. They adjust how fast enzyme molecules are synthesized, and they activate or inhibit the ones already built. In multicelled species, enzyme controls keep individual cells functioning in ways that benefit the whole body.

In some cases, a molecule that acts as an activator or inhibitor can reversibly bind to an *allosteric* site on the enzyme, not to the active site (*allo*– other; *steric*, structure). Binding alters the enzyme's shape in a way that hides or exposes the active site (Figure 6.11).

Visualize a bacterial cell making tryptophan and other amino acids—the building blocks for proteins. Even when it has made enough proteins, tryptophan synthesis continues until its increasing concentration causes **feedback inhibition**. This means a change that results from a specific activity *shuts down the activity*.

A feedback loop starts and ends at many allosteric enzymes. In this case, unused tryptophan binds to an allosteric site on the first enzyme in the tryptophan biosynthesis pathway. Binding makes the active site change shape, so less tryptophan can be made (Figure 6.12). At times when not many tryptophan molecules are around, the allosteric sites are unbound. Thus the active sites remain functional, and the synthesis rate picks up. In such ways, feedback loops quickly adjust the concentrations of substances.

EFFECTS OF TEMPERATURE, pH, AND SALINITY

What an enzyme molecule actually does also depends on conditions in the environment. Temperature, pH, and salinity all have impact on it.

Temperature is a measure of molecular motion. As it rises, it boosts reaction rates both by increasing the likelihood that a substrate will bump into an enzyme and by raising a substrate molecule's internal energy. Remember, the more energy a reactant molecule has, the closer it gets to jumping that activation energy barrier and taking part in a reaction.

Above the range of temperatures that an enzyme can tolerate, weak bonds are broken. The shape of the enzyme changes, so substrates no longer can bind to the active site. The reaction rate falls sharply (Figure 6.13). Such declines typically occur with fevers above 44°C (112°F), which people usually cannot survive.

Also, remember how the pH of solutions can vary (Section 2.6)? In the human body, most enzymes work best at pH 6–8. For instance, trypsin is active in the small intestine (pH of about 8). The enzyme pepsin is one of the exceptions. This nonspecific protease can digest any protein. It is produced in inactive form and

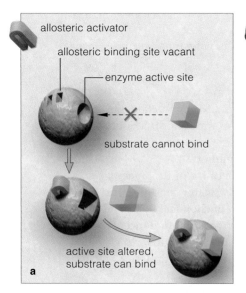

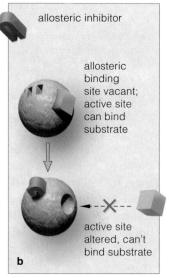

Figure 6.11 Animated! Allosteric control over enzyme activity. (**a**) An active site is unblocked when an activator binds to a vacant allosteric site. (**b**) An active site is blocked when an inhibitor binds to a vacant allosteric site.

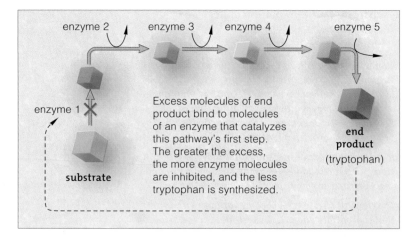

Figure 6.12 Animated! Feedback inhibition of a metabolic pathway. Five kinds of enzymes act in sequence to convert a substrate to tryptophan.

normally becomes activated only in gastric fluid, in the stomach. Gastric fluid is highly acidic, with a pH of 1–2. If activated pepsin were to leak out of the stomach, it would digest the proteins in your tissues instead of those in food. Figure 6.14 shows the effects of pH on pepsin and two other kinds of enzymes.

Also, most enzymes stop working effectively when the fluids in which they are dissolved are saltier or less salty than their range of tolerance. Too much or too little salt interferes with the hydrogen bonds that help hold an enzyme in its three-dimensional shape. By doing so, it inactivates the enzyme.

HELP FROM COFACTORS

Finally, don't forget the cofactors. These metal ions or coenzymes help at the active site of enzymes or taxi electrons, H^+, or functional groups to other reactions. **Coenzymes** are a class of organic compounds that may or may not have a vitamin component.

One or more metal ions assist nearly a third of all known enzymes. Metal ions easily give up and accept electrons. As part of coenzymes, they help products form by shifting electron arrangements in substrates or intermediates. That is what goes on at the hemes in catalase. Heme has an organic ring structure, with an iron atom at the center of the ring. As you saw in Figure 6.10, the iron atoms assist catalase in speeding the breakdown of hydrogen peroxide to water.

Like vitamin E, catalase is an **antioxidant**. It helps neutralize free radicals. *Free radicals*, or atoms with at least one unpaired electron, are leftovers of reactions. They attack the structure of DNA and other biological molecules. As we age, we make less and less catalase, so free radicals accumulate (Section 43.6).

Some coenzymes are tightly bound to an enzyme. Others, such as NAD^+ and $NADP^+$, can diffuse freely through the cytoplasm. Either way, they participate intimately in a metabolic reaction. Unlike enzymes, many become modified during the reaction, but they are regenerated elsewhere.

Controls over enzymes enhance or inhibit their activity. By doing so, they maintain, lower, or raise concentrations of many thousands of substances in coordinated ways.

Enzymes work best when the cellular environment stays within limited ranges of temperature, pH, and salinity. The actual ranges differ from one type of enzyme to the next.

Many enzymes are assisted by cofactors, which are specific metal ions or coenzymes.

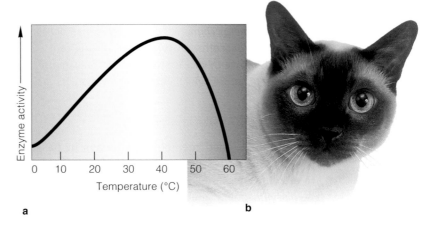

a **b**

Figure 6.13 Enzymes and the environment. (**a**) How increases in temperature affect one enzyme's activity.

(**b**) The air temperature outside the body affects the fur color of Siamese cats. Epidermal cells that give rise to the cat's fur produce a brownish-black pigment, melanin. Tyrosinase, an enzyme in the melanin production pathway, is heat-sensitive in the Siamese. It becomes less active in warmer parts of the cat's body, which end up with less melanin, and lighter fur. Put this cat's feet in booties for a few weeks and its warm feet will become light.

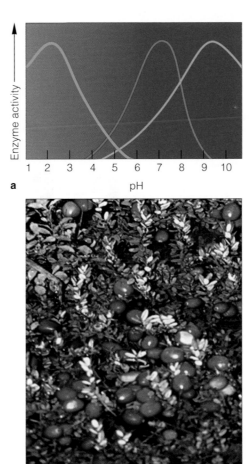

a

b

Figure 6.14 Enzymes and the environment. (**a**) How pH values affect three enzymes. The *blue* graph line shows the activity for pepsin. (**b**) Cranberry plants grow best in acidic bogs. Unlike most plants, they have no nitrate reductase. This enzyme converts nitrate (NO_3) found in most soils to metabolically useful ammonia (NH_3). In highly acidic soils, nitrogen is already in the form of ammonia (NH_4^+).

6.5 Metabolism—Organized, Enzyme-Mediated Reactions

LINK TO
SECTION
2.3

How cells use energy changes is one aspect of metabolism. Another is the concentration, conversion, and disposal of materials by energy-driven reactions. Most reactions in cells are part of stepwise metabolic pathways.

TYPES OF METABOLIC PATHWAYS

We have mentioned metabolic pathways in passing. Now let's formally define them. **Metabolic pathways** are enzyme-mediated sequences of reactions in cells. The *biosynthetic* (or anabolic) kinds require a net input of energy to produce glucose, starch, and other large molecules from small ones. Photosynthesis is the main biosynthetic pathway for the world of life (Figure 6.15).

Degradative (or catabolic) pathways are exergonic, overall, in that they end with a net release of usable energy. In degradative pathways, unstable molecules typically are broken down into smaller, more stable products, with the release of energy in forms that cells may use. Aerobic respiration is the main degradative pathway in the biosphere, and energy released during the reactions is used to form many ATP (Figure 6.15).

Many metabolic pathways are linear, a straight line from reactants to the products. In cyclic pathways, the last reaction regenerates the type of reactant molecule that is used in the first step of the reaction sequence. For instance, in the second stage of photosynthesis, a molecule known as RuBP is the entry point for cyclic reactions, and the cycle's last intermediate undergoes internal jugglings that convert it to RuBP. In branched pathways, reactants or intermediates are channeled into two or more different sequences of reactions.

THE DIRECTION OF METABOLIC REACTIONS

Bear in mind, metabolic reactions do not always run from reactants to products. They might start out in this "forward" direction. But most also run in reverse, with products being converted back to reactants.

Reversible reactions tend to run spontaneously toward **chemical equilibrium**, when the reaction rate is about the same in either direction. In most cases, the amounts of reactant and product molecules are not identical; they differ at that time (Figure 6.16). It is like a party where people drift between two rooms. The number in each room stays the same—say, thirty in one and ten in the other—even as individuals move back and forth.

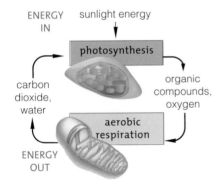

Figure 6.15 The main metabolic pathways in ecosystems. Energy from the sun drives the formation of glucose in photosynthesis, and aerobic respiration yields a great deal of usable energy from glucose breakdown.

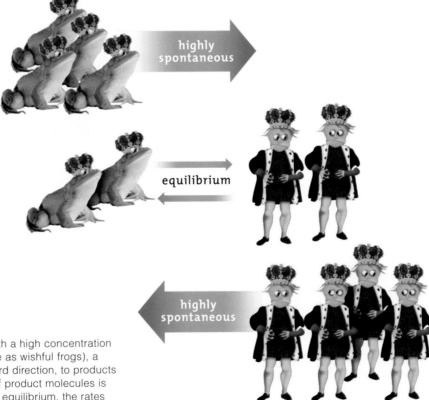

Figure 6.16 Chemical equilibrium. With a high concentration of reactant molecules (represented here as wishful frogs), a reaction runs most strongly in the forward direction, to products (the princes). When the concentration of product molecules is high, it runs most strongly in reverse. At equilibrium, the rates of the forward and reverse reactions are the same.

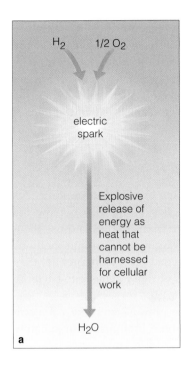

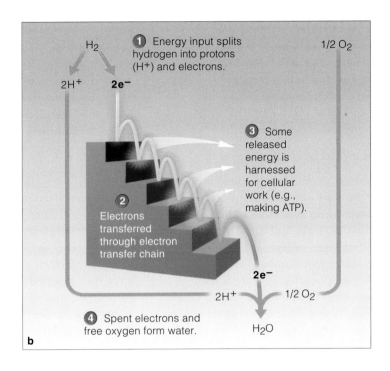

Figure 6.17 *Animated!* Uncontrolled versus controlled energy release. (**a**) Free hydrogen and oxygen exposed to an electric spark react and release energy all at once. (**b**) Electron transfer chains allow the same reaction to proceed in small, more manageable steps that can handle the released energy.

Why bother to think about this? *Each cell can bring about big changes in activities by controlling the enzymes that mediate a few steps of reversible metabolic pathways.*

For instance, when your cells need a quick bit of energy, they rapidly split glucose into two pyruvate molecules. They do so by a sequence of nine enzyme-mediated steps of a pathway called glycolysis. When glucose supplies are too low, cells quickly reverse this pathway and build glucose from pyruvate and other substances. How? Six steps of the pathway happen to be reversible, and the other three are bypassed. An input of energy from ATP drives the bypass reactions in the uphill (energetically unfavorable) direction.

What if cells did not have this reverse pathway? They would not be able to build glucose fast enough to compensate for episodes of starvation, when glucose supplies in blood become dangerously low.

REDOX REACTIONS IN THE MAIN PATHWAYS

You may be wondering: Why don't cells break down glucose all at once? Glucose, recall, is not as stable as the products of its full breakdown. Toss a cupful of glucose into a campfire, and its carbon and hydrogen atoms will explosively combine with oxygen in air. All of the released energy will be lost as heat.

Cells release energy efficiently by stepwise electron transfers, or **oxidation–reduction reactions**. In these "redox" reactions, one molecule gives up electrons (it is oxidized) and another gains them (it is reduced).

Commonly, hydrogen atoms are released at the same time. Remember, we represent free hydrogen atoms (naked protons) as H^+. Being attracted to the opposite charge of the electrons, H^+ tags along with them.

Start thinking about redox reactions, because they are central to photosynthesis and aerobic respiration. In the next two chapters, you will follow coenzymes as they pick up the electrons and H^+ stripped from substrates and then deliver them to **electron transfer chains**. Such chains are membrane-bound arrays of enzymes and other molecules that accept and give up electrons in sequence. Electrons are at a higher energy level when they enter a chain than when they leave.

Think of these electrons as descending a staircase and stingily losing a bit of energy at each step, as in Figure 6.17. In the case of photosynthesis and aerobic respiration, the stepwise electron transfers concentrate energy—in the form of H^+ electrochemical gradients—in ways that contribute to ATP formation.

Metabolic pathways are orderly, enzyme-mediated reaction sequences, some biosynthetic, others degradative.

Control over a key step of a metabolic pathway can bring about rapid shifts in cell activities.

Many aspects of metabolism involve electron transfers, or oxidation–reduction reactions. Electron transfer chains are important sites of energy exchange in both photosynthesis and aerobic respiration.

6.6 Light Up the Night—And the Lab

LINK TO
SECTION
2.3

You can always think about organisms from a "gee-whiz-ain't-nature-grand" point of view. Or you can come up with novel ways to think about what they do and how they do it. The latter way of thinking puts you squarely in the camp of biologists. Here is a case in point.

ENZYMES OF BIOLUMINESCENCE

At night, in the warm waters of tropical seas or in the summer air above gardens and fields, you may catch sight of abrupt shimmerings or flashes of light. Many species, ranging from bacteria and algae to fishes to fireflies, flash with orange, yellow, yellow-green, or blue light. In seawater, great numbers of them often flash together with startling effect (Figure 6.18*a*).

All of the flashers emit light when enzymes called luciferases transduce chemical bond energy of certain molecules into light energy. Figure 6.18*c* shows the three-dimensional structure of the luciferase found in fireflies. Remember, electrons can be excited to higher energy levels with an input of energy. In this case, reactions start when ATP, in the presence of oxygen, transfers a phosphate group to luciferin. An array of

electrons in this molecule become excited enough to enter reactions. At a certain reaction step, they release the extra energy in the form of *fluorescent* light. Any kind of destabilized molecule that reverts to a more stable configuration may emit such light. When the reactions take place in organisms, the emitted light is called **bioluminescence**.

A RESEARCH CONNECTION

People have been marveling over bioluminescence for a long time. Then biologists thought to borrow the genes for bioluminescence from fireflies and use them to make other organisms light up. Through methods of gene transfers, as sketched out in Chapter 16, they have now inserted copies of those genes into bacteria, plants, and mice (Figure 6.19).

In itself, making mice glow seems like a bizarre thing to do. However, some biologists immediately saw the potential for using bioluminescence genes as a diagnostic tool. For example, every year, 3 million people die from a lung disease caused by different strains of the bacterium *Mycobacterium tuberculosis*. No single antibiotic is effective against all the strains.

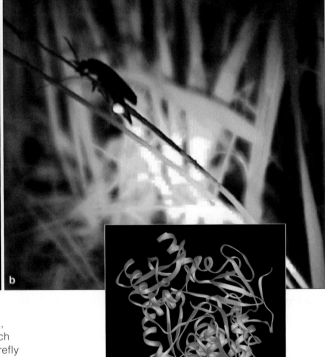

Figure 6.18 (**a**) Stirring up bioluminescent marine organisms near Vieques Island, Puerto Rico. As many as 5,000 free-living cells called dinoflagellates may be in each liter of water, and each flashes with blue light when agitated. (**b**) North American firefly (*Photinus pyralis*) emitting a flash from its light organ. Peroxisomes in this organ are packed with luciferase molecules. Firefly flashes help potential mates find each other in the dark. (**c**) Ribbon model for firefly luciferase. This enzyme catalyzes the reaction that releases light. Its single polypeptide chain is folded into multiple domains.

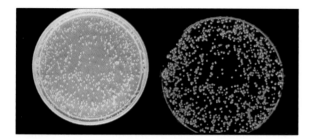

Figure 6.19 Colonies of bioluminescent bacteria in daylight (*left*) and glowing in a culture dish in the dark (*right*).

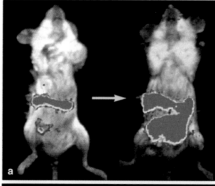

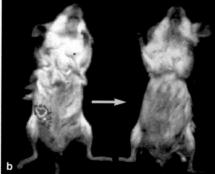

Figure 6.20 Utilizing bioluminescent bacterial cells to chart the location of infectious bacteria inside living laboratory mice and their spread through body tissues. (**a**) False-color images in this pair of photographs show how the infection spread in a control group that had not been given a dose of antibiotics. (**b**) This pair shows how antibiotics had killed most of the infectious bacterial cells.

If an infection has progressed to a dangerous stage, there is no time to waste, but there is no guarantee a treatment will be effective unless the particular strain causing a patient's infection is identified.

One way to do this is to take a sample of bacterial cells from the patient, then expose them to luciferase genes. In some cells, the genes get incorporated into the bacterial DNA. Those cells are isolated, and then colonies of their descendants are exposed to different antibiotics. When an antibiotic does *not* work, colonies glow; the cells are alive. When an antibiotic works, there is no glow; all of the bacterial cells are dead.

Christopher and Pamela Contag, two postdoctoral students at Stanford University, thought about using gene transfers to light up *Salmonella* cells in mice. As they knew, researchers who study viral or bacterial diseases had to infect dozens to hundreds of mice for experiments. Then they had to kill the mice and study their tissues to see whether infection had occurred. The practice was costly and tedious, and it meant the loss of a lot of infection-free experimental animals.

First the Contags approached a medical imaging researcher, David Benaron, with this hypothesis: If we make live, infectious bacteria bioluminescent, then flashes of light will shine through tissues of infected, living animals. As an early test of their prediction, the researchers put glowing *Salmonella* cells into a thawed chicken breast from a market. It glowed from inside.

The Contags transferred the bioluminescence genes into three *Salmonella* strains, which were injected into three experimental groups of laboratory mice. The Contags used a digital imaging camera to track and record whether infections developed in each group.

The first strain was weak; the mice fought off the infection in less than six days and did not glow. The second strain was not as weak but could not spread through the mouse body; it was localized. The third strain was dangerous. It spread rapidly through the entire mouse gut—which glowed.

Thus bioluminescent gene transfer, combined with imaging of enzyme activity, can track the course of infection. It is now used to evaluate the effectiveness of drugs in living organisms (Figure 6.20). It also may have use in gene therapy, which involves replacing one or more defective or cancer-causing genes in a patient with functional copies of the genes.

In short, people thought to use bioluminescence as visible evidence of metabolism—of the cell's capacity to acquire energy and use it to build, break apart, store, and release substances in controlled ways. Each flash reminds us that living cells are taking in energy-rich solutes, constructing membranes, storing things, replenishing enzymes, and checking out their DNA. A constant supply of energy drives these activities. The flashes remind us of how modern-day biologists are busy putting knowledge to use in practical ways.

Bioluminescence is an outcome of enzyme-mediated reactions that release energy as fluorescent light.

Biologists have transferred genes for bioluminescence, the luciferases, into a variety of organisms. The gene transfers have research applications and practical uses.

Summary

Section 6.1 Cells require energy for metabolism, or chemical work. The first law of thermodynamics tells us that energy cannot be created from scratch or destroyed. Energy can only be converted from one form to another. Examples are potential energy, such as that stored in chemical bonds, and kinetic energy (energy of motion).

The second law of thermodynamics tells us that energy tends to spread out or disperse spontaneously from concentrated to less concentrated forms, such as dispersed heat. Entropy is the measure of how much and how far a concentrated form of energy has been dispersed after an energy change.

The collective strength of the chemical bonds that hold together systems of life resists the spontaneous direction of energy flow. Those systems maintain their complex organization by being resupplied with energy lost from someplace else. Sunlight is the original source of energy for nearly all webs of life.

Section 6.2 Table 6.1 summarizes the functions of the key players in metabolism: reactants, intermediates, products, enzymes, cofactors, energy carriers, and transporters. The second law of thermodynamics lets us make this prediction about metabolism: It takes net inputs of energy to force stable molecules to combine into forms that are less stable but more concentrated sources of energy.

All reactants require a minimum amount of internal energy before they will enter into a reaction, although the amount differs among them. That amount is the activation energy for the reaction.

Some reactions are endergonic; they require a net energy input to run to completion. The formation of glucose from carbon dioxide and water is an example. Other reactions are exergonic; the net amount of energy they release is greater than the amount invested. Aerobic respiration is an example.

Cells couple reactions that require energy with other reactions that release energy. ATP is the main energy carrier between reaction sites. It jump-starts reactions by donating one or more of its phosphate groups to a reactant. It is regenerated when ADP binds to inorganic phosphate or to a phosphate group.

Biology⟨≋⟩Now
Learn about energy changes in chemical reactions and the role of ATP with the animation on BiologyNow.

Section 6.3 Enzymes are catalysts, which means they enormously enhance the rates of specific reactions. Nearly all are proteins that are much larger than their substrates (some RNAs also are catalytic). Folds in their polypeptide chains form active sites, or small clefts that create favorable microenvironments for reaction.

Enzymes speed reactions by lowering the activation energy. For each kind of reaction, that is the energy it takes to align reactive chemical groups, destabilize electric charges, and break chemical bonds. Enzymes move substrates faster to a transition state, when the reaction can proceed most easily to products.

Four mechanisms help get substrates to the transition state: Binding in an active site effectively boosts local concentrations of substrates, it orients substrates, it shuts out most or all water molecules that could interfere with the reaction, and it induces an optimum fit with the substrate that pulls it to the transition state.

Biology⟨≋⟩Now
Investigate how enzymes facilitate reactions with the animation and interaction on BiologyNow.

Section 6.4 Many factors influence enzyme action and, through it, metabolic pathways. Each type of enzyme functions best within a characteristic range of temperature, pH, and salinity. Controls over enzyme activity, including negative feedback mechanisms, influence the kinds and amounts of substances available in a given interval. Also, most enzymes require the assistance of cofactors. These metal ions and coenzymes help out at an active site by ferrying electrons, hydrogen ions, or functional groups to some other reaction site.

Biology⟨≋⟩Now
Observe mechanisms of enzyme control with the animation on BiologyNow.

Section 6.5 Cells modify concentrations of many thousands of substances, often by coordinating outputs of orderly, enzyme-mediated reaction sequences called metabolic pathways. The energy-requiring, *biosynthetic* pathways build large, unstable molecules from smaller, more stable ones. Photosynthesis is an example. The energy-releasing, *degradative* pathways break down large molecules to smaller products. Aerobic respiration is the main degradative pathway. Most metabolic reactions are reversible. Cells rapidly shift rates of metabolism by controlling a few steps of reversible pathways.

Table 6.1	Summary of the Main Participants in Metabolic Reactions
Reactant	Substance that enters a metabolic reaction or pathway; also called the substrate of a specific enzyme
Intermediate	Any substance that forms in a reaction or pathway, between the reactants and the end products
Product	Substance at the end of a reaction or pathway
Enzyme	A protein that greatly enhances reaction rates; a few RNAs also do this
Cofactor	Coenzyme (such as NAD^+) or metal ion; assists enzymes or move electrons, hydrogen, or functional groups to other reaction sites
Energy carrier	Mainly ATP; couples energy-releasing reactions with energy-requiring ones
Transport protein	Protein that passively assists or actively pumps specific solutes across a cell membrane

Cells release energy most efficiently by way of oxidation–reduction reactions, which simply are electron transfers. In both photosynthesis and aerobic respiration, redox reactions occur at electron transfer chains that are components of cell membranes.

Biology⊜Now
Compare the effects of controlled and uncontrolled energy release with the animation on BiologyNow.

Section 6.6 Bioluminescence is one outcome of enzyme-mediated reactions that release energy in the form of fluorescent light. The transfer of genes for bioluminescence into organisms has research and practical applications.

Biology⊜Now
Read the InfoTrac article, "An Enlightening Food Safety Tool," Lynn Petrak, The National Provisioner, *October 2004.*

Figure 6.21 (**a**) Superoxide dismutase. (**b**) One owner of a smattering of age spots, evidence of free radicals on the loose.

Self-Quiz

Answers in Appendix II

1. _____ is life's primary source of energy.
   a. Food   b. Water   c. Sunlight   d. ATP

2. Energy _____ .
   a. cannot be created or destroyed
   b. can change from one form to another
   c. tends to flow spontaneously in one direction
   d. all of the above

3. Entropy is a measure of _____ .
   a. order versus disorder in a system
   b. how much and how far a form of energy has been dispersed after an energy change
   c. the forerunner of an energy change
   d. all of the above

4. Enzymes _____ .
   a. are proteins, except for a few RNAs
   b. lower the activation energy of a reaction
   c. are destroyed by the reactions they catalyze
   d. a and b

5. Enzyme function is influenced by _____ .
   a. changes in temperature
   b. changes in pH
   c. changes in salinity
   d. all of the above

6. Which of the following statements is *incorrect*? A metabolic pathway _____ .
   a. has an orderly sequence of reaction steps
   b. is mediated by only one enzyme that starts it
   c. may be biosynthetic or degradative, overall
   d. all of the above

7. Match the substance with its suitable description.
   _____ coenzyme or metal ion   a. reactant
   _____ adjusts gradients at membrane b. enzyme
   _____ substance entering a reaction c. cofactor
   _____ substance formed during   d. intermediate
        a reaction   e. product
   _____ substance at end of reaction   f. energy carrier
   _____ enhances reaction rate   g. transport
   _____ mainly ATP   protein

Additional questions are available on Biology⊜Now™

Critical Thinking

1. State the law of thermodynamics that deals with the *quantity* of energy in the universe. State the law that deals with the *quality* of energy.

2. Cyanide, a toxic compound, binds irreversibly to an enzyme that is a component of electron transfer chains. The outcome is *cyanide poisoning.* Binding prevents the enzyme from donating electrons to a nearby acceptor molecule in the system. What effect will this have on ATP formation? From what you know of ATP's function, what effect will this have on a person's health?

3. Why does applying lemon juice to sliced apples keep them from turning brown?

4. One molecule of catalase can break down 6 million hydrogen peroxide molecules every minute. It is found in most organisms that live under aerobic conditions because hydrogen peroxide is toxic—cells must dispose of it fast or risk being damaged. Peroxide is catalase's substrate; but by a neat trick, catalase also can inactivate other toxins, including alcohol. Can you guess what the trick is?

5. Hydrogen peroxide bubbles if dribbled on an open cut but does not bubble on unbroken skin. Explain why.

6. *Free radicals* are unbound molecular fragments that have the wrong number of electrons. They form during many enzyme-catalyzed reactions, including the digestion of fats and amino acids. They slip out of electron transfer chains. They also form when x-rays and other kinds of ionizing radiation bombard water and other molecules. Free radicals are highly reactive. When they dock with a molecule, they alter its structure and function.

   Superoxide dismutase (Figure 6.21a) is an enzyme that works with catalase to keep free radicals and hydrogen peroxide from accumulating in cells. As we age, cells make copies of enzymes in ever diminishing numbers, in altered form, or both. When this happens to superoxide dismutase and catalase, free radicals and hydrogen peroxide build up. Like loose cannonballs, they careen through cells and blast away at the structural integrity of proteins, DNA, lipids, and other molecules. For instance, look at the "age spots" on an older person's skin (Figure 6.21b). Each spot is a mass of brownish-black pigments that have accumulated in skin cells. Do some research and identify other problems that arise when free radicals take over.

Sunlight and Survival

Think about the last bit of apple, lettuce, chicken, pizza, or any other food you put in your mouth. Where did it come from? Look past the refrigerator, the market or restaurant, and the farm. Look to plants, the starting point for nearly all of the food—the carbon-based compounds—you eat.

Plants are among the **autotrophs**, or "self-nourishing" organisms. Autotrophs get energy and carbon from the physical environment and use it to make their own food. Most bacteria, many protists, and all fungi and animals are like you; they cannot obtain energy and carbon from the physical environment. They are **heterotrophs**, which feed on autotrophs, one another, and organic wastes. *Hetero*– means other, as in "being nourished by others."

Plants are a type of *photo*autotroph. By the process of **photosynthesis**, they make sugars and other compounds by using sunlight as an energy source and carbon dioxide as their source of carbon. Each year, plants around the world produce 220 billion tons of sugar, enough to make 300 quadrillion sugar cubes. That is a LOT of sugar. They also release great amounts of oxygen (Figure 7.1).

It wasn't always this way. The first prokaryotic cells on Earth were *chemo*autotrophs. Like the existing archaeans, they did not have the enzymes for complicated metabolic magic. They extracted energy and carbon from simple organic and inorganic compounds, such as methane and hydrogen sulfide, that happened to be around. Both gases were part of the chemical brew that made up the early atmosphere. Carbon dioxide also was present, but it takes special enzymes to harness it. There was little free oxygen.

Things did not change much for about a billion years. Then light-sensitive molecules evolved in a few lineages, which became the first *photo*autotrophs. Life had tapped into an immense supply of energy. Not long afterward, parts of the photosynthetic machinery became modified

Figure 7.1 Photosynthesis—the main pathway by which energy and carbon enter the web of life. This orchard of photosynthetic autotrophs is producing apples and oxygen at the Jerzy Boyz organic farm in Chelan, Washington.

in some photoautotrophs. Water molecules could now be split apart as a source of electrons for the reactions, and supplies of water were essentially unlimited. Over time, oxygen atoms released from uncountable water molecules diffused out of uncountable numbers of cells —and the world of life would never be the same.

Free oxygen reacts fast with metals, including metal ions that help enzymes. The reactions release free radicals which, as you know, are toxic to cells. So oxygen that had accumulated in the atmosphere put selection pressure on prokaryotic populations all over the world. Prokaryotes that could not neutralize toxic oxygen radicals vanished or were marginalized in muddy sediments, deep water, and other anaerobic (oxygen-free) habitats.

As you will read in this chapter, pathways that could detoxify the oxygen radicals evolved in some lineages. One pathway, aerobic respiration, lets cells *use* oxygen's reactive properties in highly beneficial ways.

Another bonus for life: As oxygen accumulated high in the atmosphere, many atoms combined to form ozone (O_3). An ozone layer formed and became a shield against lethal ultraviolet radiation from the sun. Life could now move out of the deep ocean, out from mud, out from under rocks, and diversify under the open sky.

As you read this chapter on photosynthesis, keep in mind that its emergence and its continuity are big reasons why *you* can exist, and read this book, and think about what it takes to stay alive.

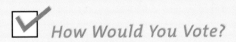

How Would You Vote?

The oxygen in Earth's atmosphere is a sure indicator that photosynthetic organisms flourish here. New technologies will allow astronomers in search of life to measure the oxygen content of the atmosphere of planets too far away for us to visit. Should public funds be used to continue this research? See BiologyNow for details, then vote online.

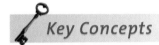

Key Concepts

THE RAINBOW CATCHERS

A one-way flow of energy through the world of life starts after chlorophylls and other pigments absorb wavelengths of visible light from the sun's rays. In plants, some bacteria, and many protists, that energy ultimately drives the synthesis of glucose and other carbohydrates. Sections 7.1, 7.2

OVERVIEW OF PHOTOSYNTHESIS

In plant cells and many protists, photosynthesis proceeds through two stages inside organelles called chloroplasts. At a membrane system in the chloroplast, the sun's energy is first converted to chemical energy. Then carbohydrates are synthesized in the chloroplast's semifluid matrix. Section 7.3

MAKING ATP AND NADPH

In the first stage of photosynthesis, sunlight energy becomes converted to chemical bond energy in ATP. NADPH forms, and free oxygen escapes into the air. Sections 7.4, 7.5

MAKING SUGARS

The second stage is the "synthesis" part of photosynthesis. Enzymes assemble sugars from atoms of carbon and oxygen obtained from carbon dioxide. The reactions use the ATP and NADPH that formed in the first stage of photosynthesis. The ATP delivers energy, and the NADPH delivers electrons and hydrogens to the reaction sites. Sections 7.6, 7.7

GLOBAL IMPACTS OF AUTOTROPHS

The emergence of the world's main energy-releasing pathway, aerobic respiration, was an evolutionary consequence of photosynthesis—the world's main energy-acquiring pathway. Collectively, photoautotrophs and chemoautotrophs make the food that sustains all of life. They also have enormous impact on the global climate. Section 7.8

Links to Earlier Concepts

Before considering the chemical basis of photosynthesis, you may wish to review the nature of electron energy levels (Section 2.3), particularly how photons and electrons interact. You will be using your knowledge of carbohydrate structure (3.4), chloroplasts (4.8), active transport proteins (5.2, 5.4), and concentration gradients (5.3).

Remember the concepts of energy flow and the underlying organization of life (6.1 and 6.2)? They help explain how energy flows through photosynthesis reactions. You also will expand your understanding of how cells harvest energy through the operation of electron transfer chains (6.5).

7.1 Sunlight as an Energy Source

LINKS TO
SECTIONS
2.3, 6.1, 6.2

Remember how energy flows in one direction through the world of life? In nearly all cases, the flow starts when photoautotrophs intercept energy, in the form of wavelengths of visible light, from the sun.

PROPERTIES OF LIGHT

An understanding of photosynthesis requires a bit of knowledge of the properties of energy that radiates from the sun. That energy undulates across space in a manner analogous to the waves moving across a sea. The term **wavelength** refers to the horizontal distance between the crests of every two successive waves of radiant energy.

Although energy travels in waves, it has a particle-like quality. When absorbed, it can be measured as if it were organized in discrete packets, or **photons**. A photon consists of a fixed amount of energy. The least energetic photons travel in longer wavelengths, and the most energetic ones travel in shorter wavelengths.

Photoautotrophs only capture light of wavelengths between 380 and 750 nanometers. Humans and other organisms see light of these wavelengths as different colors, from deep violet through blue, green, yellow, orange, and red. Figure 7.2 shows where the spectrum of visible light fits in the **electromagnetic spectrum**—the range of all wavelengths of radiant energy, from shortest (gamma rays) to longest (radio waves).

Shorter wavelengths are energetic enough to alter or break chemical bonds in DNA and proteins. That is why UV (ultraviolet) light, x-rays, and gamma rays are a threat to all organisms. That is why early life evolved away from sunlight—deep in the ocean, or in sediments, or under rocks. Life did not move onto dry land until after the ozone layer formed high above Earth. The ozone layer absorbs much of the dangerous UV light (Sections 48.1 and 48.2).

Visible light of all wavelengths combined appears white. White light separates into its individual colors when it passes through a prism or water droplets in moisture-laden air. The prism or droplets bend light of longer wavelengths (yellow to red) more than they bend shorter wavelengths (violet to blue), the result being the band of colors we see in rainbows.

FROM SUNLIGHT TO PHOTOSYNTHESIS

Pigments are a class of molecules that absorb photons in particular wavelengths only. Certain kinds are the molecular bridges from sunlight to photosynthesis.

Photons that a specific pigment does not absorb are reflected by it or continue traveling right on through it. **Chlorophyll *a***, the major pigment in all but one group of photoautotrophs, absorbs violet and red light. It reflects green and yellow light, which is why plant parts with an abundance of chlorophylls appear green. *Accessory* pigments harvest additional wavelengths. The most common accessory pigment, **chlorophyll b**, reflects green and blue light. **Carotenoids** reflect red, orange, and yellow light. Besides their photosynthetic role, carotenoids impart color to many flowers, fruits, and vegetables. **Xanthophylls** reflect yellow, brown, blue or purple light. The **anthocyanins** reflect red and purple light, as they do in cherries and many flowers. In many deciduous plants, chlorophylls in green leaves mask accessory pigments until autumn (Figure 7.3*a*).

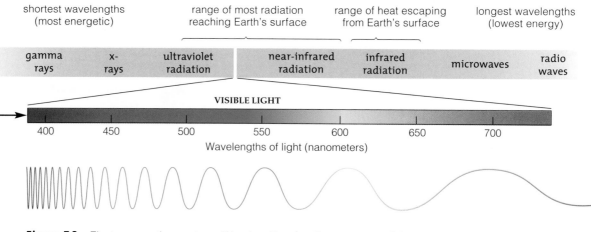

shortest wavelengths (most energetic)		range of most radiation reaching Earth's surface	range of heat escaping from Earth's surface	longest wavelengths (lowest energy)

| gamma rays | x-rays | ultraviolet radiation | near-infrared radiation | infrared radiation | microwaves | radio waves |

VISIBLE LIGHT

400   450   500   550   600   650   700

Wavelengths of light (nanometers)

Figure 7.2 Electromagnetic spectrum. Wavelengths of radiant energy undulate across space and are measured in nanometers. About 2.5 million nanometers are equal to one inch. Visible light is a very small part of the spectrum, which includes all electromagnetic waves. The shorter the wavelength, the higher the energy.

a

The **phycobilins** reflect red or blue-green light. Red algae and cyanobacteria have notable amounts of these accessory pigments. A few bacteria of ancient lineages have unique pigments. Purple bacteriorhodopsin is the main kind in the archaean *Halobacterium halobium*.

Collectively, different photosynthetic pigments can absorb nearly all wavelengths across the spectrum of visible light. What happens next? You have to zoom into a pigment for the answer. As Figure 7.3b,c shows, a pigment molecule has at least one array of atoms in which single covalent bonds alternate with double covalent bonds. Remember electron orbitals (Section 2.3)? Electrons of these atoms share one orbital that spans the entire array. That array lets the pigment act like an antenna for receiving photon energy.

Each pigment absorbs light of specific wavelengths, which correspond to photon energy. Energy inputs, remember, boost electrons to higher energy levels. A photon is absorbed by a pigment only if it has exactly enough energy to boost an electron of the pigment's antenna region to a higher energy level.

An excited electron returns to a lower energy level almost immediately and emits its extra energy as heat or as a photon. As you will see shortly, that energy bounces back and forth like a fast volleyball among a team of photosynthetic pigments. It quickly reaches the team captain—a special chlorophyll that can *give up* excited electrons and so start the reactions.

Radiation from the sun travels in waves, which differ in length and energy content. We perceive visible light of different wavelengths as different colors and measure their energy content in packets called photons.

In plants, chlorophyll a and accessory pigments absorb specific wavelengths of visible light. They are molecular bridges between the sun's energy and photosynthesis.

Pigment molecules absorb photons at their arrays of alternating single and double covalent bonds. The arrays let the pigments act like energy-receiving antennas.

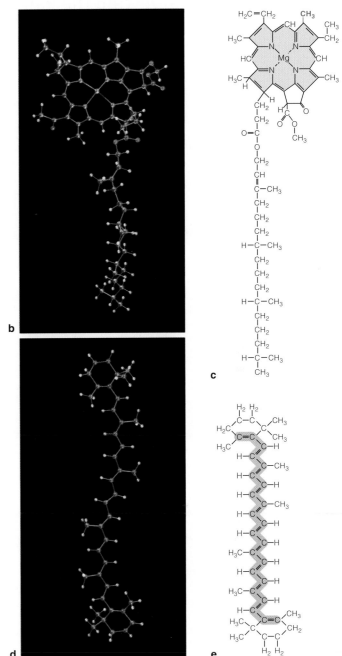

b

c

d

e

Figure 7.3 (a) Evidence of pigments in the changing leaves of autumn. In bright green leaves, photosynthetic cells continuously make chlorophyll, which masks accessory pigments. In autumn, chlorophyll synthesis lags behind its breakdown in many species. Accessory pigments then show through and give leaves characteristic red, orange, and yellow fall colors.

Ball-and-stick models and structural formulas for (b,c) chlorophyll *a* and (d,e) beta-carotene. The light-catching region of each pigment is tinted the specific color of light it transmits. Each pigment has a hydrocarbon backbone that readily dissolves in the lipid bilayer of cell membranes.

Chlorophylls *a* and *b* differ only in one functional group at the position shaded *red* (—CH_3 for chlorophyll *a* and —COO^- for chlorophyll *b*). The light-catching portion is the flattened ring structure, which is similar to a heme (Section 3.1). It holds a magnesium atom instead of iron.

7.2   Harvesting the Rainbow

LINK TO
SECTION
4,8

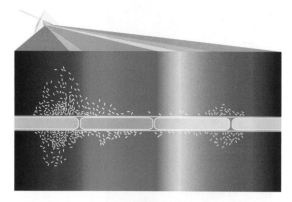

a Outcome of Engelmann's experiment

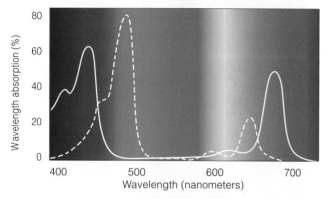

b Absorption spectra for chlorophyll *a* (solid graph line)
and chlorophyll *b* (dashed line)

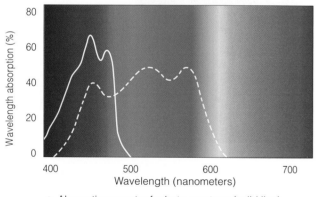

c Absorption spectra for beta-carotene (solid line)
and one of the phycobilins (dashed line)

*Different kinds of photosynthetic pigments work together.
How efficient are these pigments at harvesting light of
different wavelengths in the sun's rays?*

At one time, people thought that plants used substances
in soil to make food. By 1882, a few chemists had an idea
that plants use sunlight, water, and something in the air.
The botanist Wilhelm Theodor Engelmann wondered:
What parts of sunlight do plants favor? He already knew
that photosynthesis releases free oxygen. He came up
with a hypothesis. If photosynthesis involves certain colors
of light, then photosynthesizers will release more or less
oxygen in response to different colors.

Engelmann also knew that certain bacteria use oxygen
during aerobic respiration, and he predicted that they
would gather in places where a photosynthetic organism
was releasing the most oxygen. He directed a spectrum of
visible light across a drop of water that contained bacterial
cells (Figure 7.4a). The droplet also contained a strand of
Cladophora, a photosynthetic alga (Figure 7.5).

Most of the bacterial cells gathered where violet and
red light fell across the algal strand. More free oxygen had
to be diffusing away from parts of the strand that were
illuminated by the violet and red light—a sign that those
colors are best at driving photosynthesis.

Engelmann did identify the wavelengths. But molecular
biology was far in the future, so he did not know about the
pigments that absorb the light.

Today, an **absorption spectrum** conveys how efficiently
a given pigment absorbs light of different wavelengths. As
Figure 7.4b shows, chlorophylls are best at absorbing red
and violet light, but they transmit much of the yellow and
green light. What if you combined absorption spectra for
chlorophylls and all of the accessory pigments, including
those in Figure 7.4b,c? You would see that, collectively,
they respond to almost the full spectrum of wavelengths
from the sun. They are efficient at what they do.

Figure 7.4 *Animated!* (**a**) One of the early photosynthesis experiments.
W. T. Engelmann directed a ray of sunlight—broken into its component
colors by a crystal prism—across a water droplet on a microscope slide.
The droplet held an algal strand (*Cladophora*) and aerobic bacterial cells.
As shown here, nearly all of the cells gathered under violet and red light,
the most efficient wavelengths for photosynthesis.

(**b,c**) Later research revealed that all photosynthetic pigments combined
absorb most wavelengths in the spectrum of visible light with remarkable
efficiency. These graphs show absorption spectra for only four of many
pigments: chlorophylls *a* and *b*, beta-carotene, and a phycobilin.

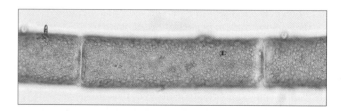

Figure 7.5 Light micrograph of the type of cells
making up one algal strand.

7.3 Overview of Photosynthesis Reactions

Plants do something you never will do. They can make their own food from no more than light, water, and carbon dioxide.

Photosynthesis proceeds in two reaction stages. In the first stage—the **light-dependent reactions**—sunlight energy is converted to chemical bond energy of ATP. Water molecules are split, and typically the coenzyme NADP⁺ accepts the released hydrogen and electrons, thus becoming NADPH. The oxygen atoms released from water molecules escape into the surroundings.

The second stage, the **light-independent reactions**, runs on energy delivered by ATP. That energy drives the synthesis of glucose and other carbohydrates. The building blocks are the hydrogen atoms and electrons from NADPH, as well as carbon and oxygen atoms stripped from carbon dioxide and water.

Photosynthesis is often summarized this way:

$$12H_2O + 6CO_2 \xrightarrow[\text{enzymes}]{\text{light energy}} 6O_2 + C_6H_{12}O_6 + 6H_2O$$

water carbon   oxygen glucose water
       dioxide

We will focus on what goes on inside **chloroplasts**, the organelles of photosynthesis in plants and many protists. A chloroplast has two outer membranes that enclose a semifluid matrix called the **stroma**. A third membrane—the **thylakoid membrane**—is folded up inside the stroma. In many cells, it looks like stacks of flattened sacs (thylakoids) connected by channels. But the space inside all the sacs and channels forms one continuous compartment, as in Figure 7.6b. Sugars are synthesized outside the compartment, in the stroma.

As you will see in the next section, the thylakoid membrane is studded with pigments. Most pigments are packed together as light-harvesting complexes. A number of **photosystems**, or reaction centers, also are embedded in the membrane, and each is surrounded by hundreds of light-harvesting complexes that pass on energy to it. With enough energy, its electrons get excited. That excitation sets in motion the first stage of reactions, as sketched out in Figure 7.6c.

In the first stage of photosynthesis, sunlight energy drives ATP and NADPH formation, and oxygen is released. In chloroplasts, this stage occurs at the thylakoid membrane.

The second stage proceeds in the stroma of chloroplasts. Energy from ATP drives the synthesis of sugars from water and carbon dioxide.

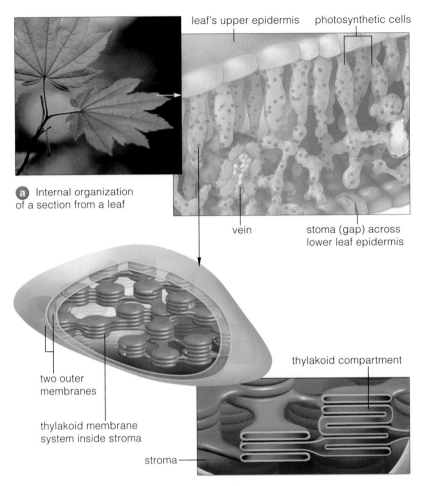

a Internal organization of a section from a leaf

leaf's upper epidermis   photosynthetic cells

vein   stoma (gap) across lower leaf epidermis

two outer membranes

thylakoid membrane system inside stroma

stroma

thylakoid compartment

b Cutaway view of a chloroplast inside the cytoplasm of one photosynthetic cell, and a close-up of its thylakoid compartment.

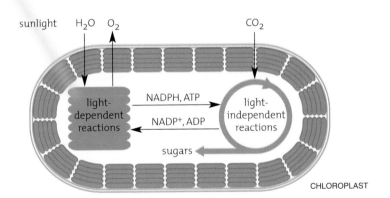

sunlight   H₂O   O₂   CO₂

light-dependent reactions   NADPH, ATP   light-independent reactions
                            NADP⁺, ADP

sugars

CHLOROPLAST

c Two stages of photosynthesis. The first stage depends on inputs of sunlight. It occurs at the thylakoid membrane system. ATP and NADPH form; free oxygen diffuses away. In the second stage, enzymes in the stroma catalyze the assembly of sugars. Energy from ATP starts the reactions. Building blocks are hydrogen atoms and electrons (from NADPH) and carbon atoms (from carbon dioxide).

Figure 7.6 *Animated!* Zooming in on sites of photosynthesis inside the leaf of a typical plant.

7.4 Light-Dependent Reactions

LINKS TO
SECTIONS
5.2, 6.5

In the first stage of photosynthesis, photons absorbed at photosystems drive ATP formation. Water molecules are split. Their oxygen diffuses away, but the coenzyme NADP+ picks up the released electrons and hydrogen.

WHAT HAPPENS TO THE ABSORBED ENERGY?

Visualize a lone photon as it collides with a pigment molecule. One of the pigment's electrons can absorb that photon's energy, which boosts the electron to a higher energy level. If nothing else were to happen, the electron would drop back to its unexcited state and lose the extra energy as a photon or as heat.

In the thylakoid membrane, however, energy that excited electrons give up is kept in play. Embedded in the membrane are many hundreds of light-harvesting complexes: circular clusterings of pigments and other proteins (Figure 7.7a). The pigments in light-harvesting complexes do not waste absorbed photons. Instead, their electrons hold on to photon energy by passing it back and forth, like a volleyball.

Energy released from one complex gets passed to another, which passes it on to another, and so on until the energy reaches a photosystem. Chloroplasts have two kinds of photosystems, *type I* and *type II*. The two have slightly different chlorophyll *a* molecules. Each contains other molecules, including different pigments. Hundreds of light-harvesting complexes surround it.

Look back on Figure 7.3, which shows the structure of chlorophyll. Two molecules of chlorophyll *a* are at the center of a photosystem. Their flat rings face each other so closely that the electrons in *both* rings are destabilized. When light-harvesting complexes pass on photon energy to a photosystem, electrons are popped right off of that special pair of chlorophylls.

The freed electrons immediately enter an electron transfer chain positioned next to the photosystem. As

you know, **electron transfer chains** are components of cell membranes. Each is an orderly array of enzymes, coenzymes, and other proteins that transfer electrons step-by-step (Section 6.5). *The entry of electrons from a photosystem into an electron transfer chain is the first step in the light-dependent reactions—in the conversion of photon energy to chemical energy for photosynthesis.*

MAKING ATP AND NADPH

Figure 7.8 tracks electrons from a type II photosystem on through an electron transfer chain in the thylakoid membrane. As certain components of the chain accept and donate the electrons, they pick up hydrogen ions (H+) from the stroma and release them into the inner thylakoid compartment. They do so again and again. Soon, concentration and electric gradients are built up across the membrane, and the combined force of the gradients attracts H+ back toward the stroma.

But H+ cannot diffuse across the membrane's lipid bilayer. It can cross only through channels inside **ATP synthases**, a type of transport protein you read about in Section 5.2. In this case, the knoblike portion of the protein projects into the stroma. Ion flow through the channel makes the knob turn, which forces inorganic phosphate to become attached to an ADP molecule. In this way, ATP forms in the stroma.

As long as electrons flow through transfer chains, the cell can keep on producing ATP. But how are the electrons from photosystem II replaced? By a process called *photolysis*, new electrons are pulled away from water molecules, which then dissociate into hydrogen ions and molecular oxygen. The free oxygen diffuses out of the chloroplast, then out of the cell and into the surroundings. Hydrogen ions remain in the thylakoid compartment. They contribute to the concentration and electric gradients that drive ATP formation.

Figure 7.7 (**a**) Ringlike array of pigment molecules that intercept rays of sunlight coming from any direction. (**b**) One of the many photosystems (represented as a *green* sphere) embedded in a chloroplast's thylakoid membrane. Each photosystem collects energy from hundreds of light-harvesting complexes that surround it; only eight complexes are shown here.

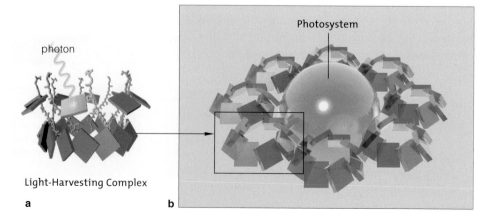

photon

Light-Harvesting Complex

a

Photosystem

b

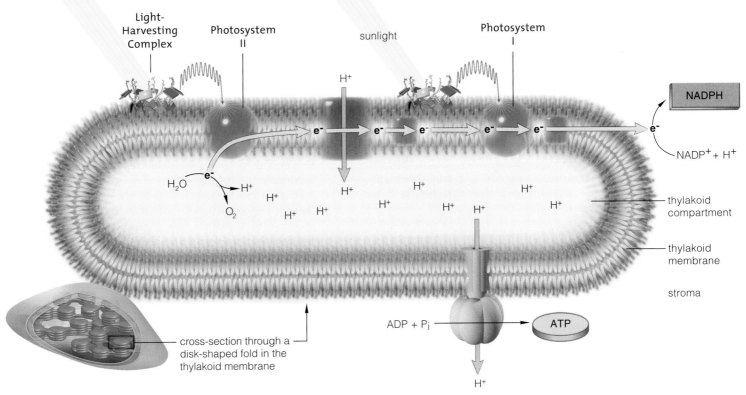

Light-Harvesting Complex

Photosystem II

sunlight

Photosystem I

NADPH

H⁺

e^-

NADP⁺ + H⁺

thylakoid compartment

thylakoid membrane

stroma

H_2O

e^-

H⁺

O_2

H⁺ H⁺ H⁺ H⁺ H⁺ H⁺ H⁺ H⁺ H⁺ H⁺ H⁺

cross-section through a disk-shaped fold in the thylakoid membrane

ADP + P$_i$

ATP

H⁺

a Light-harvesting complexes absorb photon energy (*red*), which drives electrons out of photosystem II. Replacement electrons are pulled from water molecules, which then split into oxygen and hydrogen ions (H⁺). Oxygen leaves the cell as O_2.

b The electrons released from photosystem II enter an electron transfer chain (*brown*), which also moves H⁺ from the stroma into the thylakoid compartment. The electrons continue on to photosystem I.

c H⁺ concentration and electric gradients build up across the thylakoid membrane. The force of the gradients propels H⁺ through ATP synthases. The flow causes this membrane protein to move in a way that forces the attachment of P$_i$ to ADP, thus forming ATP.

d Photon energy (*red*) drives electrons from photosystem I. An intermediary molecule (*brown*) adjacent to the photosystem in the membrane accepts and then transfers the electrons to NADP⁺, which picks up H⁺ at the same time and becomes NADPH.

Figure 7.8 *Animated!* How ATP and NADPH form during the first stage of photosynthesis. The drawing represents a cross-section through one of the disk-shaped folds of the thylakoid membrane. This entire sequence is called the *noncyclic* pathway of photosynthesis, because electrons that originally left photosystem II are not cycled back to it. They end up in NADPH.

But where do the electrons end up? After they pass through the electron transfer chain, they enter a type I photosystem where the light-harvesting complexes are volleying energy to a special pair of chlorophylls at the reaction center. The chlorophylls release electrons, which an intermediary molecule transfers to NADP⁺. When this coenzyme accepts the electrons, it attracts hydrogen ions and thereby becomes NADPH.

We have been describing the *noncyclic* pathway of ATP formation in chloroplasts, so named because the electrons that leave photosystem II do not get cycled back to it; they end up in NADPH.

When too much NADPH forms, it accumulates in the stroma, so the photosystem II pathway backs up. At such times, photosystem I may run independently so that cells can continue to make ATP. It is a *cyclic* pathway of ATP formation, because the electrons that leave photosystem I get cycled back to it. Before they

return, they pass through an electron transfer chain that moves H⁺ into the thylakoid compartment. The resulting H⁺ gradients drive ATP formation, but no NADPH forms in this shorter pathway.

Two kinds of photosystems, type I and type II, are embedded in the thylakoid membrane. Hundreds of light-harvesting complexes surround and transfer photon energy to each one.

In a noncyclic pathway of photosynthesis, photon energy forces electrons out of photosystem II and on to an electron transfer chain, which sets up H⁺ gradients that drive ATP formation. Electrons continue on through photosystem I and end up in a reduced coenzyme, NADPH.

ATP also can form by a cyclic pathway, in which electrons leave photosystem I and are cycled back to it. However, NADPH cannot form by this pathway.

7.5 Energy Flow in Photosynthesis

LINKS TO
SECTIONS
6.1–6.3

One of the recurring themes in biology is that organisms convert one form of energy to another in highly controlled ways. The energy exchanges during the light-dependent reactions, outlined in Figure 7.9, are a classic example.

The preceding section focused on a photosynthetic pathway that starts at photosystem II and ends with the formation of ATP and NADPH. However, a simpler pathway that was less energy efficient preceded it. When photoautotrophs first evolved, remember, they were anaerobic. Their light-dependent pathway of photosynthesis yielded ATP alone, and it still operates today.

Again, this set of reactions is said to be cyclic because excited electrons flow out of photosystem I, through an electron transfer chain, then back to photosystem I.

Later, the photosynthetic machinery in some kinds of photoautotrophs was remodeled. Photosystem II became part of it. That was the start of a combined sequence of reactions powerful enough to oxidize—that is, to strip hydrogen atoms from—water molecules. Free oxygen is a by-product of this pathway.

Remember, the combined pathway is noncyclic; the electrons that leave photosystem II are not returned to it. They end up in NADPH, which delivers them to the sugar factories in the stroma.

Today, some bacteria have only photosystem I. Others have only photosystem II. Cyanobacteria, plants, and all photosynthetic protists have photosystems of both types and carry out both cyclic and noncyclic pathways. Which pathway dominates depends on conditions at the time.

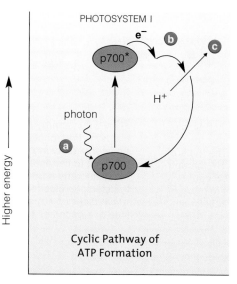

Cyclic Pathway of ATP Formation

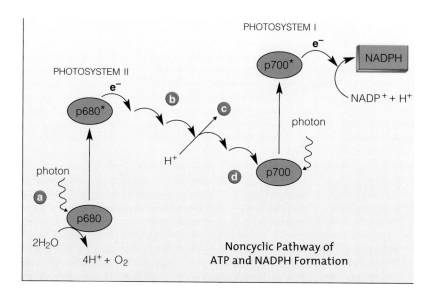

Noncyclic Pathway of ATP and NADPH Formation

a Photosystem I receives photon energy from a light-harvesting complex. It loses an electron.

b The electron passes from one molecule to another in an electron transfer chain that is embedded in the thylakoid membrane. It loses a little energy with each transfer, and ends up being reused by photosystem I (thus the pathway is considered "cyclic").

c Molecules in the transfer chain carry H⁺ across the thylakoid membrane into the inner compartment. Hydrogen ions accumulating in the compartment create an electrochemical gradient across the membrane that drives ATP synthesis, as shown in Figure 7.8.

a Photosystem II receives photon energy from a light-harvesting complex, then loses an electron. The electron moves through a different electron transfer chain. It loses a little energy with each transfer and ends up at photosystem I.

b Photosystem I receives photon energy from a light-harvesting complex, then loses an electron. Released electrons and hydrogen ions are used in the formation of NADPH from NADP⁺.

c As in the cyclic pathway, operation of the electron transfer chain pulls hydrogen ions into the thylakoid compartment. In this case, hydrogens released from dissociated water molecules also enter the compartment. The H⁺ concentration and electric gradient across the membrane are tapped for ATP formation (Figure 7.8).

d Electrons lost from photosystem I are replaced by the electrons lost from photosystem II. Electrons lost from photosystem II are replaced by electrons obtained from water. (Photolysis pulls water molecules apart into electrons, H⁺, and O₂.)

Figure 7.9 Animated! Using energy in the light-dependent reactions. The pair of chlorophyll *a* molecules at the center of photosystem I is designated p700. The pair in photosystem II is designated p680. The pairs respond most efficiently to wavelengths of 700 and 680 nanometers, respectively.

7.6 Light-Independent Reactions: The Sugar Factory

In the chloroplast's stroma, cyclic, enzyme-mediated reactions build sugars from hydrogen, carbon, and oxygen. These light-independent reactions run on energy that became conserved in ATP during the first stage of photosynthesis. NADPH that formed in the first stage donates the hydrogen and electrons. Plants get the carbon and oxygen from carbon dioxide (CO_2) in the air; algae get them from CO_2 dissolved in water.

The light-independent reactions proceed from carbon fixation, to PGAL formation, then RuBP regeneration. In **carbon fixation**, a carbon atom from CO_2 becomes attached to an organic compound. **Rubisco** (ribulose bisphosphate carboxylase/oxygenase) mediates this step in most plants. When it transfers the carbon to five-carbon RuBP (ribulose biphosphate), it opens the sugar factory—a series of enzyme-mediated reactions called the **Calvin–Benson cycle** (Figure 7.10).

The six-carbon intermediate that forms is unstable and splits at once into two PGA (phosphoglycerate) molecules, each with a three-carbon backbone (Figure 7.10*a*). Next, ATP energy and the reducing power of NADPH convert each PGA to a different three-carbon compound, PGAL (phosphoglyceraldehyde, or G3P).

How? ATP transfers a phosphate group to each PGA, and NADPH donates hydrogen and electrons to it.

Glucose, remember, has six carbon atoms. *Six CO_2 must be fixed and twelve PGAL must form to produce one glucose molecule and also to keep the Calvin–Benson cycle running.* Two PGAL combine to form one six-carbon glucose molecule with a phosphate group attached. The other ten PGAL undergo internal rearrangements in ways that regenerate RuBP (Figure 7.10*c–f*).

Most of the glucose is converted at once to sucrose or starch by other pathways that conclude the light-independent reactions. Sucrose is a transportable form of carbohydrate in plants. Excess glucose is converted to starch and briefly stored, as starch grains, in the stroma. Starch is converted to sucrose for export to leaves, stems, and roots. Plants can use photosynthetic products and intermediates as energy sources and as building blocks for all required organic compounds.

Driven by ATP energy, the light-independent reactions make sugars with hydrogen and electrons from NADPH, and with carbon and oxygen from carbon dioxide.

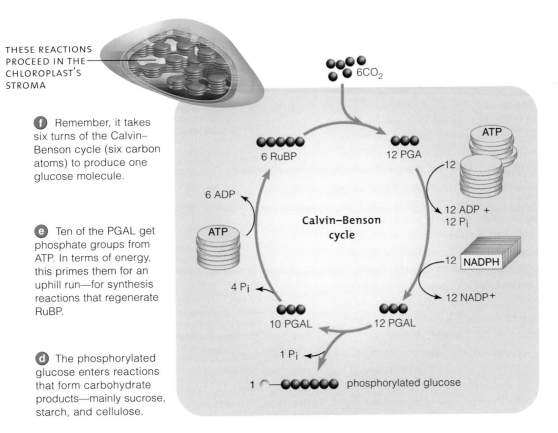

THESE REACTIONS PROCEED IN THE CHLOROPLAST'S STROMA

f Remember, it takes six turns of the Calvin–Benson cycle (six carbon atoms) to produce one glucose molecule.

e Ten of the PGAL get phosphate groups from ATP. In terms of energy, this primes them for an uphill run—for synthesis reactions that regenerate RuBP.

d The phosphorylated glucose enters reactions that form carbohydrate products—mainly sucrose, starch, and cellulose.

6 RuBP   12 PGA
6 ADP
ATP
4 P_i
Calvin–Benson cycle
10 PGAL   12 PGAL
1 P_i
1 ⟜●●●●●● phosphorylated glucose

$6CO_2$
ATP
12
12 ADP + 12 P_i
12 NADPH
12 NADP+

a CO_2 in air spaces inside a leaf diffuses into a photosynthetic cell. Six times, rubisco attaches a carbon atom of CO_2 to the RuBP that starts the Calvin–Benson cycle. Each time, the resulting intermediate splits to form two PGA molecules, for a total of twelve PGA.

b Each PGA molecule gets a phosphate group from ATP, plus hydrogen and electrons from NADPH. The resulting intermediate, PGAL, is thus primed for reaction.

c Two of the twelve PGAL molecules combine to form one molecule of glucose with an attached phosphate group.

Figure 7.10 *Animated!* Light-independent reactions of photosynthesis. The sketch is a summary of all six turns of the Calvin–Benson cycle and its product, one glucose molecule. *Brown* circles signify carbon atoms. Appendix VII details the reaction steps.

7.7   Different Plants, Different Carbon-Fixing Pathways

If sunlight intensity, air temperature, rainfall, and soil composition never varied, photosynthesis might be the same in all plants. But environments differ, and so do details of photosynthesis, as you can see by comparing what happens on hot, dry days when water is scarce.

C4 VERSUS C3 PLANTS

All plant surfaces exposed to air have a waxy, water-conserving cuticle. The only way for gases to diffuse into or out of a plant is at **stomata** (singular, stoma). These are tiny openings across the surface of leaves and green stems (Figure 7.11*a*). Stomata close on hot, dry days. Water stays inside the plant, but the CO_2 required for photosynthesis cannot diffuse in, and the O_2 by-product of photosynthesis cannot diffuse out.

That is why basswood, beans, peas, and many other plants do not grow well in hot, dry climates without steady irrigation. We call them **C3 plants**, because the *three*-carbon PGA is the first stable intermediate of the Calvin–Benson cycle. When their stomata are closed and the photosynthetic reactions are running, oxygen builds up in leaves and triggers a process that lowers a plant's sugar-making capacity. Remember rubisco, the enzyme that fixes carbon for the Calvin–Benson cycle? When O_2 levels rise, *photorespiration* dominates; rubisco attaches oxygen—not carbon—to RuBP. This reaction yields one molecule of PGA instead of two. The lower yield slows sugar production and growth of the plant. Compare Figure 7.11*a* with Figure 7.10.

C4 plants, such as corn, also close stomata on hot, dry days. But the CO_2 level does not decline as much because these plants fix carbon twice, in two types of photosynthetic cells (Figure 7.11*b*). In *mesophyll* cells, a four-carbon molecule, oxaloacetate, forms when CO_2 donates a carbon to PEP. The enzyme catalyzing this step will not use oxygen no matter how much there is. Oxaloacetate is converted to malate, which moves into *bundle-sheath cells* through plasmodesmata. The malate releases CO_2, which enters the Calvin–Benson cycle.

The C4 cycle keeps the CO_2 level near rubisco high enough to stop photorespiration. It requires one more ATP than the C3 cycle. However, less water is lost and more sugar can be made on hot, bright, dry days.

Photorespiration hampers the growth of many C3 plants. So why hasn't natural selection eliminated it? Rubisco evolved when the atmosphere held little O_2 and a great deal of CO_2. Perhaps the gene coding for rubisco's structure cannot mutate without disruptive effects on rubisco's primary role—carbon fixation.

Over the past 50 to 60 million years, the C4 cycle evolved independently in many lineages. Before then,

Leaves of basswood (*Tilia americana*)

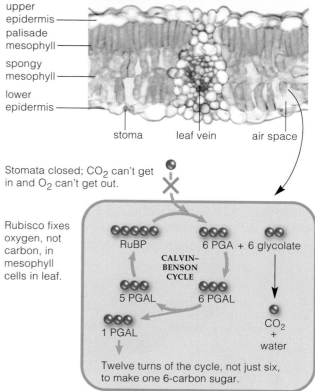

upper epidermis
palisade mesophyll
spongy mesophyll
lower epidermis

stoma   leaf vein   air space

Stomata closed; CO_2 can't get in and O_2 can't get out.

Rubisco fixes oxygen, not carbon, in mesophyll cells in leaf.

RuBP   6 PGA + 6 glycolate

CALVIN–BENSON CYCLE

5 PGAL   6 PGAL

1 PGAL

CO_2 + water

Twelve turns of the cycle, not just six, to make one 6-carbon sugar.

a Carbon fixation in C3 plants during hot, dry weather, when there is too little CO_2 and too much O_2 in leaves.

Figure 7.11 Comparison of carbon-fixing adaptations in three kinds of plants that evolved in different environments.

(**a**) The Calvin–Benson cycle, which also is called the C3 cycle, is common in evergreens and many nonwoody plants of temperate zones, such as basswood and bluegrass. (**b**) A C4 cycle is common in grasses, corn, and other plants that evolved in the tropics and that fix CO_2 twice. (**c**) Prickly pear (*Opuntia*), a CAM plant. These plants, which open stomata and fix carbon at night, include orchids, pineapples, and many succulents besides cacti.

atmospheric CO_2 levels were higher, so C3 plants had the selective advantage in hot climates. Which cycle will be most adaptive in the future? The CO_2 levels have been rising for decades and may double in the next fifty years. If so, C3 plants will yet again be at an advantage—and many vital crop plants may benefit.

Leaves of corn (*Zea mays*)

Beavertail cactus (*Opuntia basilaris*)

upper epidermis

mesophyll cell

leaf vein

bundle-sheath cell

lower epidermis

epidermis with thick cuticle

stoma

mesophyll cell

air space

Stomata closed; CO_2 can't get in and O_2 can't get out.

Stomata stay closed during day, open for CO_2 uptake at night only.

Carbon fixed in mesophyll cell, malate diffuses into adjacent bundle-sheath cell

PEP   oxaloacetate

C4 CYCLE

malate

C4 cycle operates at night, when CO_2 from aerobic respiration fixed

C4 CYCLE

In bundle-sheath cell, malate gets converted to pyruvate with release of CO_2, which enters Calvin–Benson cycle.

pyruvate

CO_2

RuBP   12 PGAL

CALVIN–BENSON CYCLE

10 PGAL   12 PGAL

2 PGAL

1 sugar

CO_2 that accumulated overnight used in C3 cycle during the day

CALVIN–BENSON CYCLE

1 sugar

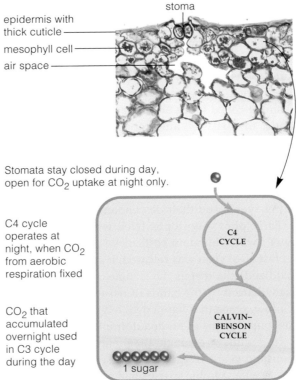

c Carbon fixation in CAM plants, adapted to hot, dry climates.

b Carbon fixation in C4 plants during hot, dry weather, when there is too little CO_2 and too much O_2 in leaves.

CAM PLANTS

We see a carbon-fixing adaptation to desert conditions in a cactus. This plant, a type of succulent, has juicy, water-storing tissues and thick surface layers that limit loss of water. It is one of many **CAM plants** (short for *Crassulacean Acid Metabolism*). A cactus will not open stomata on hot days; it opens them and fixes CO_2 *at*

night, when mesophyll cells use a C4 cycle. Each cell stores malate and other organic acids until the next day, when stomata close. Malate releases CO_2, which the cell uses in the Calvin–Benson cycle (Figure 7.11*c*).

Some CAM plants survive prolonged drought by keeping stomata shut even at night. They fix CO_2 from aerobic respiration. Not much forms, but it is enough to maintain low metabolic rates and very slow growth. Try growing cacti in mild climates, and you will see that they compete poorly with C3 and C4 plants.

C3 plants, C4 plants, and CAM plants respond differently to hot, dry conditions. At such times, stomata close to conserve water, and so photosynthetic cells must deal with too much oxygen and not enough carbon dioxide in leaves.

7.8 Autotrophs and the Biosphere

We conclude this chapter by reflecting on the mind-boggling numbers of single-celled and multicelled photosynthesizers and other autotrophs. We find them on land and in the water provinces, and they profoundly influence the biosphere.

THE ENERGY CONNECTION

This chapter opened with a brief look at the origin of photosynthesis and its impact on the world of life. All organisms require ongoing supplies of energy and carbon-based compounds for growth and survival. Autotrophs get them from the physical environment. Energy-rich carbon compounds become concentrated in single-celled kinds and in the tissues of multicelled kinds. In this way, autotrophs become concentrated stores of food tempting to heterotrophs.

Autotrophs are more than carbon-rich food baskets for the biosphere. Early practitioners of the noncyclic pathway of photosynthesis enriched the atmosphere with oxygen, and their descendants still replenish it.

Early photoautotrophs lived when iron and other metals were abundant both above and below the seas. As fast as oxygen was released, it swiftly latched onto (oxidized) the metals. Over time, it rusted them out, as evidenced by the bands of red iron deposits on the seafloor. Once that happened, oxygen bubbled out of vast populations of photoautotrophs, unimpeded.

In a wink of geologic time, maybe a few hundred thousand years, oxygen levels rose in the seas and the sky. Most anaerobic species had no means of adapting to the change, and they perished in a mass extinction. Other chemoautotrophs that could not tolerate oxygen endured in seafloor sediments, hot springs, and other anaerobic habitats. Some still live near hydrothermal vents, where superheated water spews out from big fissures in the seafloor. Archaeans near the vents get hydrogen and electrons from hydrogen sulfide in the mineral-rich water. Chemoautotrophs live in oxygen-free soils, where they extract energy from nitrogen-rich wastes and remains of other organisms.

However, among some ancient species of bacteria, metabolic pathways became modified in ways that detoxified oxygen. Later on, a pathway that released energy from organic compounds became modified in ways that allowed it to *use* oxygen as a final electron acceptor. As a direct outcome of the selection pressure exerted by oxygen—a by-product of photosynthesis—aerobic respiration had evolved (Figure 7.12).

PASTURES OF THE SEAS

Today, aerobic species are all around us. Each spring, the renewed growth of photoautotrophs is evident as trees leaf out and fields turn green. At the same time, uncountable numbers of single-celled species drifting through the ocean's surface waters make a seasonal response. You can't see them without a microscope. In some regions, a cup of seawater may hold 24 million cells of one species, and that number does not include any other aquatic species suspended in the cup.

Collectively, these cells are the "pastures of the seas." Most are bacteria and protists that ultimately feed nearly all other marine species. Their primary productivity is the start of vast aquatic food webs.

Imagine zooming in on a small patch of "pasture" in an Antarctic sea. There, tiny shrimplike crustaceans are feeding on even tinier photosynthesizers. Dense concentrations of these crustaceans, or krill, are food for other animals, such as fishes, penguins, seabirds, and immense blue whales. A single, mature whale is straining four tons of krill from the water. Before they

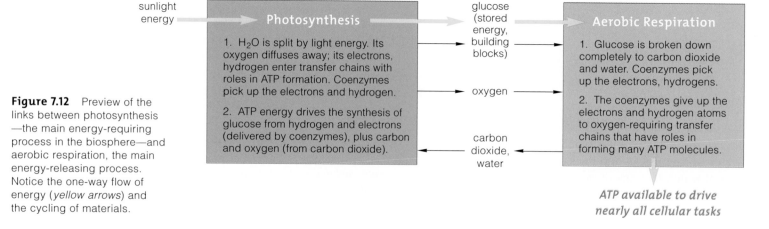

Figure 7.12 Preview of the links between photosynthesis—the main energy-requiring process in the biosphere—and aerobic respiration, the main energy-releasing process. Notice the one-way flow of energy (*yellow arrows*) and the cycling of materials.

sunlight energy → **Photosynthesis**

1. H₂O is split by light energy. Its oxygen diffuses away; its electrons, hydrogen enter transfer chains with roles in ATP formation. Coenzymes pick up the electrons and hydrogen.

2. ATP energy drives the synthesis of glucose from hydrogen and electrons (delivered by coenzymes), plus carbon and oxygen (from carbon dioxide).

glucose (stored energy, building blocks)

oxygen

carbon dioxide, water

Aerobic Respiration

1. Glucose is broken down completely to carbon dioxide and water. Coenzymes pick up the electrons, hydrogens.

2. The coenzymes give up the electrons and hydrogen atoms to oxygen-requiring transfer chains that have roles in forming many ATP molecules.

ATP available to drive nearly all cellular tasks

themselves became food for the whale, the four tons' worth of krill had munched their way through 1,200 tons of the pasture!

The pastures "bloom" in spring, when the seawater becomes warmer and greatly enriched with nutrients that currents churn up from the deep. The conditions favor huge increases in population sizes.

Until NASA gathered data from space satellites, we had no idea of the size and distribution of these marine pastures. Figure 7.13a shows the near-absence of photosynthetic activity one winter in the Atlantic Ocean. Figure 7.13b shows a springtime bloom that stretched from North Carolina all the way past Spain!

Collectively, these cells affect the global climate, because they deal with staggering numbers of gaseous reactant and product molecules. For instance, they sponge up nearly half of the carbon dioxide used in carbon fixation. Without them, atmospheric carbon dioxide would accumulate more rapidly and possibly accelerate global warming (Sections 47.9 and 47.10).

Although drastic global change is a real possibility, human activities release more carbon dioxide to the atmosphere than photoautotrophs can take up. Such activities include burning fossil fuels and setting fire to vast tracts of forests to clear land for farming.

There is more. Each day, tons of industrial wastes, raw sewage, and fertilizers in runoff from croplands enter the ocean and change its chemical composition. How long can we expect the marine photoautotrophs to function in this chemical brew? The answer may affect your life in more ways than one. It may affect populations and ecosystems throughout the world.

In sum, autotrophs exist in tremendous numbers. They nourish themselves and all other living things, and they are major players in the cycling of oxygen, nitrogen, phosphorus, and other elements all through the biosphere. Later chapters focus on their impact on the environment. In this unit, we turn next to major pathways by which all cells release the chemical bond energy that is stored in glucose and other biological molecules—the legacy of autotrophs everywhere.

Energy flow and the cycling of carbon and other nutrients through the biosphere starts with autotrophs.

Figure 7.13 Two satellite images that convey the sheer magnitude of photosynthetic activity during springtime in the surface waters of the North Atlantic Ocean. Sensors in equipment launched with the satellite recorded concentrations of chlorophyll, which were greatest in regions coded *red*.

Take a deep breath while looking at these images. You just took in free oxygen that originated with some photoautotroph, somewhere in the world. Poison the autotrophs and how long will oxygen-dependent heterotrophs last?

Summary

Section 7.1 Photosynthesis runs on energy obtained when pigment molecules absorb wavelengths of visible light from the sun. Chlorophyll *a*, the main pigment, is best at absorbing violet and red wavelengths. Diverse photosynthetic pigments are accessory pigments. They form light-harvesting complexes that capture photons of particular wavelengths. Photons not captured are reflected as the characteristic color of each pigment.

Section 7.2 Collectively, photosynthetic pigments absorb most of the full range of wavelengths in the spectrum of visible light with impressive efficiency.

Section 7.3 Photosynthesis has two stages: light-dependent and light-independent reactions. Figure 7.14 and the following equation summarize the process:

$$12H_2O + 6CO_2 \xrightarrow[\text{enzymes}]{\text{light energy}} 6O_2 + C_6H_{12}O_6 + 6H_2O$$

water carbon   oxygen glucose water
       dioxide

In chloroplasts, the light-dependent reactions occur at a thylakoid membrane that forms a single compartment in the semifluid interior (stroma).

Biology ⓔNow
View the sites where photosynthesis takes place with the animation on BiologyNow.

Sections 7.4, 7.5 Accessory pigments arrayed in clusters in the thylakoid membrane absorb photons and pass energy to many photosystems. Light-dependent reactions use electrons released from photosystems in a noncyclic or a cyclic pathway of ATP formation.

In the noncyclic pathway, electrons are released from photosystem II and enter an electron transfer chain. Their flow through the chain causes hydrogen ions to accumulate in the thylakoid compartment. They flow on to photosystem I where photon absorption also causes the release of electrons. An intermediary molecule next to photosystem I accepts the electrons and transfers them to NADP$^+$, which attracts hydrogen ions (H$^+$) at the same time and becomes a reduced coenzyme, NADPH.

The electrons lost from photosystem II are replaced by way of photolysis—a reaction that pulls electrons from water molecules, with the release of H$^+$ and O$_2$.

In the cyclic pathway, electrons from photosystem I enter an electron transfer chain, then are cycled back to the same photosystem. NADPH does not form.

In both pathways, the H$^+$ buildup in the thylakoid compartment forms concentration and electric gradients across the thylakoid membrane. H$^+$ flows in response to the gradients, through ATP synthases. The flow causes P$_i$ to be attached to ADP in the stroma, forming ATP.

Biology ⓔNow
Review the pathways by which light energy is used to form ATP with the animation on BiologyNow.

Section 7.6 The light-independent reactions proceed in the stroma. In C3 plants, the enzyme rubisco attaches carbon from CO$_2$ to RuBP to start the Calvin–Benson cycle. In this cyclic pathway, energy from ATP, carbon and oxygen from CO$_2$, and hydrogen and electrons from NADPH are used to make phosphorylated glucose, which quickly enters reactions that form the products of photosynthesis (mainly sucrose, cellulose, and starch). It takes six turns of the Calvin–Benson cycle to fix the six CO$_2$ required to make one glucose molecule.

Biology ⓔNow
Read the InfoTrac article "Robust Plants' Secret? Rubisco Activase!" Marcia Wood, Agricultural Research, November 2002.

Section 7.7 Environments differ, and so do details of sugar production. On hot, dry days, plants conserve water by closing stomata, but O$_2$ from photosynthesis cannot escape. In C3 plants, high O$_2$/low CO$_2$ levels cause the enzyme rubisco to use O$_2$ in an alternate pathway that does not make as much sugar. In C4 plants, carbon fixation occurs in one cell type, and the carbon enters the Calvin–Benson cycle in a different cell type. CAM plants close stomata in the day and fix carbon at night.

Biology ⓔNow
Read the InfoTrac article "Light of Our Lives," Norman Miller, Geographical, January 2001.

Section 7.8 Photoautotrophs and chemoautotrophs produce the food that sustains themselves and all other organisms. Also, staggering numbers of diverse aerobic and anaerobic autotrophs live in the seas as well as on land. They have impact on the global cycling of oxygen, carbon, nitrogen, and other substances, and the global climate. Human activities are having impact on them.

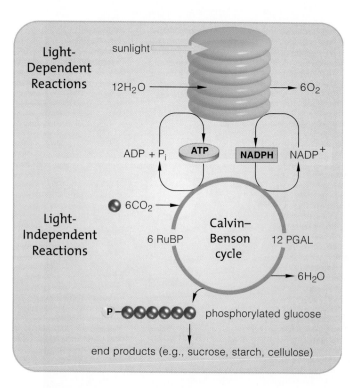

Figure 7.14 Visual summary of photosynthesis.

Figure 7.15 Leaves of *Elodea*, an aquatic plant.

Figure 7.16 (a) Red alga from a tropical reef. (b) Coastal green alga (*Codium*).

Self-Quiz

Answers in Appendix II

1. Photosynthetic autotrophs use _____ from the air as a carbon source and _____ as their energy source.

2. Chlorophyll *a* absorbs violet and red light, and it reflects light of _____ and _____ wavelengths.
   a. violet; red
   b. yellow; green
   c. green only
   d. white; orange

3. Light-*dependent* reactions in plants occur at the _____ .
   a. thylakoid membrane
   b. plasma membrane
   c. stroma
   d. cytoplasm

4. In the light-*dependent* reactions, _____ .
   a. carbon dioxide is fixed
   b. ATP and NADPH form
   c. CO_2 accepts electrons
   d. sugars form

5. What accumulates inside the thylakoid compartment during the light-*dependent* reactions?
   a. glucose   b. RuBP   c. hydrogen ions   d. CO_2

6. When a photosystem absorbs light, _____ .
   a. sugar phosphates are produced
   b. electrons are transferred to ATP
   c. RuBP accepts electrons
   d. light-dependent reactions begin

7. Light-*independent* reactions proceed in the _____ .
   a. cytoplasm   b. plasma membrane   c. stroma

8. The Calvin–Benson cycle starts when _____ .
   a. light is available
   b. carbon dioxide is attached to RuBP
   c. electrons leave photosystem II

9. What substance is *not* part of the Calvin–Benson cycle?
   a. ATP
   b. NADPH
   c. RuBP
   d. carotenoids
   e. O_2
   f. CO_2

10. Match each event with its most suitable description.
    ____ ATP formation only   a. rubisco required
    ____ CO_2 fixation   b. ATP, NADPH required
    ____ PGAL formation   c. electrons cycled back
                                     to photosystem II

Additional questions are available on **Biology⑤Now™**

Critical Thinking

1. About 200 years ago, Jan Baptista van Helmont did experiments on the nature of photosynthesis. He wanted to know where growing plants get the materials necessary for increases in size. He planted a tree seedling weighing 5 pounds in a barrel filled with 200 pounds of soil. He watered the tree regularly. Five years passed. Then van Helmont weighed the tree and the soil. The tree weighed 169 pounds, 3 ounces. The soil weighed 199 pounds, 14 ounces. Because the tree gained so much weight and the soil lost so little, he concluded the tree had gained all of its additional weight by absorbing water he had added to the barrel. Given what you know about biological molecules, why was he misguided? Knowing what you do about photosynthesis, what really happened?

2. A cat eats a bird, which earlier ate a caterpillar that chewed on a weed. Which organisms are autotrophs? Which are the heterotrophs?

3. Imagine walking through a garden of red, white, and blue petunias. Explain each of the colors in terms of which wavelengths of light the flower is absorbing.

4. Krishna exposes pea plants to a carbon radioisotope ($^{14}CO_2$), which they absorb. In which compound will the labeled carbon appear first if the plants are C3? C4?

5. While gazing into an aquarium, you observe bubbling from an aquatic plant (Figure 7.15). What is happening?

6. Only about eight classes of pigment molecules are known, but this limited group gets around in the world. For example, photoautotrophs make carotenoids, which move through food webs, as when tiny aquatic snails graze on green algae and then flamingos eat the snails. Flamingos modify the ingested carotenoids. Their cells split beta-carotene to form two molecules of vitamin A. This vitamin is the precursor of retinol, a visual pigment that transduces light into electric signals in eyes. Beta-carotene gets dissolved in fat reservoirs under the skin. Cells that give rise to bright pink feathers take it up.

   Select a similar organism and do some research to identify sources for pigments that color its surfaces.

7. Most pigments respond to only part of the rainbow of visible light. If acquiring energy is so vital, then why doesn't each kind of photosynthetic pigment absorb the whole spectrum? *Why isn't each one black?*

   If early photoautotrophs evolved in the seas, then so did their pigments. Ultraviolet and red wavelengths do not penetrate water as deeply as green and blue wavelengths do. Possibly natural selection favored the evolution of different pigments at different depths. Many relatives of the red alga in Figure 7.16*a* live deep in the sea. Some are nearly black. Green algae, such as the one in Figure 7.16*b*, live in shallow water. Their chlorophylls absorb red wavelengths, and accessory pigments harvest others. Some accessory pigments also function as shields against ultraviolet radiation.

   Speculate on how natural selection may have favored the evolution of different pigments at different depths, starting at hydrothermal vents. You might start at Richard Monasterky's article in *Science News* (September 7, 1996) on Cindy Lee Dover's work at hydrothermal vents.

When Mitochondria Spin Their Wheels

In the early 1960s, Swedish physician Rolf Luft mulled over some odd symptoms of a patient. The young woman felt weak and too hot all the time. Even on the coldest winter days she could not stop sweating, and her skin was always flushed. She was thin in spite of a huge appetite.

Luft inferred that his patient's symptoms pointed to a metabolic disorder. Her cells seemed to be spinning their wheels. They were active, but much of their activity was being dissipated as metabolic heat. So he ordered tests designed to detect her metabolic rates. The patient's oxygen consumption was the highest ever recorded!

Microscopic examination of a tissue sample from the patient's skeletal muscles revealed mitochondria, the cell's ATP-producing powerhouses. But there were far too many of them, and they were abnormally shaped. Other studies showed that the mitochondria were engaged in aerobic respiration—yet they were making very little ATP.

The disorder, now called *Luft's syndrome*, was the first to be linked directly to a defective organelle. By analogy, someone with this mitochondrial disorder functions like a city with half of its power plants shut down. Skeletal and heart muscles, the brain, and other hardworking body parts with the highest energy demands are hurt the most.

More than a hundred other mitochondrial disorders are now known. One, a heritable disease called *Friedreich's ataxia*, causes loss of coordination (ataxia), weak muscles, and visual problems. Many affected people die when they are young adults because of heart muscle irregularities. Figure 8.1 shows two affected individuals.

A mutant gene causes Friedreich's ataxia. Its abnormal protein product makes iron accumulate in mitochondria. Iron is vital for electron transfers that drive ATP formation, but too much favors the concentration of free radicals that can attack the structural integrity of all molecules of life.

Defective mitochondria also contribute to many age-related problems, including Type 1 diabetes, atherosclerosis, amyotrophic lateral sclerosis (Lou Gehrig's disease), as well as Parkinson's, Alzheimer's, and Huntington's diseases.

Clearly, human health depends on mitochondria that are structurally and functionally sound. However, zoom out from our characteristically human focus and you will find that every kind of animal, every kind of plant and fungus, and nearly every protist depend on them.

Remember how ATP forms at a membrane system inside chloroplasts? Similar events happen at a membrane system inside mitochondria. In both photosynthesis and aerobic respiration, electrons are released when an energy input breaks chemical bonds, and they are sent step-by-step through transfer chains in the membrane. Photosynthesis splits bonds in water molecules; aerobic respiration splits bonds in glucose and other organic compounds.

In mitochondria, too, energy harnessed as electrons go through transfer chains helps concentrate hydrogen ions on one side of the membrane. Here again, potential energy

Figure 8.1 Sister, brother, and broken mitochondria. Both of these individuals show symptoms of Friedreich's ataxia, a heritable genetic disorder that prevents them from making enough ATP to keep their body structurally and functionally sound. At age five, Leah started to lose her sense of balance and coordination. Six years later she was in a wheelchair and is now diabetic and partially deaf. Her brother Joshua was three when problems started. Eight years later he could not walk. He is now blind. Both sister and brother have heart problems; both had spinal fusion surgery. Special equipment allows them to attend school and work part-time. Leah is a professional model.

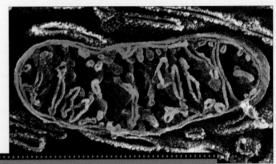

Watch the video online!
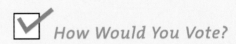

is briefly stored as concentration and electric gradients across that membrane. That stored energy is converted to the bond energy of ATP. And ATP is a transportable form of energy; it can jump-start nearly all of the metabolic reactions that keep cells, and organisms, going.

The point is, you already have a sense of how operation of electron transfer systems can concentrate energy in a form that can be tapped to make ATP. Even prokaryotic cells make ATP by operating simpler electron transfer systems, which are built into their plasma membrane. As you will see, the details differ from one group to the next. Yet even these variations do not obscure life's unity at the biochemical level.

How Would You Vote?

Developing new drugs is costly. There is little incentive for pharmaceutical companies to target ailments, such as Friedreich's ataxia, that affect relatively few individuals. Should the federal government allocate some funds to private companies that search for cures for diseases affecting a relatively small number of people? See BiologyNow for details, then vote online.

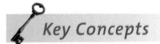

Key Concepts

THE MAIN ENERGY-RELEASING PATHWAYS

All organisms release chemical bond energy from glucose and other organic compounds to drive ATP formation. The main energy-releasing pathways all start in the cytoplasm. Only aerobic respiration, which uses free oxygen, ends in mitochondria. It has the greatest energy yield.   Section 8.1

GLYCOLYSIS—FIRST STAGE OF THE PATHWAYS

Glycolysis is the first stage of aerobic respiration. It also is the first stage of anaerobic pathways, such as alcoholic and lactate fermentation. Enzymes partially break down glucose to pyruvate, and they can do so in the presence of oxygen or in its absence.   Section 8.2

HOW AEROBIC RESPIRATION ENDS

Aerobic respiration continues through two more stages. In the second stage, pyruvate from glycolysis is broken down to carbon dioxide. Electrons and hydrogen that coenzymes picked up during the first two stages enter electron transfer chains. In the third stage, transfer chains set up conditions that favor ATP formation. Free oxygen accepts the spent electrons.   Sections 8.3, 8.4

HOW ANAEROBIC PATHWAYS END

Fermentation also starts with glycolysis, but substances other than oxygen are the final electron acceptor. The net energy yield is always small.   Section 8.5

WHAT IF GLUCOSE IS NOT AVAILABLE?

When required, molecules other than glucose can enter the aerobic pathway as alternative energy sources.   Section 8.6

PERSPECTIVE AT UNIT'S END

We see evidence of life's unity in its molecular and cellular organization and in the utter dependence of all organisms on the one-way flow of energy.   Section 8.7

Links to Earlier Concepts

This chapter expands the picture of life's dependence on energy flow by showing how all organisms tap energy stored in glucose and convert it to transportable forms, ATP especially (Sections 6.1, 6.2). You may wish to review the structure of glucose (3.1, 3.3). You will see more examples of controlled energy release at electron transfer chains (6.4). You will reflect once more on global connections between photosynthesis and aerobic respiration (7.8).

8.1 Overview of Energy-Releasing Pathways

LINKS TO
SECTIONS
6.1, 6.2

Plants make ATP during photosynthesis and use it to synthesize glucose and other carbohydrates. But all organisms, plants included, can make ATP by breaking down carbohydrates, lipids, and proteins.

Organisms stay alive only as long as they get more energy to replace the energy they use up (Section 6.1). Plants and all other photoautotrophs get energy from the sun; heterotrophs get energy by eating plants and one another. Regardless of its source, the energy must be converted to some form that can drive thousands of diverse life-sustaining reactions. Energy that becomes converted into chemical bond energy of adenosine triphosphate—ATP—serves that function.

COMPARISON OF THE MAIN TYPES OF ENERGY-RELEASING PATHWAYS

The first energy-releasing metabolic pathways were operating billions of years before Earth's oxygen-rich atmosphere evolved, so we can expect that they were *anaerobic*; the reactions did not use free oxygen. Many prokaryotes and protists still live in places where oxygen is absent or not always available. They make ATP by fermentation and other anaerobic pathways. Many eukaryotic cells still use fermentation, including skeletal muscle cells. However, the cells of nearly all eukaryotes extract energy efficiently from glucose by **aerobic respiration**, an oxygen-dependent pathway. Each breath you take provides your actively respiring cells with a fresh supply of oxygen.

Make note of this point: *In every cell, all of the main energy-releasing pathways start with the same reactions in the cytoplasm.* During the initial reactions, **glycolysis**, enzymes cleave and rearrange a glucose molecule into two molecules of **pyruvate**, an organic compound that has a three-carbon backbone.

After glycolysis, energy-releasing pathways differ. Only the aerobic pathway ends inside mitochondria. There, free oxygen accepts and removes electrons after they indirectly help the formation of ATP (Figure 8.2).

As you examine the energy-releasing pathways in sections to follow, keep in mind that enzymes catalyze each step, and intermediates formed at one step serve as substrates for the next enzyme in the pathway.

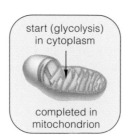

start (glycolysis) in cytoplasm

completed in cytoplasm

Anaerobic Energy-Releasing Pathways

start (glycolysis) in cytoplasm

completed in mitochondrion

Aerobic Respiration

OVERVIEW OF AEROBIC RESPIRATION

Of all energy-releasing pathways, aerobic respiration gets the most ATP for each glucose molecule. Whereas anaerobic routes have a net yield of two ATP, aerobic respiration typically yields thirty-six or more. If you were a bacterium, you would not require much ATP. Being far larger, more complex, and highly active, you

Figure 8.2 *Animated!* Where the main energy-releasing pathways of ATP formation start and end. Only aerobic respiration ends in mitochondria. This pathway alone delivers enough ATP to build and maintain big multicelled organisms, including redwoods and highly active animals, such as people and Canada geese.

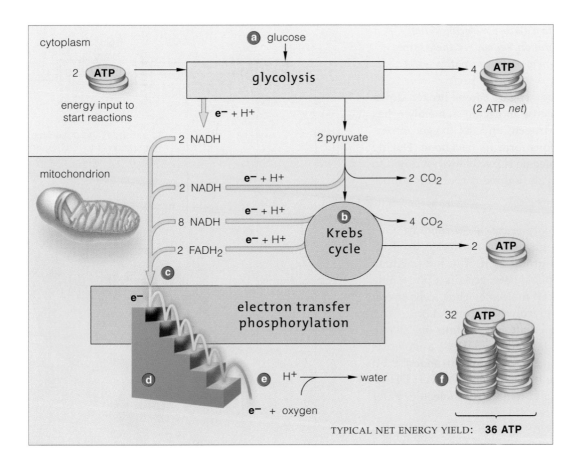

Figure 8.3 *Animated!* Overview of aerobic respiration. The reactions start in the cytoplasm, but they end inside mitochondria.

(**a**) In the first stage, glycolysis, enzymes partly break down glucose to pyruvate.

(**b**) In the second stage, enzymes break down pyruvate to carbon dioxide.

(**c**) NAD$^+$ and FAD pick up the electrons and hydrogen stripped from intermediates in both stages.

(**d**) The last stage is electron transfer phosphorylation. NADH and FADH$_2$, which are reduced coenzymes, deliver electrons to electron transfer chains. H$^+$ accompanies these electrons. Electron flow through the chains sets up H$^+$ gradients, which are tapped to make ATP.

(**e**) Oxygen accepts electrons at the end of the third stage, forming water.

(**f**) From start to finish, a typical net energy yield from a single glucose molecule is thirty-six ATP.

depend on the aerobic pathway's high yield. When a molecule of glucose is the starting material, aerobic respiration can be summarized this way:

$$C_6H_{12}O_6 \ + \ 6O_2 \ \longrightarrow \ 6CO_2 \ + \ 6H_2O$$
glucose   oxygen   carbon   water
                        dioxide

However, as you can see, the summary equation only tells us what the substances are at the start and finish of the pathway. In between are three reaction stages.

As you track the reactions, you will encounter two coenzymes, abbreviated **NAD$^+$** (nicotinamide adenine dinucleotide) and **FAD** (flavin adenine dinucleotide). Both accept electrons and hydrogen derived from intermediates that form during glucose breakdown. Unbound hydrogen atoms are hydrogen ions (H$^+$), or naked protons. When the two coenzymes are carrying electrons and hydrogen, they are in a reduced form and may be abbreviated NADH and FADH$_2$.

Figure 8.3 is your overview of aerobic respiration. Glycolysis, again, is the first stage. The second stage is a cyclic pathway, the **Krebs cycle**. Enzymes break down pyruvate to carbon dioxide and water. These reactions release many electrons and hydrogen atoms.

Few ATP form during glycolysis or the Krebs cycle. The big energy harvest comes in the third stage, after reduced coenzymes give up electrons and hydrogen to electron transfer chains—the machinery of **electron transfer phosphorylation**. Operation of these chains sets up H$^+$ concentration and electric gradients. The gradients drive ATP formation at nearby membrane transport proteins. During this final stage, many ATP molecules form. It ends when free oxygen accepts the "spent" electrons from the last portion of the transfer chain. The oxygen picks up H$^+$ at the same time and thereby forms water, a by-product of the reactions.

Nearly all metabolic reactions run on energy released from glucose and other organic compounds. The main energy-releasing pathways start in the cytoplasm with glycolysis, a series of reactions that break down glucose to pyruvate.

Anaerobic pathways have a small net energy yield, typically two ATP for each glucose molecule metabolized.

Aerobic respiration, an oxygen-dependent pathway, runs to completion in mitochondria. From start (glycolysis) to finish, it typically has a net energy yield of thirty-six ATP.

8.2 Glycolysis—Glucose Breakdown Starts

LINKS TO
SECTIONS
2.4, 3.3

Let's track what happens to a glucose molecule in the first stage of aerobic respiration. Remember, these same steps occur in anaerobic energy-releasing pathways.

Any of several six-carbon sugars can be broken down in glycolysis. Each glucose molecule, remember, has six carbon, twelve hydrogen, and six oxygen atoms (Section 3.3). The carbons form its backbone. During glycolysis, this one molecule is partly broken down to two molecules of pyruvate, a three-carbon compound:

glucose ⟶ (P)–glucose ⟶ 2 pyruvate

The initial steps of glycolysis are *energy-requiring*. One ATP molecule primes glucose to rearrange itself by donating a phosphate group to it. The intermediate that forms, fructose-6-phosphate, accepts a phosphate group from another ATP molecule. Thus, cells invest two ATP to jump-start glycolysis (Figure 8.4*a*).

The resulting intermediate is split into one PGAL (phosphoglyceraldehyde) and one DHAP, a molecule that has the same number of atoms arranged a bit differently. An enzyme can reversibly convert DHAP into PGAL. It does so, and two PGAL molecules enter the next reaction (Figure 8.4*b*).

In the first *energy-releasing* step of glycolysis, the two PGAL are converted to intermediates that give up a phosphate group to ADP, so two ATP form. In later reactions, two more intermediates do the same thing. Thus, four ATP have formed by **substrate-level phosphorylation**. We define this metabolic event as a direct transfer of a phosphate group from a substrate of a reaction to another molecule—in this case, to ADP.

Meanwhile, the two PGAL give up electrons and hydrogens to two NAD$^+$, which becomes NADH.

Even though four ATP are now formed, remember that two ATP were invested to start the reactions. The *net* yield of glycolysis is two ATP and two NADH.

To summarize, glycolysis converts bond energy of glucose to bond energy of ATP—a transportable form of energy. The electrons and hydrogen stripped from glucose and picked up by NAD$^+$ can enter the next stage of reactions. So can the products of glycolysis—two pyruvate molecules.

> *Glycolysis is a series of reactions that partially break down glucose or other six-carbon sugars to two molecules of pyruvate. It takes two ATP to jump-start the reactions.*
>
> *Two NADH and four ATP form. However, when we subtract the two ATP required to start the reactions, the net energy yield of glycolysis is two ATP from one glucose molecule.*

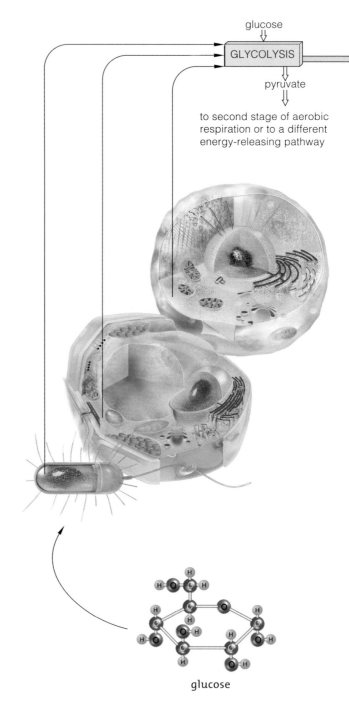

glucose

Figure 8.4 *Animated!* Glycolysis. This first stage of the main energy-releasing pathways occurs in the cytoplasm of all prokaryotic and eukaryotic cells. Glucose is the reactant in this example. Appendix VII gives the structural formulas of reaction intermediates and products. Two pyruvate, two NADH, and four ATP form in glycolysis. Cells invest two ATP to start the reactions, however, so the *net* energy yield is two ATP.

Depending on the type of cell and environmental conditions, the pyruvate may enter the second set of reactions of the aerobic pathway, including the Krebs cycle. Or it may be used in other reactions, such as those of fermentation.

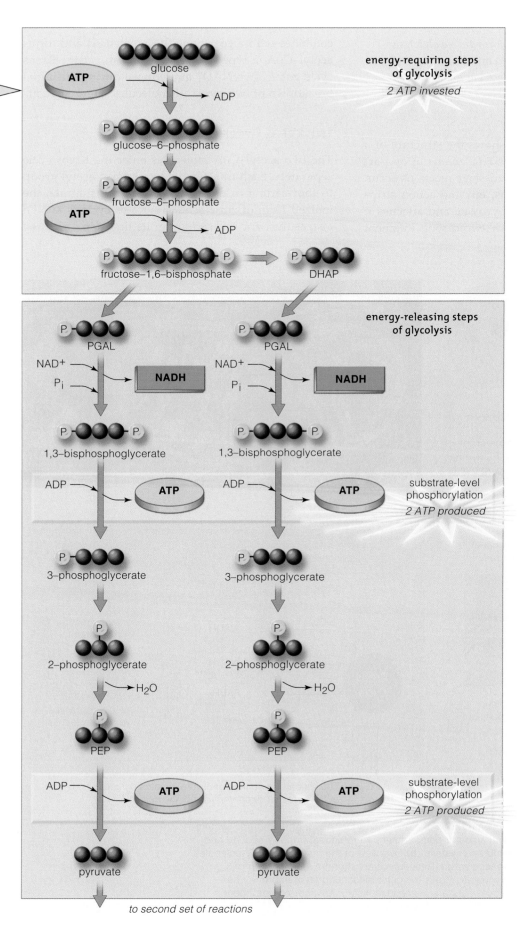

Track the six carbon atoms (*brown circles*) of glucose. Glycolysis requires an energy investment of two ATP:

a One ATP transfers a phosphate group to glucose, jump-starting the reactions.

b Another ATP transfers a phosphate group to an intermediate, causing it to split into two three-carbon compounds: PGAL and DHAP (dihydroxyacetone phosphate). Both have the same atoms, arranged differently, and they are interconvertible. But only PGAL can continue on in glycolysis. DHAP gets converted, so two PGAL are available for the next reaction.

c Two NADH form when each PGAL gives up two electrons and a hydrogen atom to NAD⁺.

d Two intermediates each transfer a phosphate group to ADP. *Thus, two ATP have formed by direct phosphate group transfers.* The original energy investment of two ATP is now paid off.

e Two more intermediates form. Each gives up one hydrogen atom and an —OH group. These combine as water. Two molecules called PEP form by these reactions.

f Each PEP transfers a phosphate group to ADP. *Once again, two ATP have formed by substrate-level phosphorylation.*

In sum, glycolysis has a net energy yield of two ATP for each glucose molecule. Two NADH also form during the reactions, and two molecules of pyruvate are the end products.

8.3 Second Stage of Aerobic Respiration

LINKS TO
SECTIONS
1.1, 2.4

The two pyruvate molecules that form during glycolysis may be completely dismantled in a mitochondrion. Many coenzymes pick up the released electrons and hydrogens.

ACETYL–CoA FORMATION

Start with Figure 8.5, which shows the structure of a typical mitochondrion. Figure 8.6a zooms in on part of the interior where the second-stage reactions occur. At the start of these reactions, enzyme action strips one carbon atom from each pyruvate and attaches it to oxygen, forming CO_2. Each two-carbon fragment combines with a coenzyme (designated A) and forms acetyl–CoA, a type of cofactor that can get the Krebs cycle going. Two NAD^+ are reduced during the initial breakdown of two pyruvate molecules (Figure 8.7a,b).

THE KREBS CYCLE

The two acetyl–CoA molecules enter the Krebs cycle separately. Each transfers its two-carbon acetyl group to four-carbon oxaloacetate, which forms citrate, the ionized form of citric acid. The Krebs cycle is known also as the *citric acid cycle*, after its first step.

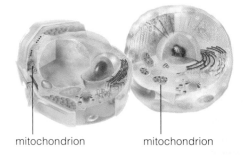

mitochondrion   mitochondrion

Figure 8.5 Scanning electron micrograph of a mitochondrion, sliced crosswise. Remember, nearly all eukaryotic species, including plants and animals, contain these organelles.

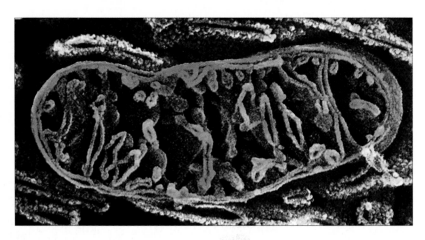

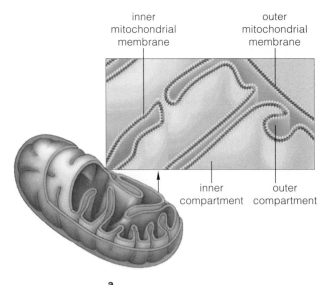

inner mitochondrial membrane   outer mitochondrial membrane

inner compartment   outer compartment

Two pyruvates cross the inner mitochondrial membrane   outer mitochondrial compartment

Krebs Cycle

2 NADH   inner mitochondrial compartment

6 NADH

2 FADH₂   *Eight NADH, two FADH₂, and two ATP are the payoff from the complete breakdown of two pyruvates in the second-stage reactions.*

2 ATP

6 CO_2 The six carbon atoms from two pyruvates diffuse out of the mitochondrion, then out of the cell, in six CO_2.

a   b

Figure 8.6 *Animated!* (**a**) Functional zones of a mitochondrion. An inner membrane system divides the interior into an inner and an outer compartment. Aerobic respiration's second and third stages take place at this membrane system. (**b**) Overview of the number of ATP molecules and coenzymes that form in the second stage. Reactions start after the membrane proteins transport the two pyruvate from glycolysis across the outer mitochondrial membrane, then across the inner membrane. Both pyruvates are dismantled inside the inner compartment.

a One carbon atom is stripped from each pyruvate and is released as CO_2. The remaining fragment binds with coenzyme A, forming acetyl–CoA.

Acetyl-CoA Formation

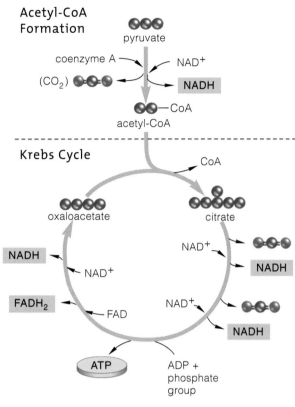

pyruvate

coenzyme A

NAD$^+$

(CO_2)

NADH

—CoA

acetyl-CoA

b NAD$^+$ picks up hydrogen and electrons, forming one NADH.

Krebs Cycle

CoA

oxaloacetate

citrate

h The final steps regenerate oxaloacetate. NAD$^+$ picks up hydrogen and electrons, forming NADH. *At this point in the cycle, three NADH and one FADH$_2$ have formed.*

NADH

NAD$^+$

NAD$^+$

NADH

FADH$_2$

FAD

NAD$^+$

NADH

c In the first step of the Krebs cycle, acetyl–CoA transfers two carbons to oxaloacetate, forming citrate.

d In rearrangements of intermediates, another carbon atom is released as CO_2, and NADH forms as NAD$^+$ picks up hydrogen and electrons.

e Another carbon atom is released as CO_2. Another NADH forms. *The three carbon atoms that entered the second-stage reactions in each pyruvate have now been released.*

g FADH$_2$ forms as the coenzyme FAD picks up electrons and hydrogen.

ATP

ADP + phosphate group

f A phosphate group is attached to ADP. At this point, one ATP has formed by substrate-level phosphorylation.

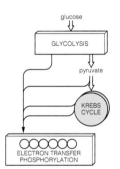

glucose

GLYCOLYSIS

pyruvate

KREBS CYCLE

ELECTRON TRANSFER PHOSPHORYLATION

Figure 8.7 *Animated!* Aerobic respiration's second stage: formation of acetyl–CoA and the Krebs cycle. The reactions proceed in a mitochondrion's inner compartment. *It takes two turns of the cycle to break down the two pyruvates from glucose.* A total of two ATP, eight NADH, two FADH$_2$, and six CO_2 molecules form. Organisms release the CO_2 from the reactions into their surroundings. For details, see Appendix VII.

The second-stage reactions run twice, one for each pyruvate molecule. The remaining carbon atoms are released in the form of CO_2 (Figure 8.7d,e). Only two ATP form during the two turns, which does not add much to the small net yield from glycolysis. However, in addition to the two NAD$^+$ that were reduced when the acetyl–CoA formed, six more NAD$^+$ and two FAD molecules are reduced. With their cargo of electrons and hydrogen, these coenzymes—*eight NADH and two FADH$_2$*—are a big potential payoff for the cell. All of the electrons have potential energy, which coenzymes can deliver to the final reaction sites.

As you can see from Figure 8.7, a total of *six* carbon atoms (from two pyruvates) depart during the second stage of aerobic respiration, in six molecules of CO_2. Therefore, the glucose from glycolysis has lost all of its carbons; it has become fully oxidized.

A final note: Figure 8.7 is a simplified version of these second-stage reactions. For interested students, Figure B in Appendix VII offers more details.

Aerobic respiration's second stage starts after two pyruvate molecules from glycolysis move from the cytoplasm, across the outer and inner mitochondrial membranes, and then into the inner mitochondrial compartment.

During these reactions, pyruvate is converted to acetyl–CoA, which starts the Krebs cycle. Two ATP and ten coenzymes (eight NADH, two FADH$_2$) form. All of pyruvate's carbons depart, in the form of carbon dioxide.

Together with two coenzymes (NADH) that formed during glycolysis, the ten coenzymes from the second stage will deliver electrons and hydrogen to the third and final stage.

8.4 Third Stage of Aerobic Respiration—A Big Energy Payoff

LINKS TO
SECTIONS
5.2, 6.4

In the aerobic pathway's third stage, energy release goes into high gear. Coenzymes from the first two stages provide the hydrogen and electrons that drive the formation of many ATP. Electron transfer chains and ATP synthases function as the machinery.

ELECTRON TRANSFER PHOSPHORYLATION

The third stage starts as coenzymes donate electrons to electron transfer chains in the inner mitochondrial membrane (Figure 8.8a). The flow of electrons through the chains drives the attachment of phosphate to ADP molecules. That is what the name of the event, *electron transfer phosphorylation*, means.

Incremental energy release, recall, is more efficient than one burst of energy, nearly all of which would be lost as unusable heat (Section 6.4). As electrons flow through the chains, they transfer energy bit by bit, in tiny amounts, to molecules that briefly store it. The two NADH that formed in the cytoplasm (by glycolysis) cannot directly reach the ATP-forming machinery. They must give up their electrons and hydrogen to transport proteins, which shuttle them across the inner membrane, into the innermost compartment. There, NAD$^+$ or FAD

pick them up. The eight NADH and two FADH$_2$ from the second stage are already inside.

All of the coenzymes turn over electrons to transfer chains and at the same time give up hydrogen, which now has a positive charge (H$^+$). Again, electrons lose a bit of energy at each transfer through the chain. At three transfers, that energy drives the pumping of H$^+$ into the outer compartment (Figure 8.8b). There, many ions accumulate—which sets up concentration and electric gradients across the inner membrane.

H$^+$ cannot cross a lipid bilayer. Instead, it follows its gradients by flowing through ATP synthases (Section 5.2 and Figure 8.8c). The ion flow causes parts of the ATP synthase molecules to change their shape in a reversible way. The change promotes the attachment of unbound phosphate to ADP, thus forming ATP.

The last components of the electron transfer chains pass electrons to oxygen, which combines with H$^+$ and thereby forms water. *Oxygen is the final acceptor of electrons originally stripped from glucose.*

In oxygen-starved cells, the electrons have nowhere to go. The whole chain backs up with electrons all the way to NADPH, so no H$^+$ gradients form, and no ATP forms, either. Without oxygen, cells of complex organisms do not survive long. They cannot produce enough ATP to sustain life processes.

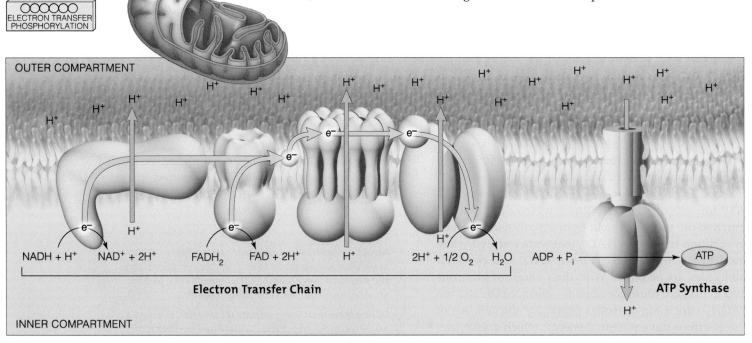

(a) At the inner mitochondrial membrane, NADH and FADH$_2$ give up electrons to transfer chains. When electrons are transferred through the chains, unbound hydrogen (H$^+$) is shuttled across the membrane to the outer compartment.

(b) Free oxygen is the final acceptor of electrons at the end of the transfer chain.

(c) H$^+$ concentration and electric gradients now exist across the membrane. H$^+$ follows the gradients through the interior of ATP synthases, to the inner compartment. The flow drives the formation of ATP from ADP and unbound phosphate (P$_i$).

Figure 8.8 Electron transfer phosphorylation, the third and final stage of aerobic respiration.

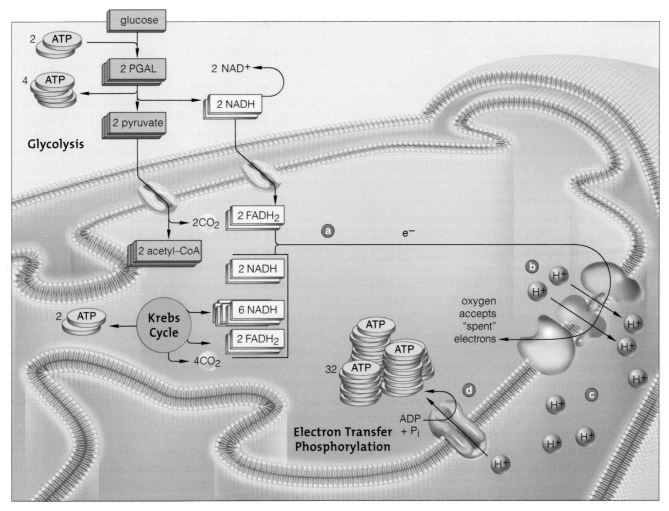

(a) Electrons and hydrogen from NADH and FADH$_2$ that formed during the first and second stages enter electron transfer chains.

(b) As electrons are being transferred through these chains, H$^+$ ions are shuttled across the inner membrane, into the outer compartment.

(c) More H$^+$ accumulates in the outer compartment than in the inner one. Chemical and electric gradients have been established across the inner membrane.

(d) Hydrogen ions follow the gradients through the interior of ATP synthases, driving ATP formation from ADP and phosphate (P$_i$).

Figure 8.9 *Animated!* Summary of the transfers of electrons and hydrogen from coenzymes involved in ATP formation in mitochondria.

SUMMING UP: THE ENERGY HARVEST

Thirty-two ATP typically form in the third stage. Add the four ATP from the earlier stages, and the net yield from a glucose molecule is *thirty-six ATP* (Figure 8.9). By contrast, anaerobic pathways may use up eighteen glucose molecules to get the same net yield.

The yield varies. First, reactant, intermediate, and product concentrations can change it. Second, the two NADH from glycolysis cannot enter a mitochondrion. They give up electrons and hydrogen to transport proteins in the outer mitochondrial membrane, which shuttle them across. NAD$^+$ or FAD that are already inside accept them, forming NADH or FADH$_2$.

When NADH inside delivers electrons to a certain entry point into a transfer chain, enough H$^+$ gets pumped across the membrane to make three ATP.

FADH$_2$ delivers them to a different entry point. Less H$^+$ is pumped across, and only two ATP form.

In liver, heart, and kidney cells, the electrons and hydrogen enter the first entry point, and the energy harvest is thirty-eight ATP. More commonly, as in skeletal muscle and brain cells, they are transferred to FAD, so the harvest is thirty-six ATP.

In aerobic respiration's third stage, electrons from NADH and FADH$_2$ flow through transfer chains and H$^+$ is shuttled across the mitochondrion's inner membrane, into the outer compartment. H$^+$ concentration and electric gradients form across the inner membrane. H$^+$ flows back outside through ATP synthases, which drives the formation of many ATP.

8.5   Anaerobic Energy-Releasing Pathways

LINK TO
SECTION
7.8

Unlike aerobic respiration, the anaerobic pathways do not use oxygen as the final acceptor of electrons. Their final steps have an important function: they regenerate NAD⁺.

FERMENTATION PATHWAYS

Fermenters are diverse. Many are protists and bacteria in marshes, bogs, mud, deep sea sediments, the animal gut, canned foods, sewage treatment ponds, and other oxygen-free places. When exposed to oxygen, some die. Bacteria that cause botulism are examples. Other fermenters are indifferent to oxygen's presence. Still other kinds use oxygen, but they also can switch to fermentation when oxygen becomes scarce.

Glycolysis is the first stage of fermentation, just as it is in aerobic respiration (Figure 8.4). Here again, two pyruvate, two NADH, and two ATP form. But the last steps do not degrade glucose to carbon dioxide and water. They get no more ATP beyond the small yield from glycolysis. *The final steps in fermentation pathways regenerate the essential coenzyme NAD⁺.*

Fermentation yields enough energy to sustain many single-celled anaerobic species. It helps some aerobic cells when oxygen levels are stressfully low. But it isn't enough to sustain large, multicelled organisms, this being why you will never see anaerobic elephants.

Alcoholic Fermentation   In **alcoholic fermentation**, the three-carbon backbone of two pyruvate molecules from glycolysis is split. The reactions result in two molecules of acetaldehyde (an intermediate with a two-carbon backbone), and two of carbon dioxide. Acetaldehyde accepts electrons and hydrogen from NADH to form ethyl alcohol, or ethanol (Figure 8.10).

Some yeasts are famous fermenters. Bakers mix *Saccharomyces cerevisiae* cells and sugar into dough. The bubbles of carbon dioxide that form expand the dough (make it rise). Oven heat forces bubbles out of the dough, and the alcohol product evaporates away.

Wild and cultivated strains of *Saccharomyces* help produce alcohol in wine. Crushed grapes are left in vats along with the yeast, which converts sugar in the juice to ethanol. Ethanol is toxic to microbes. When a fermenting brew's ethanol content nears 10 percent, yeast cells start to die, and fermentation ends.

Lactate Fermentation   With **lactate fermentation**, the NADH gives up electrons and hydrogen to the pyruvate. The transfer converts pyruvate to lactate, a three-carbon compound (Figure 8.11). You probably have heard of lactic acid, the non-ionized form of this compound. But lactate is by far the most common form in living cells, which is our focus here.

Lactate fermentation by *Lactobacillus* and certain other bacteria can spoil food, yet some species have commercial uses. Huge populations in vats of milk give us cheeses, yogurt, buttermilk, and other dairy products. Fermenters also help in curing meats and in pickling some fruits and vegetables, such as sauerkraut. Lactate is an acid; it gives these foods a sour taste.

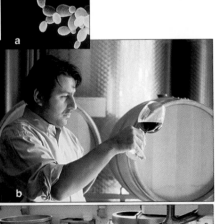

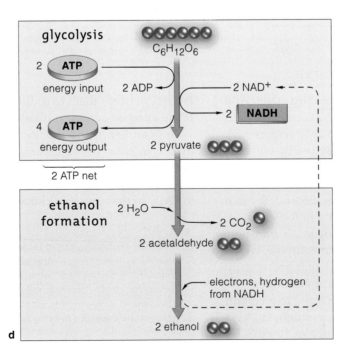

Figure 8.10 *Animated!* Steps of alcoholic fermentation. (**a**) Yeasts, which are single-celled fungi, use this anaerobic pathway to make ATP.

(**b**) A vintner examining the color and clarity of one fermentation product of *Saccharomyces*. Strains of this yeast live on the sugar-rich tissues of ripe grapes.

(**c**) Carbon dioxide released from cells of *S. cerevisiae* is making bread dough rise in this bakery.

(**d**) Alcoholic fermentation. The intermediate acetaldehyde functions as the final electron acceptor. The end product of the reactions is ethanol (ethyl alcohol).

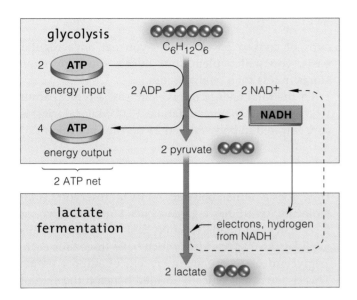

Figure 8.11 *Animated!* Steps of lactate fermentation. In this anaerobic pathway, the product (lactate) is the final acceptor of electrons originally stripped from glucose. These fermentation reactions have a net energy yield of two ATP (from glycolysis).

Figure 8.12 Sprinters, calling upon lactate fermentation in their muscles. The micrograph, a cross-section through part of a muscle, reveals three types of fibers. The lighter fibers sustain short, intense bursts of speed; they make ATP by lactate fermentation. The darker fibers contribute to endurance; they make ATP by aerobic respiration.

Lactate fermentation as well as aerobic respiration yields ATP for muscles that are partnered with bones. These skeletal muscles contain a mixture of cell types. Cells fused together inside *slow-twitch* muscle fibers support light, steady, prolonged activity, as during marathon runs or bird migrations. Cells of slow-twitch muscle fibers make ATP only by aerobic respiration, and they have many mitochondria. They are dark red because they hold large amounts of myoglobin. This pigment, which is related to hemoglobin, binds and stores oxygen for aerobic respiration (Figure 8.12).

By contrast, cells of pale *fast-twitch* muscle fibers have few mitochondria and no myoglobin. They use lactate fermentation to make ATP. They function when energy needs are immediate and intense, as during weight lifting or sprints (Figure 8.12). The pathway produces ATP quickly but not for very long; it does not support sustained activity. That is one reason you do not see chickens migrating. The flight muscles in a chicken contain mostly fast-twitch fibers, which make up the "white" breast meat.

Short bursts of flight evolved in the ancestors of chickens, perhaps as a way to flee from predators or improve agility during territorial battles. Chickens do walk and run; hence the "dark meat" (slow-twitch muscle) in their thighs and legs. Would you expect to see light or dark breast muscles in a migratory duck or in an albatross that skims ocean waves for months?

Section 37.5 offers more information on alternative energy pathways in skeletal and cardiac muscle cells.

ANAEROBIC ELECTRON TRANSFERS

Especially among prokaryotes, we see less common but more diverse energy-releasing pathways, some of which are topics of later chapters. Many assist in the global cycling of sulfur, nitrogen, and other elements. Collectively, the practitioners of these pathways affect nutrient availability in ecosystems everywhere.

Some bacteria and archaeans engage in **anaerobic electron transfers**. Electrons from organic compounds flow through transfer chains in the plasma membrane and H^+ flows out of the cell through ATP synthases. Inorganic compounds are often used as final electron acceptors. The net energy yield is variable but small.

Some anaerobic species in waterlogged soil give up electrons to sulfate, forming a putrid gas (hydrogen sulfide). Other anaerobes live in the nutrient-rich mud of some aquatic habitats. Still others are the basis of food webs at hydrothermal vents (Sections 7.8, 48.15).

In alcoholic fermentation, the final acceptor of electrons from glucose is acetaldehyde, a reaction intermediate. In lactate fermentation, it is pyruvate.

Both pathways have a net energy yield of two ATP, which forms in glycolysis. The remaining reactions regenerate the coenzyme NAD^+, without which glycolysis would stop.

Some bacteria and archaeans generate ATP by anaerobic electron transfers across the plasma membrane.

8.6 Alternative Energy Sources in the Body

LINKS TO
SECTIONS
3.3, 3.4

So far, you have looked at what happens after glucose molecules enter an energy-releasing pathway. Now start thinking about what cells can do when they have too much or too little glucose.

THE FATE OF GLUCOSE AT MEALTIME AND BETWEEN MEALS

What happens to glucose at mealtime? While you and all other mammals are eating, glucose and other small organic molecules are being absorbed across the gut lining, and your blood is transporting them through the body. The rising glucose concentration in blood prompts an organ, the pancreas, to secrete insulin. This hormone makes cells take up glucose faster.

Cells trap the incoming glucose by converting it to glucose–6–phosphate. This intermediate of glycolysis forms as ATP transfers a phosphate group to glucose (Figures 8.4 and 8.13). Phosphorylated glucose cannot be transported out of the cell.

When a cell takes in more glucose than it requires for energy, its ATP-forming machinery goes into high gear. Unless it is using ATP rapidly, the cytoplasmic concentration of ATP rises, and glucose–6–phosphate is diverted into a biosynthesis pathway. The glucose gets converted to glycogen, a storage polysaccharide in animal cells (Section 3.3). Liver cells and muscle cells especially favor the conversion. Together, these two types of cells maintain the body's largest stores of glycogen molecules.

Between meals, the blood level of glucose declines. If the decline were not countered, that would be bad news for the brain, your body's glucose hog. At any time, your brain is taking up more than two-thirds of the freely circulating glucose. Why? The brain's many hundreds of millions of nerve cells (neurons) use this sugar as their preferred energy source.

The pancreas responds to low glucose levels by secreting glucagon. This hormone causes liver cells to convert stored glycogen to glucose and send it back to the blood. Only liver cells do this; muscle cells will not give it up. The glucose level in blood rises, so brain cells keep on functioning. Thus, *hormones control whether your cells use free glucose as an energy source or tuck it away.*

Don't let this explanation lead you to believe that your cells store enormous amounts of glycogen. Glycogen makes up only 1 percent or so of the total energy reserves of the average adult's body—the energy equivalent of two cups of cooked pasta. Unless you eat on a regular basis, you will deplete your liver's small glycogen stores in less than twelve hours.

Of the total energy reserves in, say, a typical adult who eats well, 78 percent (about 10,000 kilocalories) is concentrated in body fat and 21 percent in proteins.

ENERGY FROM FATS

How does a human body access its fat reservoir? A fat molecule, recall, has a glycerol head and one, two, or three fatty acid tails (Section 3.4). The body stores most fats as triglycerides, which have three tails each. Triglycerides accumulate in fat cells of adipose tissue. This tissue is strategically located beneath the skin of buttocks and other body regions.

When the blood glucose level falls, triglycerides are tapped as an energy alternative. Enzymes in fat cells cleave bonds between glycerol and fatty acids, which both enter the blood. Enzymes in the liver convert the glycerol to PGAL. As you know, PGAL is one of the key intermediates in glycolysis (Figure 8.4). Nearly all cells of your body take up circulating fatty acids, and enzymes inside them cleave the fatty acid backbones. The fragments become converted to acetyl–CoA, which can enter the Krebs cycle.

Compared with glucose, a fatty acid tail has more carbon-bound hydrogen atoms, so it yields more ATP. Between meals or during steady, prolonged exercise, fatty acid conversions supply about half of the ATP that muscle, liver, and kidney cells require.

What happens if you eat too many carbohydrates? Aerobic respiration converts the glucose subunits to pyruvate, then to acetyl–CoA, which enters the Krebs cycle. When too much glucose is circulating through the body, acetyl–CoA is diverted to a pathway that synthesizes fatty acids. *Too much glucose ends up as fat.*

ENERGY FROM PROTEINS

Some enzymes in your digestive system split dietary proteins into their amino acid subunits, which are then absorbed into the bloodstream. Cells use amino acids to make proteins or other nitrogen-containing compounds. Even so, when you eat more protein than your body needs, amino acids are further degraded. Their $-NH_3^+$ group is pulled off, so ammonia (NH_3) forms. Depending on the types of amino acids, the leftover carbon backbones are split, and acetyl–CoA, pyruvate, or an intermediate of the Krebs cycle forms. Your cells can divert any of these organic compounds into the Krebs cycle (Figure 8.13).

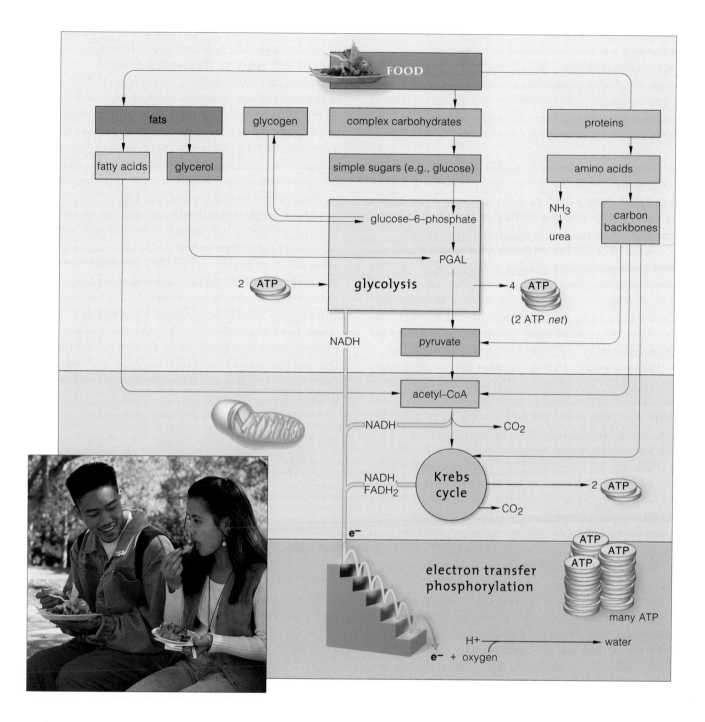

Figure 8.13 **Animated!** Reaction sites where different organic compounds can enter the stages of aerobic respiration. The compounds shown are alternative energy sources in humans and other mammals. Notice how complex carbohydrates, fats, and proteins cannot enter the aerobic pathway directly. First the digestive system, then individual cells, must break apart these molecules to simpler compounds.

As you can see, maintaining and accessing energy reserves is complicated business. Controlling the use of glucose is special because it is the fuel of choice for the brain. However, providing all of your cells with energy starts with the kinds of food you eat.

In humans and other mammals, the entrance of glucose or other organic compounds into an energy-releasing pathway depends on the kinds and proportions of carbohydrates, fats, and proteins in the diet.

8.7 Reflections on Life's Unity

In this unit, you traveled through many levels of life's organization. You have a sense that all organisms tap into a one-way flow of energy, that they alone make certain organic molecules, and that life emerges when molecules become organized and interact as units called cells. In short, you have a sense of life's molecular unity.

At this point in the book, you still may have difficulty sensing the connections between yourself—a highly intelligent being—and such remote-sounding events as energy flow and the cycling of carbon, hydrogen, and oxygen. Is this really the stuff of humanity?

Think back on the structure of a water molecule. Two hydrogen atoms sharing electrons with oxygen may not seem close to your daily life. Yet, through that sharing, water molecules show a polarity that invites them to hydrogen-bond with one another. The chemical behavior of three simple atoms is a start for the organization of lifeless matter into living things.

For now you can visualize other diverse molecules interspersed through water. The nonpolar kinds resist interaction with water; polar kinds dissolve in it. On their own, the phospholipids among them assemble into a two-layered film. Such lipid bilayers, recall, are the framework of cell membranes, hence all cells.

From the very beginning, the cell has been the basic *living* unit. The essence of life is not some mysterious force. It is organization and metabolic control. With a membrane to contain them, metabolic reactions *can* be controlled. With molecular mechanisms built into their membranes, cells can respond to energy changes and to shifts in solute concentrations in the environment. Response mechanisms operate by "telling" proteins —enzymes—when and what to build or tear down.

And it is not some mysterious force that creates proteins. DNA, the double-stranded treasurehouse of inheritance, has the chemical structure—*the chemical message*—that helps molecule reproduce molecule, one generation after the next. In your body, DNA strands

tell trillions of cells how countless molecules must be built or torn apart for their stored energy.

So yes, carbon, hydrogen, oxygen, and other atoms of organic molecules are the stuff of you, and us, and all of life. Yet it takes far more than organic molecules to complete the picture. Life continues only as long as a continuous flow of energy sustains its organization. It takes energy to assemble molecules into cells, cells into organisms, organisms into communities, and so on through the biosphere (Section 1.1).

Reflect on photosynthesis and aerobic respiration. Photosynthesizers use energy from the sun and raw materials to feed themselves and, indirectly, nearly all other forms of life. Long ago they enriched the whole atmosphere with oxygen, a leftover from a noncyclic pathway. That atmosphere exerted selection pressure and favored aerobic respiration, a novel way to break down food molecules by using the free oxygen. And photosynthesizers made more food with leftovers of the aerobic pathway—carbon dioxide and water. In this way the cycling of carbon, hydrogen, and oxygen through living things came full circle.

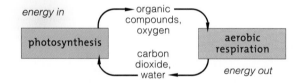

With few exceptions, infusions of energy from the sun sustain life's sweeping organization. And energy, remember, flows through time in one direction—from concentrated to less concentrated forms (Section 6.1). Only as long as more energy flows into the great web of life can life continue in all its rich expressions.

So life is no more *and no less* than a marvelously complex system for prolonging order. Sustained with energy transfusions from the sun, life continues by a capacity for self-reproduction. With energy and the hereditary codes of DNA, matter becomes organized, generation after generation. Even with the death of individuals, life elsewhere is prolonged. With each death, molecules are released and may be cycled as raw materials for new generations.

With this flow of energy and cycling of materials through time, each birth is affirmation of our ongoing capacity for organization, each death a renewal.

The diversity of life, and its continuity through time, arises from unity at the bioenergetic and molecular levels.

Summary

Section 8.1   All organisms, including photosynthetic types, access the chemical bond energy of glucose and other organic compounds, then use it to make ATP. They do so because ATP is a transportable form of energy that can jump-start nearly all metabolic reactions. They break apart compounds and release electrons and hydrogen, both of which have roles in ATP formation.

Glycolysis, the partial breakdown of one glucose to two pyruvate molecules, takes place in the cytoplasm of all cells. It is the first stage of all the main energy-releasing pathways, and it does not use free oxygen.

Anaerobic pathways end in the cytoplasm, and the net yield of ATP is small. An oxygen-requiring pathway called aerobic respiration continues in mitochondria. It releases far more usable energy from glucose.

Biology ❸ Now
Get an overview of aerobic respiration with the animation on BiologyNow.

Section 8.2   It takes two ATP molecules to jump-start glycolysis. One activates six-carbon glucose; the other primes an intermediate to split into two three-carbon molecules (PGAL). The two PGAL give up electrons and hydrogen to coenzymes (NAD^+), forming two NADH. Two intermediates each give up a phosphate group to ADP, forming two ATP. Two more intermediates do the same.

Thus, during glycolysis, four ATP form by substrate-level phosphorylation, but the *net* yield is two ATP (because two ATP had to be invested up front). The end products of glycolysis are two molecules of pyruvate, each with a three-carbon backbone.

Biology ❸ Now
Take a step-by-step journey through glycolysis with the animation on BiologyNow.

Section 8.3   The second and third stages of aerobic respiration get under way when the two pyruvates from glycolysis enter a mitochondrion.

The second stage consists of the Krebs cycle and a few preparatory steps before it. It takes two full turns of these cyclic reactions to break down both of the pyruvates.

Before the cycle, each pyruvate is stripped of a carbon atom (which departs in CO_2) as well as electrons and hydrogen (which NAD^+ picks up, forming NADH). A coenzyme (acetyl–CoA) picks up the two-carbon leftover.

The Krebs cycle starts when acetyl–CoA transfers the leftover to oxaloacetate, forming citrate. During stepwise rearrangements, intermediates give up two carbon atoms (which depart in CO_2) and electrons and hydrogen (which three NAD^+ and one FAD accept). An ATP forms.

In total, then, the second stage of aerobic respiration—including preparatory steps and two full turns of the Krebs cycle—results in the formation of six CO_2, two ATP, eight NADH, and two $FADH_2$.

Biology ❸ Now
Explore a mitochondrion and observe the reactions inside with the animation on BiologyNow.

Section 8.4   The third stage of aerobic respiration takes place at electron transfer chains and ATP synthases in the inner mitochondrial membrane.

The electron transfer chains accept electrons and hydrogen from NADH and $FADH_2$ that formed during the first two stages of the aerobic pathway. Electron flow through the chains causes H^+ to accumulate in the inner mitochondrial compartment, so H^+ concentration and electric gradients build up across the inner membrane.

H^+ follows its gradients and flows back to the outer mitochondrial compartment, through the interior of ATP synthases. This ion flow causes reversible changes in the shape of parts of the ATP synthase. Those changes force ADP to combine with unbound phosphate, thus forming ATP. This happens repeatedly, so many ATP molecules are produced.

Free oxygen picks up the electrons at the end of the transfer chains and combines with H^+, forming water.

Aerobic respiration has a typical net energy yield of thirty-six ATP for each glucose molecule metabolized.

Biology ❸ Now
Study how each step in aerobic respiration contributes to energy harvests with the animation on BiologyNow.

Section 8.5   Anaerobic energy-releasing pathways do not use free oxygen and occur only in cytoplasm. The net energy yield is small.

Fermentation pathways follow glycolysis and add no more to the two ATP net yield from glycolysis. They function only to regenerate NAD^+.

In alcoholic fermentation, the two pyruvates from glycolysis are converted to two acetaldehyde and two CO_2 molecules. When NADH transfers electrons and hydrogen to acetaldehyde, two ethanol (ethyl alcohol) molecules form and NAD^+ is regenerated.

In lactate fermentation, NAD^+ is regenerated when NADH gives up electrons and hydrogen to the two pyruvates. Two lactate molecules are end products.

Slow-twitch and fast-twitch skeletal muscle fibers support different levels of activity. Aerobic respiration and lactate fermentation occur in different fibers that make up these muscles.

Many prokaryotes make ATP by anaerobic electron transfers. They have electron transfer chains and ATP synthases in their plasma membrane. Sulfate or some other inorganic substance in the environment is often the final electron acceptor.

Biology ❸ Now
Compare alcoholic and lactate fermentation with the animation on BiologyNow.

Section 8.6   In the human body, simple sugars from carbohydrates, glycerol and fatty acids from fats, and carbon backbones of amino acids from proteins can enter the aerobic pathway as alternative energy sources.

Biology ❸ Now
Follow the breakdown of different organic molecules with the interaction on BiologyNow.

Section 8.7   Life's diversity and continuity arise from its unity at the bioenergetic and molecular levels.

Self-Quiz

Answers in Appendix II

1. Glycolysis starts and ends in the _____ .
   - a. nucleus
   - b. mitochondrion
   - c. plasma membrane
   - d. cytoplasm

2. Which of the following molecules does not form during glycolysis?
   - a. NADH
   - b. pyruvate
   - c. $FADH_2$
   - d. ATP

3. Aerobic respiration is completed in the _____ .
   - a. nucleus
   - b. mitochondrion
   - c. plasma membrane
   - d. cytoplasm

4. In the third stage of aerobic respiration, _____ is the final acceptor of electrons from glucose.
   - a. water
   - b. hydrogen
   - c. oxygen
   - d. NADH

5. Fill in the blanks in the diagram below.

6. In alcoholic fermentation, _____ is the final acceptor of electrons stripped from glucose.
   - a. oxygen
   - b. pyruvate
   - c. acetaldehyde
   - d. sulfate

7. Fermentation makes no more ATP beyond the small yield from glycolysis. The remaining reactions _____ .
   - a. regenerate FAD
   - b. regenerate NAD^+
   - c. regenerate NADH
   - d. regenerate $FADH_2$

8. In certain organisms and under certain conditions, _____ can be used as an energy alternative to glucose.
   - a. fatty acids
   - b. glycerol
   - c. amino acids
   - d. all of the above

9. Match the event with its most suitable description.
   - ____ glycolysis
   - ____ fermentation
   - ____ Krebs cycle
   - ____ electron transfer phosphorylation
   - a. ATP, NADH, $FADH_2$, CO_2, and water form
   - b. glucose to two pyruvates
   - c. NAD^+ regenerated, two ATP net
   - d. H^+ flows through ATP synthases

Additional questions are available on Biology🌀Now™

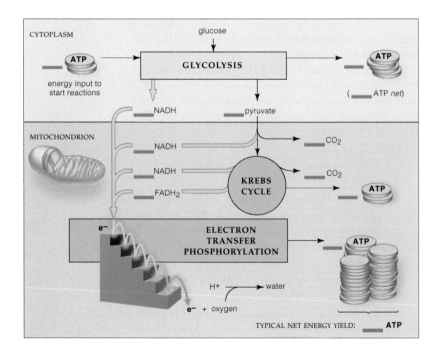

Critical Thinking

1. Living cells of your body absolutely do not use their nucleic acids as alternative energy sources. Suggest why.

2. Suppose you start a body-building program. You are already eating plenty of carbohydrates. Now a qualified nutritionist recommends that you start a protein-rich diet that includes protein supplements. Speculate on how extra dietary proteins will be put to use, and in which tissues.

3. Each year, Canada geese lift off in precise formation from their northern breeding grounds. They head south to spend the winter months in warmer climates, then make the return trip in spring. As is the case for other migratory birds, their flight muscle cells are efficient at using fatty acids as an energy source. Remember, the carbon backbone of fatty acids can be cleaved into small fragments that can be converted to acetyl–CoA for entry into the Krebs cycle.

   Suppose a lesser Canada goose from Alaska's Point Barrow has been steadily flapping along for about three thousand kilometers and is approaching Klamath Falls, Oregon. It looks down and notices a rabbit sprinting from a coyote with a taste for rabbit.

   With a stunning burst of speed, the rabbit reaches the safety of its burrow.

   Which energy-releasing pathway predominated in muscle cells in the rabbit's legs? Why was the Canada goose relying on a different pathway for most of its journey? And why wouldn't the pathway of choice in goose flight muscle cells be much good for a rabbit making a mad dash from its enemy?

4. At high altitudes, oxygen levels are low. Mountain climbers risk altitude sickness, which is characterized by shortness of breath, weakness, dizziness, and confusion.

   Oddly, early symptoms of *cyanide poisoning* resemble altitude sickness. This highly toxic poison binds tightly to a cytochrome, the last molecule in mitochondrial electron transfer chains. When cyanide becomes bound to it, the cytochrome can't transfer electrons to the next component of the chain. Explain why cytochrome shutdown might cause the same symptoms as altitude sickness.

5. ATP forms in mitochondria. In warm-blooded animals, so does a lot of heat, which is circulated in ways that help control body temperature. Cells of *brown adipose tissue* make a protein that disrupts the formation of electron transfer chains in mitochondrial membranes. H^+ gradients are affected, so fewer ATP form; electrons in the transfer chains give up more of their energy as heat. Because of this, some researchers hypothesize that brown adipose tissue may not be like white adipose tissue, which is an energy (fat) reservoir. Brown adipose tissue may function in thermogenesis, or heat production.

   Mitochondria, recall, contain their own DNA, which may have mutated independently in human populations that evolved in the Arctic and in the hot tropics. If that is so, then mitochondrial function may be adapted to climate.

   How do you suppose such a mitochondrial adaptation might affect people living where the temperature range no longer correlates with their ancestral heritage? Would you expect people whose ancestors evolved in the Arctic to be more or less likely to put on a lot of weight than those whose ancestors lived in the tropics? See *Science*, January 9, 2004: 223–226 for more information.

II Principles of Inheritance

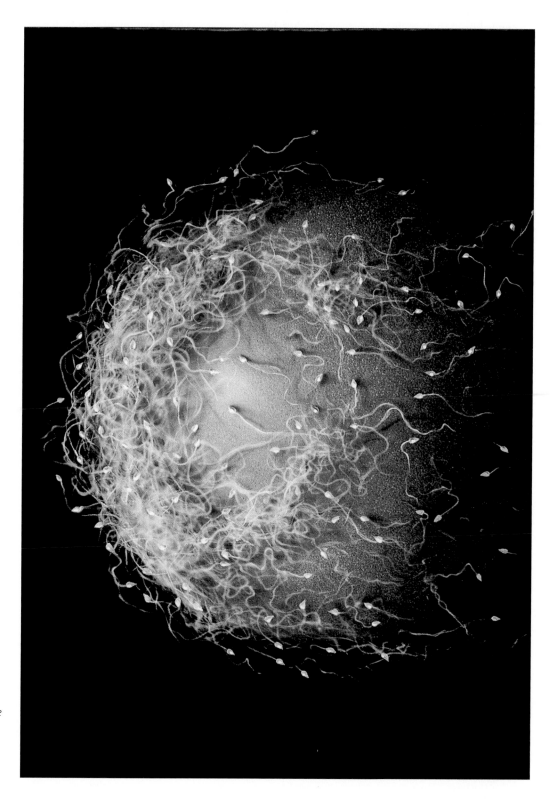

Human sperm, one of which will penetrate this mature egg and so set the stage for the development of a new individual in the image of its parents. This exquisite art is based on a scanning electron micrograph.

Henrietta's Immortal Cells

Each human starts out as a fertilized egg. By the time of birth, the body consists of about a trillion cells, all descended from that single cell. Even in an adult, billions of cells still divide every day and replace their damaged or worn-out predecessors.

In 1951, George and Margaret Gey of Johns Hopkins University were trying to develop a way to keep human cells dividing outside the body. An "immortal" cell lineage could help researchers study basic life processes as well as cancer and other diseases. Using cells to study cancer would be a better alternative than experimenting with patients and risking their already vulnerable lives.

For almost thirty years, the Geys tried to grow normal and diseased human cells. But they could not stop the cellular descendants from dying within a few weeks.

Mary Kubicek, a lab assistant, tried again and again to establish a self-perpetuating lineage of cultured human cancer cells. She was about to give up, but she prepared one last sample. She named them *HeLa* cells, a code for the first two letters of the patient's first and last names.

The HeLa cells began to divide. They divided again and again. Four days later, the researchers had to subdivide the cells into more culture tubes. The cell populations increased at a phenomenal rate; cells were dividing every twenty-four hours and coating the inside of the tubes within days.

Sadly, cancer cells in the patient were dividing just as frequently. Six months after she had been diagnosed with cancer, malignant cells had invaded tissues throughout her body. Two months after that, Henrietta Lacks, a young woman from Baltimore, was dead.

Although Henrietta passed away, her cells lived on in the Geys' laboratory (Figure 9.1). In time, HeLa cells were shipped to research laboratories all over the world. The Geys used HeLa cells to identify viral strains that cause polio, which at the time was epidemic. They developed the tissue culture techniques that were used to grow a vaccine. Other researchers used HeLa cells to investigate cancer, viral growth, protein synthesis, the effects of radiation on cells, and more. Some HeLa cells even traveled into space for experiments on the *Discoverer XVII* satellite. Even now, hundreds of important research projects move forward annually, thanks to Henrietta's immortal cells.

Figure 9.2 shows a photograph of Henrietta. She was only thirty-one when the runaway cell divisions killed her. Decades later, her legacy continues to help humans everywhere, through cellular descendants that are still dividing day after day.

Understanding cell division—and, ultimately, how new individuals are put together in the image of their parents—starts with answers to three questions. *First*, what kind of

Figure 9.1 The terrifying beauty of dividing HeLa cells—a legacy of Henrietta Lacks, who was a young casualty of cancer. Her cellular contribution to science is still helping others every day.

Figure 9.2 Henrietta Lacks.

information guides inheritance? *Second*, how is information copied in a parent cell before being distributed to each daughter cell? *Third*, what kinds of mechanisms actually parcel out the information to daughter cells?

We will require more than one chapter to survey the nature of cell reproduction and other mechanisms of inheritance. In this chapter, we introduce the structures and mechanisms that cells use to reproduce.

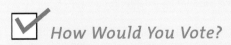
Watch the video online!

 How Would You Vote?

It is illegal to sell your organs, but you can sell your cells, including eggs, sperm, and blood cells. HeLa cells are still being sold all over the world by cell culture firms. Should the family of Henrietta Lacks share in the profits? See BiologyNow for details, then vote online.

 **Key Concepts**

CHROMOSOMES AND DIVIDING CELLS

Individuals of a species have a characteristic number of chromosomes in their cells. Those chromosomes differ in length and shape, and they carry different parts of the cell's hereditary information. Division mechanisms parcel out the information to each daughter cell, along with enough cytoplasm to start up its own operation. Section 9.1

WHERE MITOSIS FITS IN THE CELL CYCLE

A cell cycle starts when a daughter cell forms and ends when that cell completes its own division. A typical cycle goes through interphase, mitosis, and cytoplasmic division. In interphase, a cell increases its mass and number of components, and copies its DNA. Section 9.2

STAGES OF MITOSIS

Mitosis divides the nucleus, not the cytoplasm. It has four continuous stages: prophase, metaphase, anaphase, and telophase. A microtubular spindle forms. It moves the cell's duplicated chromosomes into two parcels, which end up in two genetically identical nuclei. Section 9.3

HOW THE CYTOPLASM DIVIDES

After nuclear division, the cytoplasm divides and typically puts a nucleus in each daughter cell. The cytoplasm of an animal cell is simply pinched in two. In plant cells, a cross-wall forms in the cytoplasm and divides it. Section 9.4

THE CELL CYCLE AND CANCER

Built-in mechanisms monitor and control the timing and rate of cell division. On rare occasions, the surveillance mechanisms fail, and cell division becomes uncontrollable. Tumor formation and cancer are the outcome. Section 9.5

Links to Earlier Concepts

Before reading, review the description about the changing appearance of chromosomes in the nucleus of eukaryotic cells (Section 4.6). You may wish to review the introduction to microtubules and the motor proteins associated with them (4.10, 4.11). Doing so will help you understand the nature of the mitotic spindle and the potential value of cancer research that is zeroing in on it. A look back at the walls of plant cells (4.9) will help give you a sense of why they cannot divide by pinching their cytoplasm in two parcels, as animal cells do.

9.1 Overview of Cell Division Mechanisms

LINK TO
SECTION
4.5

*The continuity of life depends on **reproduction**. By this process, parents produce a new generation of cells or multicelled individuals like themselves. Cell division is the bridge between generations.*

A dividing cell faces a challenge. Each of its daughter cells must get information encoded in the parental DNA and enough cytoplasm to start up its own operation. DNA "tells" it which proteins to build. Some of the proteins are structural materials; others are enzymes that speed construction of organic compounds. If the cell does not inherit all of the required information, it will not be able to grow or function properly.

In addition, the parent cell's cytoplasm already has enzymes, organelles, and other metabolic machinery. When a daughter cell inherits what looks like a blob of cytoplasm, it actually is getting start-up machinery that will keep it running until it can use information in DNA for growing on its own.

MITOSIS, MEIOSIS, AND THE PROKARYOTES

Eukaryotic cells cannot simply split in two, because their DNA is housed in a single nucleus. They do split the cytoplasm into daughter cells, but not until *after* DNA has been copied and packaged into more than a single nucleus by way of mitosis or meiosis.

Mitosis is a nuclear division mechanism that occurs in *somatic* cells (body cells) of multicelled eukaryotes. It is the basis of increases in body size during growth, replacements of worn-out or dead cells, and tissue repair. Many plants, animals, fungi, and single-celled protists also reproduce asexually, or make copies of themselves, by way of mitosis (Table 9.1).

Meiosis is a different nuclear division mechanism. It precedes the formation of gametes or spores, and it is the basis of sexual reproduction. The type of gametes known as sperm and eggs develop from *germ* cells, or immature reproductive cells. Spores form in the life cycle of many protists as well as plants and fungi.

As you will discover in this chapter and the next, meiosis and mitosis have much in common. Even so, their outcomes differ.

What about prokaryotes—bacteria and archaeans? All of these cells reproduce asexually by prokaryotic fission, an entirely different mechanism. We consider prokaryotic fission later, in Section 21.2.

KEY POINTS ABOUT CHROMOSOME STRUCTURE

In Section 4.5, you read that a eukaryotic chromosome is one double-stranded DNA molecule with a lot of proteins attached to it. Each eukaryotic species has a characteristic number of chromosomes inside its cells, and before a cell enters nuclear division, it duplicates every one of them. Each chromosome and its copy stay attached to each other as **sister chromatids** until late in the division process. Figure 9.3 is a simple way to think about what an unduplicated chromosome and a duplicated chromosome look like.

During an early stage of mitosis or meiosis, each duplicated chromosome coils back on itself again and

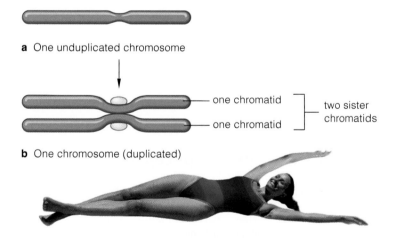

a One unduplicated chromosome

one chromatid — two sister
one chromatid — chromatids

b One chromosome (duplicated)

Figure 9.3 A simple way to visualize a eukaryotic chromosome in the unduplicated state and duplicated state. Eukaryotic cells are duplicated before mitosis or meiosis. Each becomes two sister chromatids. Students sometimes have trouble visualizing which is which. Think of a chromatid as one arm and leg of a sunbather.

Table 9.1	Comparison of Cell Division Mechanisms
Mechanisms	Functions
Mitosis, cytoplasmic division	In *all* multicelled eukaryotes, the basis of the following three processes: 1. Increases in body size during growth 2. Replacement of dead or worn-out cells 3. Repair of damaged tissues In single-celled and many multicelled species, *also* the basis of asexual reproduction
Meiosis, cytoplasmic division	In single-celled and multicelled eukaryotes, the basis of sexual reproduction; precedes gamete formation or spore formation (Chapter 10)
Prokaryotic fission	In bacteria and archaeans, the basis of asexual reproduction (Section 21.2)

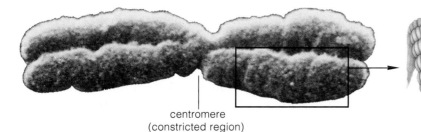

centromere
(constricted region)

a A duplicated chromosome in its most condensed form.

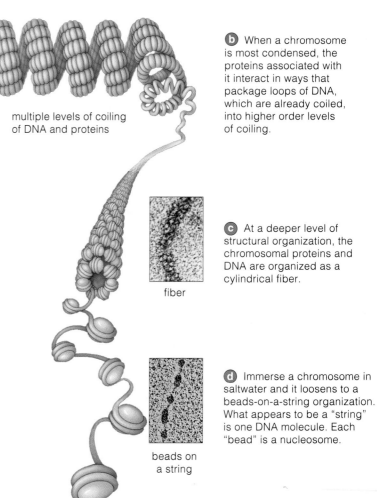

multiple levels of coiling
of DNA and proteins

b When a chromosome is most condensed, the proteins associated with it interact in ways that package loops of DNA, which are already coiled, into higher order levels of coiling.

fiber

c At a deeper level of structural organization, the chromosomal proteins and DNA are organized as a cylindrical fiber.

beads on
a string

d Immerse a chromosome in saltwater and it loosens to a beads-on-a-string organization. What appears to be a "string" is one DNA molecule. Each "bead" is a nucleosome.

DNA
double
helix

core of
histones

nucleosome

e A nucleosome consists of part of a DNA molecule looped twice around a core of histone proteins.

again, into a highly condensed form. Figure 9.4*a* gives an example of a duplicated human chromosome when it is most condensed. The orderly coiling arises from interactions between the DNA molecule and proteins associated with it. At regular intervals, the double-stranded DNA winds twice around tiny "spools" of proteins called **histones**. The repeating histone–DNA spools look like beads on a string when microscopists loosen them up (Figure 9.4*d*). Each of the "beads" is a **nucleosome**, which is the smallest unit of structural organization in eukaryotic chromosomes (Figure 9.4*e*).

While each duplicated chromosome is condensing, it becomes constricted at a predictable location along its length in a pronounced way. This constriction is a **centromere** (Figure 9.4*a*). The centromere's location is different for each type of chromosome and is one of its defining characteristics. During nuclear division, we find a kinetochore at the centromere of each chromatid. A kinetochore is a docking site for microtubules that will help move the chromatids to prescribed locations during nuclear division.

What is the point of this structural organization? The tight packaging probably keeps the chromosomes from getting tangled up while they are moved and sorted out into parcels *during* nuclear division. Also, *between* cell divisions, nucleosome packaging can be loosened in specific regions. In many cases, enzymes thereby gain access to the precise bits of hereditary information that a cell requires at that time.

When a cell divides, each daughter cell receives a required number of chromosomes and some cytoplasm. Eukaryotic cells divide their nucleus first, then the cytoplasm.

In eukaryotes, a nuclear division mechanism called mitosis is the basis of bodily growth, cell replacements, tissue repair, and often asexual reproduction.

Also in eukaryotes, a nuclear division mechanism called meiosis precedes the formation of gametes and, in many species, spores. Meiosis is the basis of sexual reproduction.

Figure 9.4 *Animated!* (**a**) Scanning electron micrograph of a duplicated human chromosome in its most condensed form. (**b,c**) Proteins package many loops of coiled DNA into the coiled array of a cylindrical fiber. (**d,e**) The smallest unit of structural organization is the nucleosome: part of a DNA molecule looped twice around a core of histone molecules. The transmission electron micrographs correspond to organizational levels (**c**) and (**d**).

9.2 Introducing the Cell Cycle

Start by thinking of reproduction in terms of a cell's life, from the time it forms until it divides. This interval is not the same as a life cycle—the sequence of stages through which individuals of a species pass during their lifetime.

A **cell cycle** is a series of events from one cell division to the next (Figure 9.5). It starts when a new daughter cell forms by mitosis and cytoplasmic division. It ends when that cell divides. Mitosis, cytoplasmic division, and interphase constitute one turn of the cycle.

THE WONDER OF INTERPHASE

During **interphase**, the cell increases in mass, roughly doubles the number of components in its cytoplasm, and duplicates its DNA. For most cells, interphase is the longest portion of the cell cycle. Biologists divide it into three stages:

G1   Interval ("*Gap*") of cell growth and functioning before the onset of DNA replication

S   Time of "*Synthesis*" (DNA replication)

G2   Second interval (*Gap*), after DNA replication when the cell prepares for division

G1, S, and G2 are code names for some events that are just amazing, considering how much DNA is packed in a nucleus. Remember, if you could stretch out all the DNA molecules from one of your somatic cells in a single line, they would extend past the fingertips of both outstretched arms. A line of all the DNA from one salamander cell would stretch about 540 feet.

The wonder is, enzymes and other proteins in cells *selectively* access, activate, and silence information in all that DNA. They also make base-by-base copies of every DNA molecule before cells divide. Most of this cellular work is completed in interphase.

G1, S, and G2 of interphase have distinct patterns of biosynthesis. Most of your cells remain in G1 while they are making proteins, carbohydrates, and lipids. Cells destined to divide enter S, when they copy their DNA as well as the proteins attached to it. During G2, they make proteins that will drive mitosis.

The length of the cycle is about the same for cells of the same type. However, it can differ from one cell type to the next. All of the neurons (nerve cells) inside your brain are stuck in G1 of interphase and normally will not divide again. Yet 2 million to 3 million cells in your red bone marrow are dividing every second. They give rise to new red blood cells, which will last only a few months before they wear out. As another example, when a new sea urchin is developing, the number of cells doubles every two hours.

Once S begins, DNA replication usually proceeds at a predictable rate and ends before a cell prepares to divide. The rate holds for all cells of a species, so you may well wonder if the cell cycle has built-in molecular brakes. It does. Apply the brakes that are supposed to work in G1, and the cycle stalls in G1. Lift the brakes, and the cell cycle runs to completion. Said another way, *control mechanisms govern the rate of cell division.*

Imagine a car losing its brakes just as it starts down a steep mountain road. As you will read later in the chapter, cancer begins this way. The crucial controls over the cell cycle are lost.

One more point: Stressful conditions often interrupt the cell cycle. For instance, when deprived of a vital nutrient, the free-living cells known as amoebas remain in interphase. They will remain there as long as they have not moved past a certain checkpoint. However, past that point, the cycle normally continues regardless of outside conditions, because built-in control mechanisms have lifted the brakes.

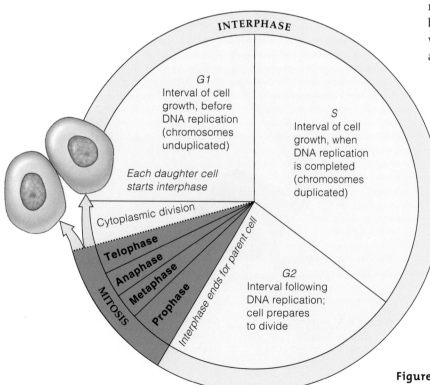

INTERPHASE

G1
Interval of cell growth, before DNA replication (chromosomes unduplicated)

S
Interval of cell growth, when DNA replication is completed (chromosomes duplicated)

Each daughter cell starts interphase

Cytoplasmic division

Telophase
Anaphase
Metaphase
Prophase

MITOSIS

Interphase ends for parent cell

G2
Interval following DNA replication; cell prepares to divide

Figure 9.5 *Animated!* Eukaryotic cell cycle, generalized. The length of each interval differs among different cell types.

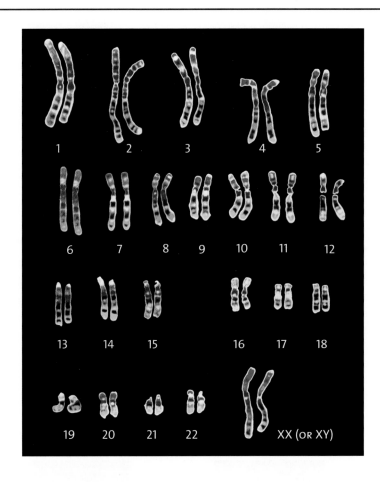

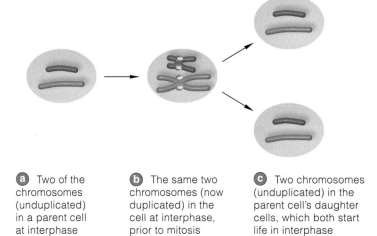

a Two of the chromosomes (unduplicated) in a parent cell at interphase

b The same two chromosomes (now duplicated) in the cell at interphase, prior to mitosis

c Two chromosomes (unduplicated) in the parent cell's daughter cells, which both start life in interphase

Figure 9.6 Preview of how mitosis maintains a parental chromosome number, one generation to the next. Human diploid cells have twenty-three pairs of metaphase chromosomes, for a total of forty-six (*left*). (The last ones in the lineup are a pair of sex chromosomes. In human females, they are two X chromosomes; in males, they are XY.)

The sketches above track what happens to just two of the forty-six. When all goes well, each time a human somatic cell undergoes mitosis and cytoplasmic division, daughter cells end up with an unduplicated set of twenty-three pairs of chromosomes. The icon below shows the bipolar mitotic spindle that helps bring about this outcome.

MITOSIS AND THE CHROMOSOME NUMBER

Mitosis follows G2, and it maintains the parent cell's chromosome number. The **chromosome number** is the sum of all chromosomes in cells of a given type. Body cells of gorillas and chimpanzees have 48, pea plants have 14, and humans have 46 (Figure 9.6).

Actually, your cells have a **diploid number** (*2n*) of chromosomes; there are two of each type. Those 46 are like volumes of two sets of books numbered from 1 to 23. You have two volumes of, say, chromosome 22—*a pair of them*. Except for one sex chromosome pairing (XY), both have the same length and shape, and carry the same hereditary information about the same traits.

Think of them as two sets of books on how to build a house. Your father gave you one set. Your mother had her own ideas about wiring, plumbing, and so on. She gave you an alternate edition on the same topics, but it says slightly different things about many of them.

With mitosis, a diploid parent cell can produce two diploid daughter cells. This doesn't mean each merely gets forty-six or forty-eight or fourteen chromosomes. If only the total mattered, then one cell might get, say, two pairs of chromosome 22 and no pairs whatsoever of chromosome 9. But neither cell could function like its parent *without two of each type of chromosome.*

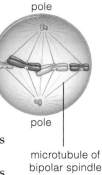

pole

pole

microtubule of bipolar spindle

Mitosis has four stages—*prophase*, *metaphase*, *anaphase*, and *telophase*—which require a **bipolar mitotic spindle**. This dynamic structure consists of microtubules that grow or shrink as tubulin subunits are added to or lost from their ends. Its microtubules extend from both spindle poles. Some overlap midway between the two poles, and others tether the duplicated chromosomes.

The next section will explain how the microtubules extending from one pole connect to one chromatid of each chromosome, and microtubules from the other pole connect to its sister. As you will see, the spindle moves the two chromatids apart, to opposite poles. The result is two complete sets of now-unduplicated chromosomes, one for each forthcoming daughter cell. Figure 9.6 is a preview of how mitosis maintains the parental chromosome number.

Interphase, mitosis, and cytoplasmic division constitute one turn of the cell cycle. During interphase, a new cell increases its mass, doubles the number of its components, and duplicates its chromosomes. The cycle ends after the cell undergoes mitosis and then divides its cytoplasm.

9.3 A Closer Look at Mitosis

LINKS TO
SECTIONS
4.10, 4.11

Focus now on a "typical" animal cell to see how mitosis can keep the chromosome number constant, division after division, from one cell generation to the next.

We know that a cell is in **prophase**, the first stage of mitosis, when its chromosomes become visible in light microscopes as threadlike forms. ("Mitosis" is from the Greek *mitos,* meaning thread.) Each chromosome was duplicated earlier, in interphase; each is two sister chromatids joined at the centromere. Now they twist and fold. By late prophase, they will be condensed in thicker, compact, rod-shaped forms (Figure 9.7*a–c*).

Also before prophase, two barrel-shaped centrioles and two centrosomes started duplicating themselves next to the nucleus. A centriole, recall, gives rise to a flagellum or cilium (Section 4.11). If you observe this structure, you can bet that flagellated or ciliated cells develop during the organism's life cycle.

In animal cells, each centriole helps organize one **centrosome**, a center where microtubules originate. In prophase, one set of the duplicated centrioles and centrosomes moves to the other side of the nucleus,

then microtubules grow out of each centrosome. *These are the microtubules that form the bipolar spindle.*

During the transition from prophase to metaphase, the nuclear envelope breaks up completely into many tiny, flattened vesicles. The microtubules are now free to interact with chromosomes and with one another. Many dock at kinetochores; others keep on growing from centrosomes until they overlap midway between the two spindle poles. Remember the motor proteins associated with microtubules (Section 4.11)? Energy from ATP activates dyneins and kinesins, which then generate the force to assemble the mitotic spindle, and to bind and move the chromosomes.

Again, some microtubules extending from one pole tether one chromatid of each chromosome, and some from the opposite pole tether the sister chromatid. The opposing sets of microtubules engage in a tug-of-war. They add and lose tubulin subunits, so they grow and shrink until they are the same length. At that point, **metaphase**, all duplicated chromosomes are aligned midway between the spindle poles (*meta–*, midway). The alignment is crucial for the next stage of mitosis.

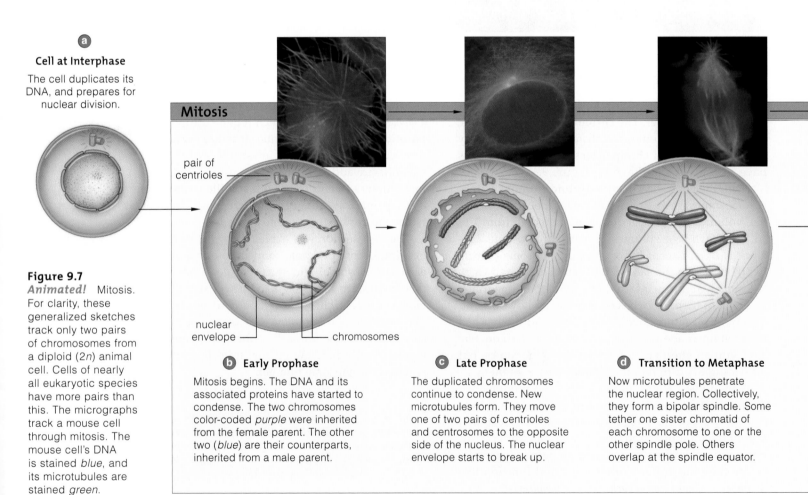

a

Cell at Interphase

The cell duplicates its DNA, and prepares for nuclear division.

Mitosis

pair of centrioles

Figure 9.7
Animated! Mitosis. For clarity, these generalized sketches track only two pairs of chromosomes from a diploid (2*n*) animal cell. Cells of nearly all eukaryotic species have more pairs than this. The micrographs track a mouse cell through mitosis. The mouse cell's DNA is stained *blue*, and its microtubules are stained *green*.

nuclear envelope — chromosomes

b Early Prophase

Mitosis begins. The DNA and its associated proteins have started to condense. The two chromosomes color-coded *purple* were inherited from the female parent. The other two (*blue*) are their counterparts, inherited from a male parent.

c Late Prophase

The duplicated chromosomes continue to condense. New microtubules form. They move one of two pairs of centrioles and centrosomes to the opposite side of the nucleus. The nuclear envelope starts to break up.

d Transition to Metaphase

Now microtubules penetrate the nuclear region. Collectively, they form a bipolar spindle. Some tether one sister chromatid of each chromosome to one or the other spindle pole. Others overlap at the spindle equator.

At **anaphase**, sister chromatids of each chromosome are moved toward opposite spindle poles. How? Motor proteins attached to each chromatid's kinetochore are inching along microtubular tracks that lead to one or the other spindle pole. The microtubules themselves are shrinking at both ends even as motor proteins are dragging the chromatids with them. And so the sister chromatids end up at opposite poles (Figure 9.7f).

At the same time, the spindle poles themselves are being pushed farther apart! Different microtubules, the ones that overlap midway between the spindle poles, are ratcheting past one another. Motor proteins drive this interaction and push the poles apart.

One sister chromatid is a duplicate of the other. So once they detach from each other at anaphase, there are two separate chromosomes, one at each pole.

Telophase gets under way when one of each type of chromosome reaches a spindle pole. Two genetically identical clusters of chromosomes are now located at opposite "ends" of the cell. All of the chromosomes decondense and become threadlike. Vesicles derived from old nuclear envelope fuse and form patches of membrane around each cluster. Patch joins with patch until a new nuclear envelope encloses each cluster. And so two nuclei form (Figure 9.7g). In our example, the parent cell had a diploid number of chromosomes. So does each nucleus. Once two nuclei have formed, telophase is over—and so is mitosis.

Prior to mitosis, each chromosome in a cell's nucleus is duplicated, so it consists of two sister chromatids.

In prophase, chromosomes condense to rodlike forms, and microtubules form a bipolar spindle. The nuclear envelope breaks up. Some microtubules harness the chromosomes.

At metaphase, all duplicated chromosomes are aligned midway between the spindle's poles, at its equator.

At anaphase, microtubules move the sister chromatids of each chromosome apart, to opposite spindle poles.

At telophase, a new nuclear envelope forms around each of two clusters of decondensing chromosomes.

Thus two daughter nuclei have formed. Each has the same chromosome number as the parent cell's nucleus.

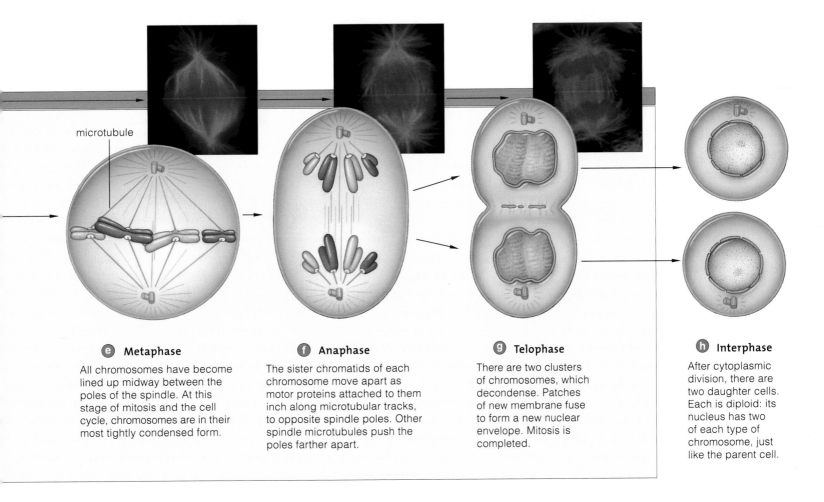

e **Metaphase**

All chromosomes have become lined up midway between the poles of the spindle. At this stage of mitosis and the cell cycle, chromosomes are in their most tightly condensed form.

f **Anaphase**

The sister chromatids of each chromosome move apart as motor proteins attached to them inch along microtubular tracks, to opposite spindle poles. Other spindle microtubules push the poles farther apart.

g **Telophase**

There are two clusters of chromosomes, which decondense. Patches of new membrane fuse to form a new nuclear envelope. Mitosis is completed.

h **Interphase**

After cytoplasmic division, there are two daughter cells. Each is diploid: its nucleus has two of each type of chromosome, just like the parent cell.

microtubule

9.4 Cytoplasmic Division Mechanisms

LINKS TO
SECTIONS
4.9, 4.11

In most cell types, the cytoplasm usually divides at some time between late anaphase and the end of telophase. The mechanism of cytoplasmic division—or, more formally, cytokinesis—differs among species.

HOW DO ANIMAL CELLS DIVIDE?

Dividing animal cells partition their cytoplasm by a **contractile ring mechanism**. Most often, the plasma membrane starts to sink inward as a thin indentation about halfway between the cell's poles (Figure 9.8*a*). This is a cleavage furrow, the first visible sign that the cytoplasm is dividing. It advances until it extends all around the cell. As it does so, it deepens along a plane corresponding to the former spindle's equator.

What is going on? Part of the cell cortex, that mesh of cytoskeletal elements under the plasma membrane,

is a ring of actin filaments organized as a thin band around the cell's midsection. The band is anchored to the plasma membrane. When energized by ATP, all the filaments contract and slide past one another in a way that shrinks the band diameter (compare Section 4.11). Being attached to the plasma membrane, the band drags it inward until the cytoplasm is pinched in two (Figures 9.8*a* and 9.9). Two daughter cells form this way. Each ends up with a nucleus and cytoplasm, enclosed within a plasma membrane.

HOW DO PLANT CELLS DIVIDE?

The contractile ring mechanism that works for animal cells could not work for plant cells. The contractile force could not pinch through plant cell walls, which are stiff with cellulose and often lignin. Microtubules

 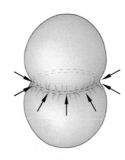 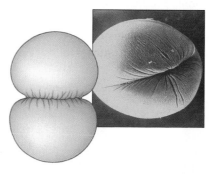

1 Mitosis is completed, and the bipolar spindle is starting to disassemble.

2 At the former spindle equator, a ring of actin filaments attached to the plasma membrane contracts.

3 The diameter of the contractile ring continues to shrink and pull the cell surface inward.

4 The contractile mechanism continues to operate until the cytoplasm is partitioned.

a Contractile Ring Formation

 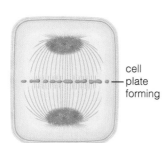 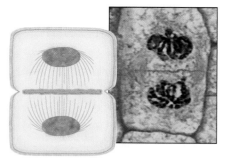 

cell plate forming

1 The plane of division and of a future cross-wall was established by a band of microtubules and actin filaments that formed and broke up before mitosis. Vesicles cluster here when mitosis ends.

2 Vesicle membranes fuse. The wall material is sandwiched between two new membranes that lengthen along the plane of a newly forming cell plate.

3 Cellulose is deposited inside the sandwich. In time, these deposits will form two cell walls. Others will form the middle lamella between the walls and cement them together.

4 A cell plate grows at its margins until it fuses with the parent cell plasma membrane. The primary wall of growing plant cells is still thin. New material is deposited on it.

b Cell Plate Formation

Figure 9.8 *Animated!* Cytoplasmic division of an animal cell (**a**) and a plant cell (**b**).

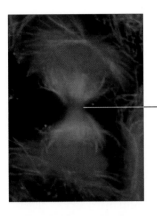

ring of microfilaments midway between the two spindle poles, in the same plane as the spindle equator

Figure 9.9 Micrograph capturing the contractile ring action inside an animal cell.

just beneath the plasma membrane help orient the fibers of cellulose in the wall. Before prophase, these microtubules disassemble and new ones assemble in a narrow band around the nucleus—a band that also includes actin filaments. As other microtubules form the bipolar spindle, the narrow band disappears, and an actin-depleted zone is left behind. *The zone marks the plane of cytoplasmic division* (Figure 9.8*b*).

Along that plane, tiny vesicles packed with wall-building materials from Golgi bodies fuse with one another. Together, deposits of these materials form a disk-shaped structure known as a cell plate. Deposits of cellulose accumulate at the plate. In time, they are thick enough to form a cross-wall through the cell. New plasma membrane extends across both sides of it. The wall grows until it bridges the cytoplasm and partitions the parent cell. This cytoplasmic division mechanism is known as **cell plate formation**.

APPRECIATE THE PROCESS!

Take a moment to look closely at your hands. Visualize the cells making up your palms, thumbs, and fingers. Now imagine the mitotic divisions that produced all of the cell generations that preceded them while you were developing, early on, inside your mother (Figure 9.10). And be grateful for the astonishing precision of mechanisms that led to their formation at prescribed times, in prescribed numbers, for the alternatives can be terrible indeed.

Why? Good health and survival itself depend on the proper timing and completion of cell cycle events. Some genetic disorders arise as a result of mistakes during the duplication or distribution of even one chromosome. In other cases, unchecked cell divisions often destroy surrounding tissues and, ultimately, the

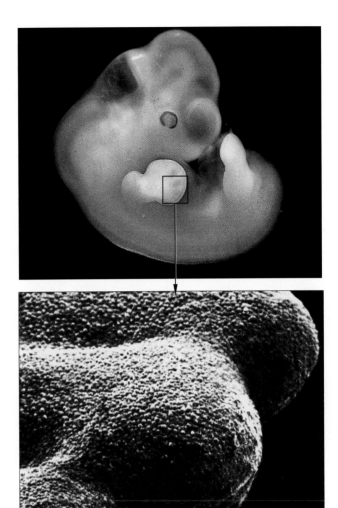

Figure 9.10 The paddlelike structure of a human embryo that develops into a hand by mitosis, cytoplasmic divisions, and other processes. The scanning electron micrograph shows individual cells.

individual. Such losses can start in body cells. They can start in the germ cells that give rise to sperm and eggs, although rarely. The last section of this chapter can give you a sense of the consequences.

After mitosis, a separate mechanism partitions the cytoplasm into two daughter cells, each with a nucleus.

A contractile ring mechanism partitions a dividing animal cell. A band of actin filaments around the cell midsection contracts and pinches the cytoplasm in two.

A mechanism called cell plate formation partitions plant cells. Golgi-derived vesicles deposit material at a plane of cytoplasmic division to form a cross-wall, which connects to the parent cell wall.

LINKS TO
SECTIONS
4.10, 6.5

9.5 When Control Is Lost

Controls over cell division affect growth and reproduction. On rare occasions, something goes wrong in a somatic cell or reproductive cell. Cancer may be the outcome.

THE CELL CYCLE REVISITED

Every second, millions of cells in your skin, bone marrow, gut lining, liver, and elsewhere are dividing and replacing their worn-out, dead, and dying predecessors. They do not divide willy-nilly. Many mechanisms control cell growth, DNA replication, and division. They also control when the division machinery is put to rest.

What happens when something goes wrong? Suppose, for instance, that sister chromatids do not separate as they should during mitosis. As a result, one daughter cell ends up with too many chromosomes and the other with too few. Or suppose the wrong nucleotide gets added to a growing strand of DNA during the replication process. Suppose free radicals, peroxides, or ultraviolet radiation attack and disrupt the structure of chromosomal DNA (Section 6.5). Such problems are frequent but inevitable, and a cell may not function properly unless they are quickly countered.

The cell cycle has built-in checkpoints where specific proteins monitor the structure of chromosomal DNA. At different points, these proteins monitor whether the preceding phase of the cycle was successfully completed. Other kinds sense whether conditions favor cell division. All of these proteins—the products of checkpoint genes—form the mechanisms that can advance, delay, or block the cell cycle.

For example, **kinases** are a class of enzymes that can activate other molecules by transferring a phosphate group to them. When DNA is broken or incomplete, they activate certain proteins in a cascade of signaling events that ultimately stop the cell cycle or induce cell death. As another example, the checkpoint proteins called **growth factors** promote transcription of genes that have roles in the body's growth. One of these, epidermal growth factor, activates a kinase by binding to receptors on target cells in epithelial tissues. Binding is a signal to start mitosis.

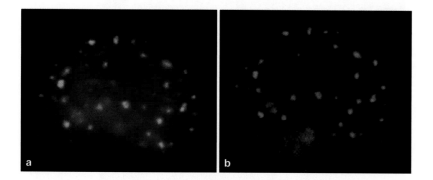

Figure 9.11 Protein products of checkpoint genes in action. A form of radiation damaged the DNA inside this nucleus. (**a**) *Green* dots pinpoint the location of *53BP1*, and (**b**) *red* dots pinpoint the location of *BRCA1*. Both proteins have clustered around the same chromosome breaks in the same nucleus. The integrated action of these proteins and others blocks mitosis until the DNA breaks are fixed.

CHECKPOINT FAILURE AND TUMORS

Sometimes a checkpoint gene mutates and its protein product no longer functions properly. When all checkpoint mechanisms fail, the cell loses control over the cell cycle. Figures 9.11 through 9.14 show a few of the outcomes.

In some cases, the cycle gets stuck in mitosis. Mitotic cell divisions occur over and over again, with no transition into or out of interphase. In other cases, chromosomal DNA that has been damaged is replicated. In still other cases, signaling mechanisms that can make an abnormal cell commit suicide are disabled. You will read more about this mechanism in Section 28.5. Regardless of the cause, the cell's continually dividing descendants form an abnormal mass—a **tumor**—in the surrounding tissue.

Usually, several checkpoint proteins are absent in tumor cells. That is why checkpoint gene products that inhibit mitosis are called *tumor suppressors*. Checkpoint genes encoding proteins that stimulate mitosis are known as *proto-oncogenes*. Mutations that alter their products or the rate at which they are synthesized help transform a normal cell into a tumor cell. Mutant checkpoint genes are linked with an increased risk of tumor formation, and sometimes they run in families.

Moles and other tumors are **neoplasms**, or abnormal masses of cells that lost controls over how they grow and divide. Ordinary skin moles are among the noncancerous, or *benign*, neoplasms. They grow very slowly, and their cells retain the surface recognition proteins that keep them in

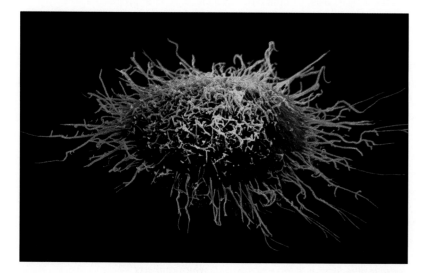

Figure 9.12 Scanning electron micrograph of the surface of a cervical cancer cell, the kind of malignant cell that killed Henrietta Lacks.

benign
tumor

malignant
tumor

a Cancer cells break away from their home tissue.

b The metastasizing cells become attached to the wall of a blood vessel or lymph vessel. They release digestive enzymes onto it. Then they cross the wall at the breach.

c Cancer cells creep or tumble along inside blood vessels, then leave the bloodstream the same way they got in. They start new tumors in new tissues.

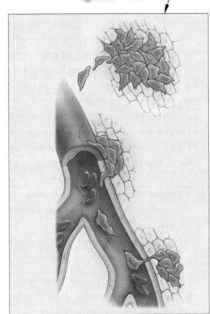

Figure 9.13 *Animated!* Comparison of benign and malignant tumors. Benign tumors typically are slow-growing and stay put in their home tissue. Cells of a malignant tumor migrate abnormally through the body and establish colonies even in distant tissues.

Figure 9.14 Skin cancers. (**a**) A *basal cell carcinoma* is the most common type. This slow-growing, raised lump is typically uncolored, reddish-brown, or black.

(**b**) The second most common form of skin cancer is called a *squamous cell carcinoma*. This pink growth, firm to the touch, grows fast under the surface of skin exposed to the sun.

(**c**) *Malignant melanoma* spreads fastest. Cells form dark, encrusted lumps. They may itch like an insect bite or bleed easily.

their home tissue (Figure 9.13). Unless a benign neoplasm grows too large or becomes irritating, it poses no threat to the body.

CHARACTERISTICS OF CANCER

Cancers are the abnormally growing and dividing cells of a *malignant* neoplasm. They disrupt surrounding tissues, both physically and metabolically. Cancer cells are grossly disfigured. They can break loose from home tissues, slip into and out of blood vessels and lymph vessels, and invade other tissues where they do not belong (Figure 9.13).

Cancer cells typically display four characteristics. *First*, they grow and divide abnormally. Controls on overcrowding in tissues are lost and cell populations reach extremely high densities. The number of small blood vessels, or capillaries, that transport gases and other substances to and from the growing cell mass also increases abnormally.

Second, the cytoplasm and plasma membrane of cancer cells become grossly altered. The membrane becomes leaky and has altered or missing proteins. The whole cytoskeleton shrinks, becomes disorganized, or both. Enzyme action shifts, as in amplified reliance on ATP formation by glycolysis.

Third, cancer cells often have a weakened capacity for adhesion. Because their recognition proteins are altered or lost, they cannot stay anchored in proper tissues. They break away and may establish growing colonies in distant tissues. *Metastasis* is the name for this process of abnormal cell migration and tissue invasion.

Fourth, cancer cells may have lethal effects. Unless they are eradicated by surgery, chemotherapy, or some other procedures, their uncontrollable divisions can put the individual on a painful road to death.

Each year in the developed countries alone, cancers cause 15 to 20 percent of all deaths. And cancer is not just a human problem. Cancers are known to occur in most of the animal species studied to date.

Cancer is a multistep process. Researchers have already identified many of the mutant genes that contribute to it. They also are working to identify drugs that specifically target and destroy cancer cells or stop them from dividing.

HeLa cells, for instance, were used in early tests of taxol, an anticancer drug that stops spindles from disassembling. With this kind of research, we may one day have drugs that can put the brakes on cancer cells. We return to this topic in later chapters.

Summary

Section 9.1 By processes of reproduction, parents produce a new generation of individuals like themselves. Cell division is a bridge between generations. When a cell divides, its daughter cells each receive a required number of DNA molecules and some cytoplasm.

Only eukaryotic cells undergo mitosis, meiosis, or both. These nuclear division mechanisms partition the duplicated chromosomes of a parent cell into daughter nuclei. A separate mechanism divides the cytoplasm. Prokaryotic cells divide by a different mechanism.

Mitosis is the basis of multicellular growth, cell replacements, and tissue repair. Also, many singled-celled and multicelled species reproduce asexually by mitosis.

Meiosis, the basis of sexual reproduction, precedes the formation of gametes or spores.

A eukaryotic chromosome is a molecule of DNA and many histones and other proteins associated with it. The proteins structurally organize the chromosome and affect access to genes. The smallest unit of organization, the nucleosome, consists of a stretch of double-stranded DNA looped twice around a spool of histones.

When duplicated, the chromosome consists of two sister chromatids, each with a kinetochore (a docking site for microtubules). Until late in mitosis (or meiosis), the two remain attached at their centromere region.

Biology⑤Now
Explore the structure of a chromosome with the animation on BiologyNow.

Section 9.2 Each cell cycle starts when a new cell forms, runs through interphase, and ends when that cell reproduces by nuclear and cytoplasmic division. A cell carries out most functions in interphase: it increases in mass, roughly doubles the number of its cytoplasmic components, then duplicates each of its chromosomes.

Biology⑤Now
Investigate the stages of the cell cycle with the interaction on BiologyNow.

Section 9.3 The sum of all chromosomes in cells of a given type is the chromosome number. Human body cells have a diploid chromosome number of 46 (pairs of 23 types of chromosomes). Mitosis, which maintains the chromosome number, has four continuous stages:

Prophase. The duplicated, threadlike chromosomes start to condense. With the help of motor proteins, new microtubules start forming a bipolar mitotic spindle. The nuclear envelope starts to break apart. Some of the microtubules extending from one spindle pole tether one chromatid of each chromosome; others extending from the opposite pole tether the sister chromatid. Microtubules extending from both poles grow until they overlap at the spindle's midpoint.

Metaphase. At metaphase, all chromosomes have become aligned at the spindle's midpoint.

Anaphase. Sister chromatids detach from each other. The kinetochore of each drags it along microtubules, which are shortening at both ends. The microtubules that overlap ratchet past each other, pushing the spindle poles farther apart. Different motor proteins drive the movements. One of each type of parental chromosome ends up clustered together at each spindle pole.

Telophase. The chromosomes decondense to threadlike form. A new nuclear envelope forms around each cluster. Both nuclei have the parental chromosome number.

Fill in the blanks of the diagram below to check your understanding of the four stages of mitosis, and how it maintains the chromosome number.

Biology⑤Now
Observe how mitosis occurs with the animation on BiologyNow.

Section 9.4 The mechanisms of cytoplasmic division differ. In animal cells, a microfilament ring that is part of the cell cortex contracts and pulls the cell surface inward until the cytoplasm is partitioned. In plant cells, a band of microtubules and microfilaments forms around the nucleus before mitosis starts. It marks the site where a cell plate will form from Golgi-derived material. The cell plate will enlarge and become a cross-wall that will partition the cytoplasm.

Biology⑤Now
Compare the cytoplasmic division of plant and animal cells with the animation on BiologyNow.

Section 9.5 Checkpoint gene products are part of the controls over the cell cycle. Mutant checkpoint genes cause tumors by disrupting normal controls. Cancer is a multistep process involving altered cells that grow and divide abnormally. Malignant cells may metastasize, or break loose and colonize distant tissues.

Biology⑤Now
See how cancers spread throughout a body with the animation on BiologyNow.

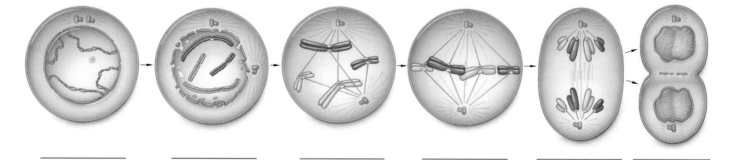

_____ _____ _____   _____ _____ _____

Self-Quiz

Answers in Appendix II

1. Mitosis and cytoplasmic division function in _____ .
   a. asexual reproduction of single-celled eukaryotes
   b. growth, tissue repair, often asexual reproduction
   c. gamete formation in prokaryotes
   d. both a and b

2. A duplicated chromosome has _____ chromatid(s).
   a. one   b. two   c. three   d. four

3. The basic unit that structurally organizes a eukaryotic chromosome is the _____ .
   a. higher order coiling   c. nucleosome
   b. bipolar mitotic spindle   d. microfilament

4. The chromosome number is _____ .
   a. the sum of all chromosomes in cells of a given type
   b. an identifiable feature of each species
   c. maintained by mitosis
   d. all of the above

5. A somatic cell having two of each type of chromosome has a(n) _____ chromosome number.
   a. diploid   b. haploid   c. tetraploid   d. abnormal

6. Interphase is the part of the cell cycle when _____ .
   a. a cell ceases to function
   b. a germ cell forms its spindle apparatus
   c. a cell grows and duplicates its DNA
   d. mitosis proceeds

7. After mitosis, the chromosome number of a daughter cell is _____ the parent cell's.
   a. the same as   c. rearranged compared to
   b. one-half   d. doubled compared to

8. Only _____ is not a stage of mitosis.
   a. prophase b. interphase c. metaphase d. anaphase

9. Match each stage with the events listed.
   ____ metaphase   a. sister chromatids move apart
   ____ prophase   b. chromosomes start to condense
   ____ telophase   c. daughter nuclei form
   ____ anaphase   d. all duplicated chromosomes are
                        aligned at the spindle equator

Additional questions are available on Biology🌀Now™

Critical Thinking

1. Figure 9.15 shows a cell going through stages of mitosis. Notice the barrel-shaped spindle that is quite evident at anaphase. Also notice the dense array of short microtubules midway between the two clusters of chromosomes at telophase. From these clues, would you say that this is a plant cell or an animal cell?

2. Pacific yews (*Taxus brevifolius*) are among the slowest growing trees, which makes them vulnerable to extinction. People started stripping their bark and killing them when they heard that *taxol,* a chemical extracted from the bark, may work against breast and ovarian cancer. It takes bark from about six trees to treat one patient. Do some research and find out why taxol has potential as an anticancer drug and what has been done to protect the trees.

3. X-rays emitted from some radioisotopes damage DNA, especially in cells undergoing DNA replication. Humans

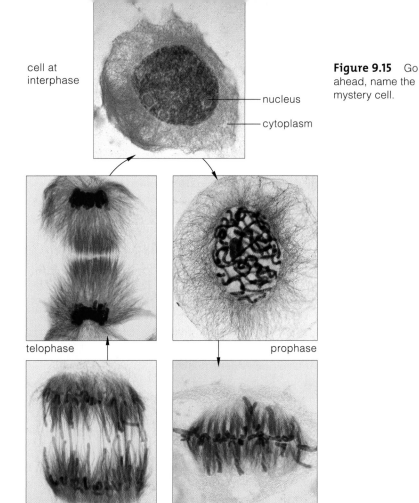

cell at interphase

nucleus

cytoplasm

telophase

prophase

anaphase

metaphase

Figure 9.15 Go ahead, name the mystery cell.

exposed to high levels of x-rays face *radiation poisoning.* Hair loss and a damaged gut lining are early symptoms. Speculate why. Also speculate on why radiation exposure is used as a therapy to treat some cancers.

4. Suppose you have a way to measure the amount of DNA in a single cell during the cell cycle. You first measure the amount at the G1 phase. At what points during the remainder of the cycle would you predict changes in the amount of DNA per cell?

5. The cervix is part of the uterus, a chamber in which embryos develop. The *Pap smear* is a screening procedure that can detect *cervical cancer* in its earliest stages.

   Treatments range from freezing precancerous cells or killing them with a laser beam to removal of the uterus (a hysterectomy). The treatments are more than 90 percent effective when this cancer is detected early. Survival chances plummet to less than 9 percent after it spreads.

   Most cervical cancers develop slowly. Unsafe sex increases the risk. A key risk factor is infection by human papillomaviruses (HPV), which cause genital warts. Viral genes coding for the tumor-inducing proteins get inserted into the DNA of cervical cells. Of one group of cervical cancer patients, 91 percent had been infected with HPV.

   Not all women request Pap smears. Many wrongly believe the procedure is costly. Many do not recognize the importance of abstinence or "safe" sex. Others don't want to think about whether they have cancer. Knowing about the cell cycle and cancer, what would you say to a woman who falls in one or more of these groups?

Why Sex?

Single-celled eukaryotes started to engage in sex many hundreds of millions of years ago, although no one knows how. An unsolved puzzle is why they did it at all.

Asexual reproduction by way of mitotic cell division is easier and faster. Evolutionarily speaking, one individual alone parcels out its DNA to offspring, which are just like their parent. Without sex, that one individual has all of its DNA represented in the new generation. The advantages are evident among most of the protists and fungi, which reproduce asexually most of the time. They quickly give rise to huge populations of cells just like themselves. The advantages are evident among many plants and many invertebrates, including corals, sea stars, and flatworms. Even after bits of these organisms bud or break off, or if the body splits in two, the parts grow into complete copies of the parent. How can the costs of sexual reproduction—such as all of the energy required to construct and use special mate-attracting body parts—beat that?

Sexual reproduction can be an alternative adaptation in changing environments. Consider the plant-sucking insects called aphids. In spring and summer, when plant juices are plentiful, a female aphid reproduces by parthenogenesis. In one day she can give birth to as many as five females, all from unfertilized eggs (Figure 10.1a). Aphid population sizes soar until autumn, when food dwindles. Males now form from eggs, aphids engage in sex, and large fertilized eggs are laid that can withstand winter conditions. Next spring, the eggs develop into asexually oriented females.

Alternative adaptations to the environment also may be why we find a few all-female species of fishes, reptiles, and birds—not mammals—in nature. Not content to let it go at that, University of Tokyo researchers recently fused two mouse eggs in a test tube and made an embryo with no DNA from a male. The embryo developed into Kaguya, the world's first fatherless mammal (Figure 10.1b). The female mouse grew up, engaged in sex with a male mouse, and gave birth to offspring. But back to the big picture:

Sexual reproduction also has advantages when other organisms change. This is especially apparent when we consider the interactions between predators and prey, or between hosts and the parasites or pathogens that infect them. An intriguing idea, the Red Queen hypothesis, may explain the connection between these interactions and sexual reproduction.

In Lewis Carroll's book *Through the Looking Glass*, the Queen of Hearts tells Alice, "Now here, you see, it takes all the running you can do, to keep in the same place." When mutation introduces a better defense against a predator, parasite, or pathogen, we can comfortably predict that natural selection will favor it. However, we also can predict that selection will favor individual predators, parasites, or pathogens that have a novel means to overcome the new defense. The interacting species coevolve; each is running as fast as it can to keep up with the ongoing changes in the other. Talk about an evolutionary treadmill.

Applying the Red Queen hypothesis to our questions, sexual reproduction endures because individuals that practice it can come up with far more variety in heritable defenses compared to the ones that do not. Remember the chromosomes? Sexual reproducers typically have a diploid chromosome number; they inherit two of each type, from two parents. Their two sets of chromosomes

Figure 10.1 Reproductive moments. (**a**) Aphid giving birth. Like females of some other sexually reproducing species, this one reproduces asexually in spring but engages in sex before winter. (**b**) A fatherless mouse. (**c**) Poppy plant being helped by a beetle, which makes pollen deliveries for it. (**d**) Mealybugs mating.

generally hold information about the same traits, but the information about a given trait is not always *exactly* the same on both of them. Some of it might even be bad under prevailing conditions but might be useful in the future. As you will see shortly, meiosis and fertilization mix up information, so that a tremendous variety of novel traits is tried out among the offspring of each new generation. The capacity for rapid, adaptive responses to abiotic and biotic conditions may well be present somewhere in the expressed range of variation.

Asexual reproduction cannot shuffle information into novel combinations. It puts out the same versions of traits again and again into the environmental testing ground. Doing so works well enough—as long as the organism is already equipped to handle change.

With this chapter, we turn to mechanisms of sexual reproduction. Three interconnected events—meiosis, the formation of gametes, and fertilization—are hallmarks of this reproductive mode. The outcome is the production of offspring that display novel combinations of traits. As you will see throughout the book, that outcome has contributed immensely to the range of diversity, past and present.

Watch the video online!

 How Would You Vote?

Japanese researchers have successfully created a "fatherless" mouse that contains the genetic material from the eggs of two females. The mouse is healthy and fully fertile. Do you think researchers should be allowed to try the same process with human eggs? See BiologyNow for details, then vote online.

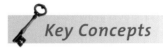 Key Concepts

SEXUAL VERSUS ASEXUAL REPRODUCTION

By asexual reproduction, one parent alone transmits genetic information to offspring. By sexual reproduction, offspring inherit novel combinations of information from more than one parent, because those parents typically differ in their alleles. Alleles are slightly different molecular forms of a gene that specify different versions of the same trait. Section 10.1

OVERVIEW OF MEIOSIS

Meiosis, a nuclear division mechanism, divides the parental chromosome number by half. It occurs only in cells set aside for sexual reproduction. Section 10.2

STAGES OF MEIOSIS

Meiosis sorts out a reproductive cell's chromosomes into four new nuclei. After it ends, gametes form by way of cytoplasmic division and other events. Section 10.3

CHROMOSOME RECOMBINATIONS AND SHUFFLINGS

During meiosis, each pair of chromosomes swaps segments and exchanges alleles. Also, one of each pair is randomly aligned for distribution into a new nucleus. Which ends up in a given gamete is a matter of chance. Chromosomes are shuffled again at fertilization. These events contribute to variation in traits among offspring. Section 10.4

SEXUAL REPRODUCTION IN THE LIFE CYCLES

In animals, gametes form by different mechanisms in males and females. In most plants, spore formation and other events intervene between meiosis and gamete formation. Spores store and protect hereditary information through times that predictably do not favor survival of offspring. Section 10.5

MITOSIS AND MEIOSIS COMPARED

Recent molecular evidence suggests that meiosis originated through mechanisms that already existed for mitosis and, before that, for repairing damaged DNA. Section 10.6

Links to Earlier Concepts

For this chapter, you will be drawing on your sense of the dynamic nature of microtubule assembly and disassembly (Sections 4.10, 4.11, 9.2). Be sure you have a clear picture of the structural organization of chromosomes (9.1) and that you can define chromosome number (9.2). Reflect on how a bipolar spindle made of microtubules moves chromosomes during nuclear division (9.3), and how the cytoplasm gets divided following nuclear division (9.4). You will be revisiting the checkpoint gene products that monitor and repair chromosomal DNA during the cell cycle (9.5).

10.1 Introducing Alleles

LINKS TO
SECTIONS
9.1, 9.2

Asexual reproduction produces genetically identical copies of a parent. Sexual reproduction introduces variation in the details of traits among offspring.

When an orchid or aphid reproduces by itself, what sort of offspring does it get? By the process of **asexual reproduction**, all offspring inherit the same number and kinds of genes from a single parent. **Genes** are sequences of chromosomal DNA. The genes for each species contain all the heritable information necessary to make new individuals. Rare mutations aside, then, asexually produced individuals can only be *clones*, or genetically identical copies of the parent.

Inheritance gets far more interesting with **sexual reproduction**, a process involving meiosis, formation of gametes, and fertilization—a union of two gametes. In most sexual reproducers, such as humans, the first cell of a new individual holds *pairs of genes* on pairs of chromosomes. Usually, one of each pair is maternal and the other paternal in origin (Figure 10.2).

If information in all pairs of genes were identical down to the last detail, sexual reproduction would also produce clones. Just imagine—you, every person you know, the entire human population might be a clone, with everybody looking alike. But the two genes of a pair might *not* be identical. Why not? The molecular structure of any gene can change permanently; it can mutate. So two genes that happen to be paired in an individual's cells may "say" slightly different things about a trait. Each unique molecular form of the same gene is called an **allele**.

Such tiny differences affect thousands of traits. For instance, whether your chin has a dimple depends on which pair of alleles you inherited at one chromosome location. One kind of allele at that location says "put a dimple in the chin." Another kind says "no dimple." Alleles are one reason why the individuals of sexually reproducing species do not all look alike. *With sexual reproduction, offspring inherit new combinations of alleles, which lead to variations in the details of their traits.*

This chapter gets into the cellular basis of sexual reproduction. More importantly, it starts you thinking about far-reaching effects of gene shufflings at certain stages of the process. The process introduces variations in traits among offspring that are typically acted upon by agents of natural selection. Thus, *variation in traits is a foundation for evolution.*

Figure 10.2 A maternal and a paternal chromosome. Any gene on one might be slightly different structurally than the same gene on the other.

> *Sexual reproduction introduces variation in traits by bestowing novel combinations of alleles on offspring.*

10.2 What Meiosis Does

Meiosis is a nuclear division process that divides a parental chromosome number by half in specialized reproductive cells. Sexual reproduction will not work without it.

THINK "HOMOLOGUES"

Think back to the preceding chapter and its focus on mitotic cell division. Unlike mitosis, meiosis sorts out chromosomes into parcels not once but *twice*. Unlike mitosis, it is the first step leading to the formation of gametes. Male and female gametes—such as sperm and eggs—fuse to form a new individual. In most multicelled eukaryotes, cells that form in specialized reproductive structures or organs are the forerunners of gametes. Figure 10.3 gives three examples of where cells that give rise to gametes originate.

As you know, the **chromosome number** is the sum total of chromosomes in cells of a given type. If a cell has a **diploid number** (2n), it has a *pair* of each type of chromosome, often from two parents. Except for a pairing of nonidentical sex chromosomes, each pair has the same length, shape, and assortment of genes, and they line up with each other at meiosis. We call them **homologous chromosomes** (*hom–* means alike).

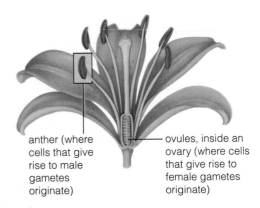

anther (where cells that give rise to male gametes originate)

ovules, inside an ovary (where cells that give rise to female gametes originate)

a Flowering plant

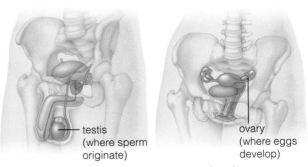

testis (where sperm originate)

ovary (where eggs develop)

b Human male

c Human female

Figure 10.3 Examples of reproductive organs, where cells that give rise to gametes originate.

The body cells of humans are diploid, with 23 + 23 homologous chromosomes (Figure 10.4). So are human germ cells that give rise to gametes. Following meiosis, every gamete normally gets 23 chromosomes—one of each type. Meiosis reduced the parental chromosome number by half, to a **haploid number** (*n*).

TWO DIVISIONS, NOT ONE

Bear in mind, meiosis *is* similar to mitosis in certain respects. As in mitosis, a germ cell duplicates its DNA in interphase. The two DNA molecules and associated proteins stay attached at the centromere, the notably constricted region along their length. For as long as they remain attached, we call them **sister chromatids**:

one chromosome in the duplicated state

As in mitosis, the microtubules of a spindle apparatus move the chromosomes in prescribed directions.

With meiosis, however, *chromosomes go through two consecutive divisions that end with the formation of four haploid nuclei.* The germ cell does not enter interphase between the two nuclear divisions, which are known as meiosis I and meiosis II:

	Meiosis I		Meiosis II
Interphase (DNA is replicated prior to meiosis I)	Prophase I Metaphase I Anaphase I Telophase I	*No* interphase (DNA is *not* replicated prior to meiosis II)	Prophase II Metaphase II Anaphase II Telophase II

In meiosis I, each duplicated chromosome aligns with its partner, *homologue to homologue*. After the two chromosomes of every pair have lined up with each other, they are moved apart:

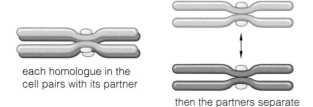

each homologue in the cell pairs with its partner

then the partners separate

The cytoplasm typically starts to divide at some point after each homologue detaches from its partner. The two daughter cells formed this way are haploid, with *one* of each type of chromosome. Don't forget, these chromosomes are still in the duplicated state.

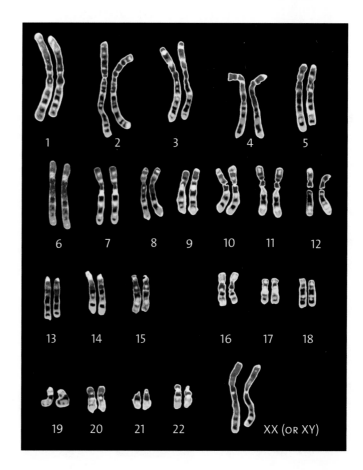

Figure 10.4 Another look at the twenty-three pairs of homologous human chromosomes. This example is from a human female, with two X chromosomes. Human males have a different pairing of sex chromosomes (XY). These chromosomes have been labeled with fluorescent markers.

Next, during meiosis II, *the two sister chromatids of each chromosome are separated from each other:*

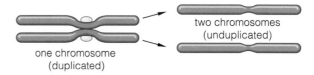

There are now four parcels of 23 chromosomes, and each has one chromosome of each type. New nuclear envelopes enclose them, as four nuclei. Typically the cytoplasm divides once more, so the outcome is four haploid (*n*) cells. Figure 10.5 on the next two pages puts these chromosomal movements in the context of the sequential stages of meiosis.

Meiosis, a nuclear division mechanism, reduces a parental cell's chromosome number by half—to a haploid number (n).

10.3 Visual Tour of Meiosis

Meiosis I

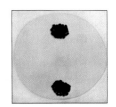

plasma
membrane

newly forming
microtubules in
the cytoplasm

spindle equator
(midway between
the two poles)

one pair of
homologous
chromosomes

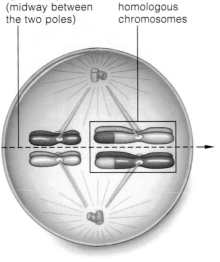
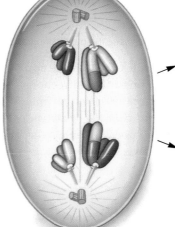
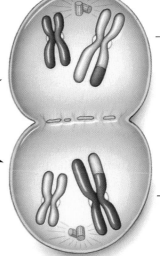

breakup
of nuclear
envelope

pair of centrioles,
and a centrosome,
moving to opposite
sides of nucleus

ⓐ Prophase I

As prophase I begins, chromosomes become
visible as threadlike forms. Each pairs with its
homologue and usually swaps segments with it,
as indicated by the breaks in color in the large
chromosomes. Microtubules are forming a
bipolar spindle (Section 9.3). If two pairs of
centrioles are present, one pair is moved to
the opposite side of the nuclear envelope,
which is starting to break up.

ⓑ Metaphase I

Microtubules from one spindle
pole have tethered one of each type
of chromosome; microtubules from
the other pole have tethered its
homologue. By metaphase I, a
tug-of-war between the two sets
of microtubules has aligned all
chromosomes midway between
the poles.

ⓒ Anaphase I

Microtubules attached to each
chromosome shorten and move
it toward a spindle pole. Other
microtubules, which extend
from the poles and overlap at
the spindle equator, ratchet
past each other and push the
two poles farther apart. Motor
proteins drive the ratcheting.

ⓓ Telophase I

One of each type of
chromosome has
now arrived at the
spindle poles. For
most species, the
cytoplasm divides at
some point, forming
two haploid cells.
All chromosomes
are still duplicated.

Figure 10.5 *Animated!* Meiosis in one type of animal cell. This is a nuclear division mechanism.
It reduces the parental chromosome number in immature reproductive cells by half, to the haploid
number, for forthcoming gametes. To keep things simple, we track only two pairs of homologous
chromosomes. Maternal chromosomes are shaded *purple* and paternal chromosomes *blue*.

Of the four haploid cells that form by meiosis and cytoplasmic divisions, one or all may develop into
gametes and function in sexual reproduction. In plants, the cells that form may develop into spores,
a stage that precedes gamete formation in the life cycle.

Meiosis II

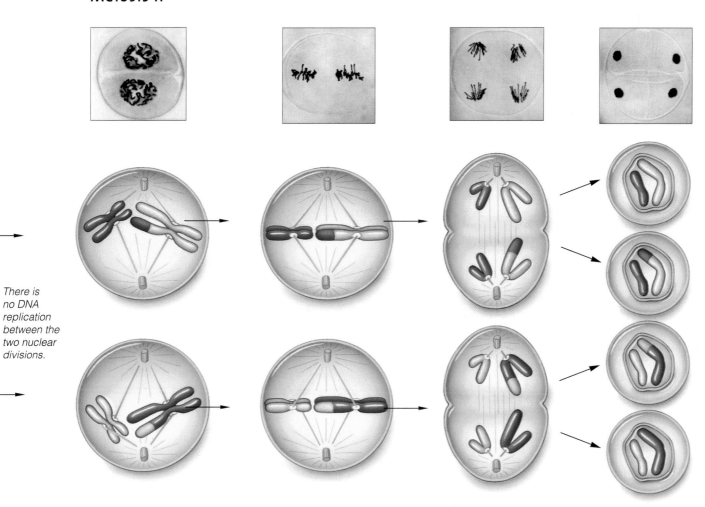

There is no DNA replication between the two nuclear divisions.

ⓔ Prophase II

A new bipolar spindle forms in each haploid cell. Microtubules have moved one member of the pair of centrioles to the opposite end of each cell. One chromatid of each chromosome becomes tethered to one spindle pole, and its sister chromatid becomes tethered to the opposite pole.

ⓕ Metaphase II

Microtubules from both spindle poles have assembled and disassembled in a tug-of-war that ended at metaphase II, when all chromosomes are positioned midway between the poles.

ⓖ Anaphase II

The attachment between sister chromatids of each chromosome breaks. Each is now a separate chromosome but is still tethered to microtubules, which move it toward a spindle pole. Other microtubules push the poles apart. A parcel of unduplicated chromosomes ends up near each pole. One of each type of chromosome is present in each parcel.

ⓗ Telophase II

In telophase II, four nuclei form as a new nuclear envelope encloses each cluster of chromosomes. After cytoplasmic division, each of the resulting daughter cells has a haploid (*n*) number of chromosomes.

10.4   How Meiosis Introduces Variations in Traits

As Sections 10.2 and 10.3 make clear, the basic function of meiosis is the reduction of a parental chromosome number by half. In evolutionary terms, two other functions are as important: Prophase I crossovers and the random alignment of chromosomes at metaphase I contribute greatly to the variation in traits among offspring.

The preceding section mentioned in passing that pairs of homologous chromosomes swap parts of themselves during prophase I. It also showed how a homologous chromosome becomes aligned with its partner during prophase I. Both events introduce new combinations of alleles into the gametes that form at some point *after* meiosis. Along with the chromosome shufflings that occur during fertilization, they contribute to variation in traits that occur among new generations of offspring in sexually reproducing species. Later in the book, you will explore how variation in traits has evolutionary

and ecological consequences. We suggest that you read this section closely. It will serve you well later on.

CROSSING OVER IN PROPHASE I

Figure 10.6*a* is a simple sketch of a pair of duplicated chromosomes, early in prophase I of meiosis. Notice how they are in threadlike form. All chromosomes in a germ cell condense this way. When they do, each is drawn close to its homologue. The chromatids of one become stitched point by point along their length to the chromatids of the other, with little space between them. This tight, parallel orientation favors **crossing over**, a molecular interaction between a chromatid of one chromosome and a chromatid of the homologous partner. DNA strands break and seal in complex ways, but the outcome is that the two "nonsister" chromatids exchange corresponding segments; they swap genes.

a This maternal chromosome (*purple*) and paternal chromosome (*blue*) were duplicated in interphase. They appear in microscopes early in prophase I, when they are starting to condense to threadlike form. Sister chromatids of each chromosome are positioned so close together that they look like a single thread. (We pulled them apart a bit in this sketch so you can distinguish between them.)

b Each chromosome now becomes zippered up with its homologous partner, so all four chromatids are tightly aligned. If the two sex chromosomes have different forms (such as X paired with Y), they still get tightly aligned, but only in a tiny region at their ends.

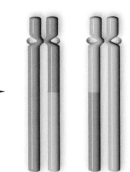

c Here is the simplest way to think about crossing over. However, don't forget that the chromosomes are not really rod-shaped during early prophase I. They are still condensing to threadlike form, and each is tightly aligned with its homologous partner.

d The intimate contact encourages crossovers at various intervals along the length of nonsister chromatids. Here we show the location of just one crossover.

e Nonsister chromatids exchange segments at the crossover site. They keep on condensing into thicker, rodlike forms. They will be fully unzipped from each other by metaphase I.

f Crossing over breaks up the old combinations of alleles and puts new ones together in homologous chromosomes. It mixes up maternal and paternal information about traits.

Figure 10.6 *Animated!*   Key events of prophase I, the first stage of meiosis. For clarity, we show only one pair of homologous chromosomes and one crossover. More than one crossover usually occurs in each chromosome pair. *Blue* signifies a paternal chromosome, and *purple*, its maternal homologue.

Gene swapping would be pointless if each type of gene never varied. But remember, a gene can come in slightly different forms—alleles. You can predict that a number of the alleles on one chromosome will *not* be identical to their partner alleles on the homologous chromosome. Each crossover event is a chance to swap slightly different versions of heritable information on gene products.

We will look at the mechanism of crossing over in later chapters. For now, just remember this: *Crossing over leads to recombinations among genes of homologous chromosomes, and eventually to variation in traits among offspring.*

METAPHASE I ALIGNMENTS

Major shufflings of intact chromosomes start during the transition from prophase I to metaphase I. Suppose this is happening right now in one of your germ cells. Crossovers have already made genetic mosaics of the chromosomes, but put this aside in order to simplify tracking. Just call the twenty-three chromosomes you inherited from your mother the *maternal* chromosomes and the twenty-three you inherited from your father the *paternal* chromosomes.

At metaphase I, microtubules from both poles have now aligned all of the duplicated chromosomes at the spindle equator (Figure 10.5b). Have they tethered all maternal chromosomes to one pole and all paternal chromosomes to the other? Maybe, but probably not. When the microtubules were growing, they latched on to the first chromosome they contacted. Because the tethering was random, there is no particular pattern to the metaphase I positions of maternal and paternal chromosomes.

Now carry this thought one step further. During anaphase I, when a duplicated chromosome is moved away from its homologous partner, *either partner* can end up at either spindle pole.

Think of the possibilities while tracking just three pairs of homologues. By metaphase I, these three pairs may be arranged in any one of four possible positions (Figure 10.7). This means that eight combinations (2^3) are possible for forthcoming gametes.

Cells that give rise to human gametes have twenty-three pairs of homologous chromosomes, not three. Thus, every time a human sperm or egg forms, there is a total of *8,388,608* (or 2^{23}) possible combinations of maternal and paternal chromosomes! Moreover, in a sperm or an egg, many hundreds of alleles inherited from the mother might not "say" the exact same thing about hundreds of different traits as alleles inherited

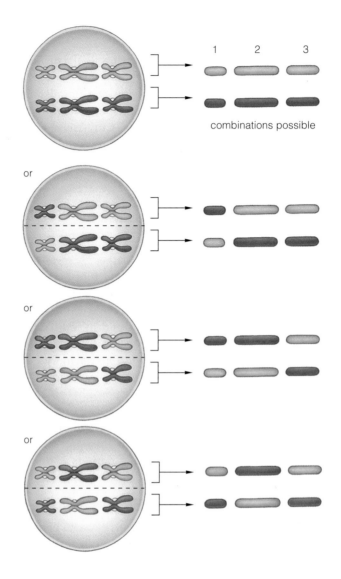

combinations possible

Figure 10.7 *Animated!* Possible outcomes for the random alignment of merely three pairs of homologous chromosomes at metaphase I. The three types of chromosomes are labeled 1, 2, and 3. With four alignments, eight combinations of maternal chromosomes (*purple*) and paternal chromosomes (*blue*) are possible in gametes.

from the father. Are you getting an idea of why such fascinating combinations of traits show up among the generations of your own family tree?

Crossing over, an interaction between a pair of homologous chromosomes, breaks up old combinations of alleles and puts new ones together during prophase I of meiosis.

The random tethering and subsequent positioning of each pair of maternal and paternal chromosomes at metaphase I lead to different combinations of maternal and paternal traits in each new generation.

10.5 From Gametes to Offspring

LINK TO
SECTION
9.4

*What happens to the gametes that form after meiosis?
Later chapters have specific examples. Here, simply focus
on where they fit in the life cycles of plants and animals.*

Gametes are not all the same in their details. Human sperm have one tail, opossum sperm have two, and roundworm sperm have none. Crayfish sperm look like pinwheels. Most eggs are microscopic in size, yet an ostrich egg inside its shell is as big as a football. A flowering plant's male gamete is just a sperm nucleus.

GAMETE FORMATION IN PLANTS

The life cycle of most plant species alternates between sporophyte and gametophyte stages. A *sporophyte* is a multicelled spore-producing body that makes sexual spores by way of meiosis (Figure 10.8*a*). In plants, each **spore** is a haploid reproductive cell that is not a gamete and that does not take part in fertilization. At some point, the spore undergoes mitotic cell divisions that give rise to a *gametophyte*. One or more gametes do form inside this multicelled haploid body.

Pine trees are examples of sporophytes, and their female gametophytes form on the scales of pinecones. Rose bushes and fuschias also are sporophytes, and gametophytes form inside their flowers. You will be focusing on plant life cycles in Chapters 23 and 32.

GAMETE FORMATION IN ANIMALS

In animals, diploid germ cells give rise to gametes. In a male reproductive system, a germ cell develops into a primary spermatocyte. This large, immature cell enters meiosis and cytoplasmic divisions. Four haploid cells result and develop into spermatids (Figure 10.9). Each

cell undergoes changes, such as the formation of a tail, and becomes a **sperm**, a type of mature male gamete.

In female animals, a germ cell becomes a primary oocyte, which is an immature egg. Unlike sperm, the primary oocyte increases in size and stockpiles many cytoplasmic components. In addition, its four daughter cells differ in size and function (Figure 10.10).

When the primary oocyte divides after meiosis I, one daughter cell—the secondary oocyte—gets nearly all of the cytoplasm. The other cell, a first polar body, is exceedingly small. Later, both of these haploid cells enter meiosis II, then cytoplasmic division. One of the secondary oocyte's daughter cells becomes the second polar body. The other daughter cell gets most of the cytoplasm and develops into a gamete. The mature female gamete is an ovum (plural, ova). An ovum also is known informally as an **egg**.

And so we have one egg. The three polar bodies that formed don't function as gametes and aren't rich in nutrients or plump with cytoplasm. In time they degenerate. But their formation assures that the egg will have a haploid chromosome number. Also, by getting most of the cytoplasm, the egg holds enough metabolic machinery to support early cell divisions of the new individual, as Chapters 43 and 44 explain.

MORE SHUFFLINGS AT FERTILIZATION

The chromosome number characteristic of the parents is restored at **fertilization**, a time when a female and male gamete unite and their haploid nuclei fuse. If meiosis did not precede fertilization, the chromosome number would double in each generation. Doublings would disrupt hereditary information, usually for the worse. Why? That information is like a fine-tuned set

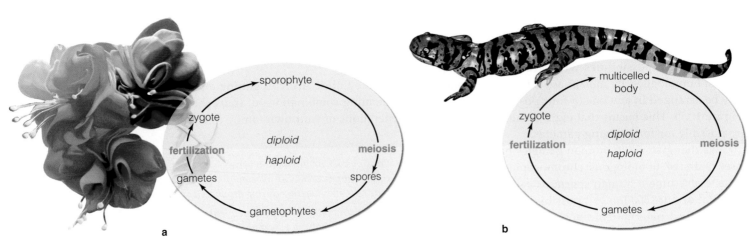

Figure 10.8 (**a**) Generalized life cycle for most plants. (**b**) Generalized life cycle for animals. The zygote is the first cell to form when the nuclei of two gametes fuse at fertilization.

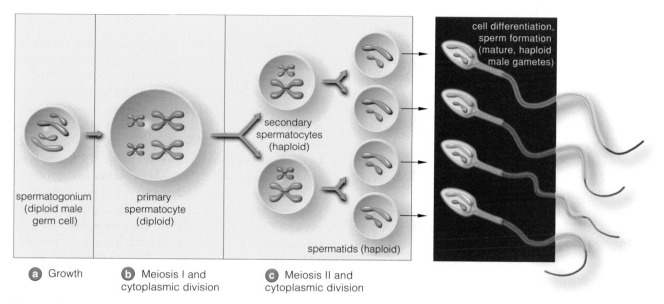

(a) Growth   (b) Meiosis I and cytoplasmic division   (c) Meiosis II and cytoplasmic division

Figure 10.9 *Animated!* Generalized sketch of sperm formation in animals. Figure 44.4 shows a specific example (how sperm form in human males).

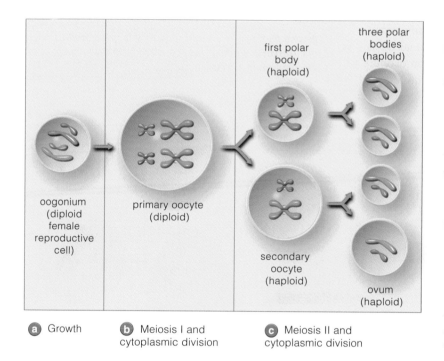

(a) Growth   (b) Meiosis I and cytoplasmic division   (c) Meiosis II and cytoplasmic division

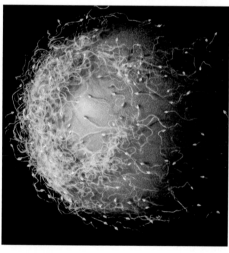

Figure 10.10 *Animated!* Animal egg formation. Eggs are far larger than sperm and larger than the three polar bodies. The painting above, based on a scanning electron micrograph, depicts human sperm surrounding an ovum.

of blueprints that must be followed exactly, page after page, to build a normal individual.

Fertilization also adds to variation among offspring. Reflect on the possibilities for humans alone. During prophase I, every human chromosome undergoes an average of two or three crossovers. In addition to the crossovers, random positioning of pairs of paternal and maternal chromosomes at metaphase I results in one of millions of possible chromosome combinations in each gamete. And of all male and female gametes that form, *which* two actually get together is a matter of chance. The sheer number of combinations that can exist at fertilization is staggering!

The distribution of random mixes of chromosomes into gametes, random metaphase chromosome alignments, and fertilization contribute to variation in traits of offspring.

10.6 Mitosis and Meiosis—An Ancestral Connection?

LINKS TO
SECTIONS
9.2, 9.5

This chapter opened with hypotheses about the survival advantages of asexual and sexual reproduction. It seems like a giant evolutionary step from producing clones to producing genetically varied offspring. But was it?

Figure 10.11 shows an obvious parallel between the four stages of mitosis and meiosis II. The same kind of bipolar spindle assorts duplicated chromosomes into parcels in very similar ways. Recent studies also reveal striking similarities at the molecular level.

In all organisms, from prokaryotes to mammals, certain genes code for proteins that can recognize and repair breaks in the double-stranded DNA molecules of chromosomes. Such damage, recall, is monitored by products of checkpoint genes while DNA is being replicated during the cell cycle (Sections 9.2 and 9.5). If they detect a problem, there is a pause in the cycle until it is repaired. Even in bacteria—the most ancient lineages on Earth—a mechanism exists that may well have been recruited for mitosis and meiosis.

Some highly conserved gene products often repair breaks and odd rearrangements in chromosomal DNA that occur during mitosis. They also put chromosomal DNA back together in prophase I, after homologous chromosomes exchange segments. This outcome—a form of genetic recombination—could have been part of the evolution of sexual reproduction.

Is *Giardia intestinalis* one model? This descendent of one of the earliest eukaryotic lineages does not have mitochondria, and it does not form a bipolar spindle during mitosis. This single-celled parasite has never been observed to reproduce sexually. Yet it has gene products that serve in meiosis in higher eukaryotes.

We invite you to think about these possibilities as you read later chapters in the book. We invite you to explore likely connections on your own. For instance, when you look at *Chlamydomonas*, a single-celled alga of freshwater habitats, mull over the fact that haploid *Chlamydomonas* cells reproduce asexually by mitotic cell division. But two cells of different mating strains also can function as *gametes*; they can fuse and form a diploid individual. Do such cells offer more clues to the origin of sexual reproduction? Maybe.

Recombination mechanisms that are vital for reproduction of eukaryotic cells might have evolved from DNA repair mechanisms in prokaryotic ancestors.

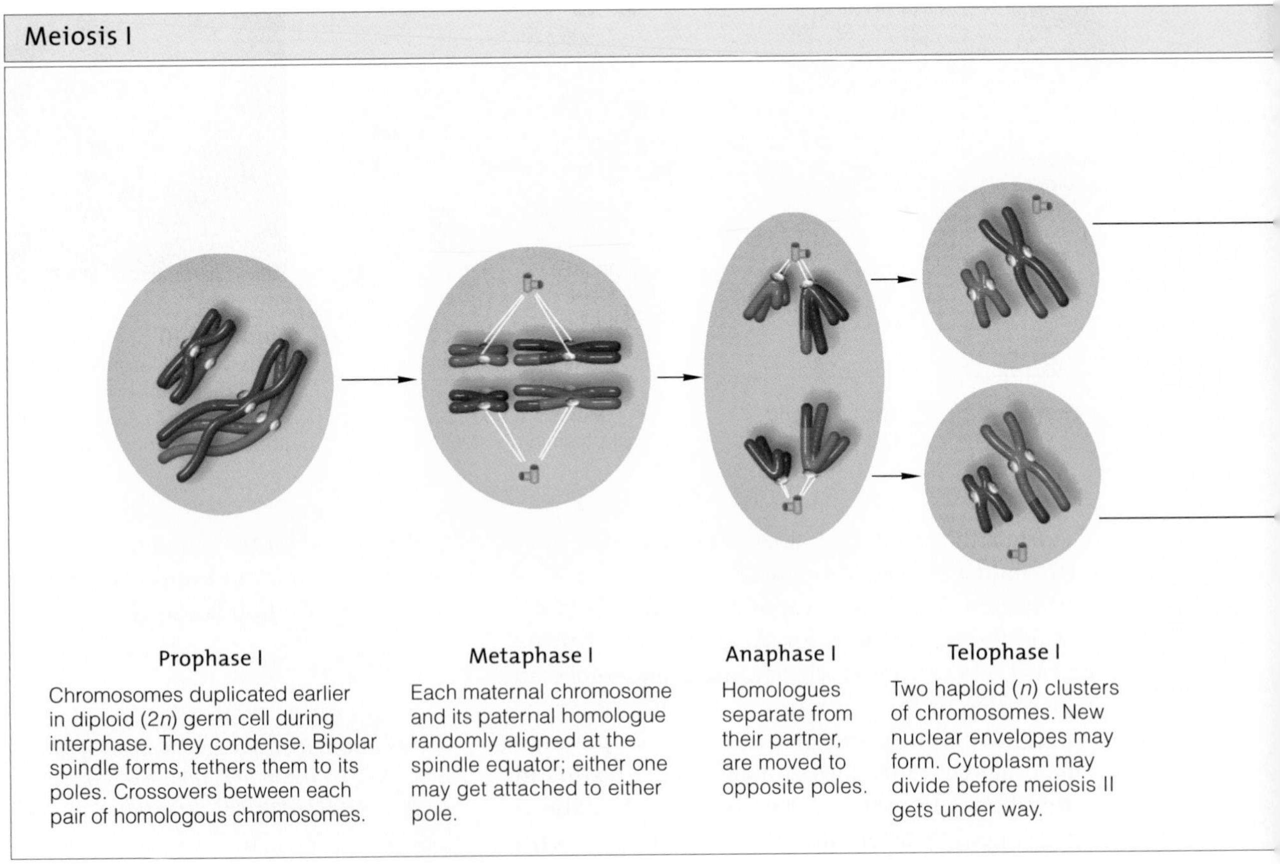

Meiosis I

Figure 10.11 Comparative summary of key features of mitosis and meiosis, starting with a diploid cell. Only two paternal and two maternal chromosomes are shown. Both were duplicated in interphase, prior to nuclear division. Both use a bipolar spindle made of microtubules to sort out and move the chromosomes.

Mitosis maintains the parental chromosome number. Meiosis halves it, to the haploid number.

Mitotic cell division is the basis of asexual reproduction among eukaryotes. It is the basis of growth and tissue repair of multicelled eukaryotic species.

Meiotic cell division is a required step before the formation of gametes or sexual spores.

Prophase I

Chromosomes duplicated earlier in diploid (2*n*) germ cell during interphase. They condense. Bipolar spindle forms, tethers them to its poles. Crossovers between each pair of homologous chromosomes.

Metaphase I

Each maternal chromosome and its paternal homologue randomly aligned at the spindle equator; either one may get attached to either pole.

Anaphase I

Homologues separate from their partner, are moved to opposite poles.

Telophase I

Two haploid (*n*) clusters of chromosomes. New nuclear envelopes may form. Cytoplasm may divide before meiosis II gets under way.

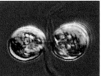

Giardia intestinalis   *Chlamydomonas* cells mating

Mitosis

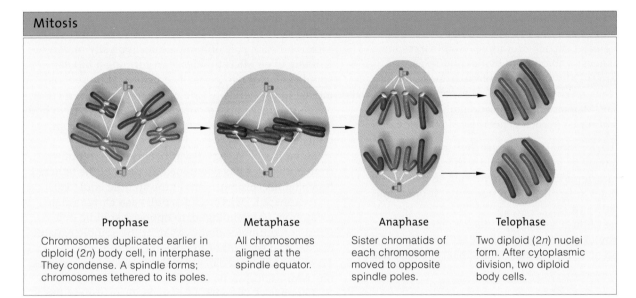

Prophase

Chromosomes duplicated earlier in diploid (2*n*) body cell, in interphase. They condense. A spindle forms; chromosomes tethered to its poles.

Metaphase

All chromosomes aligned at the spindle equator.

Anaphase

Sister chromatids of each chromosome moved to opposite spindle poles.

Telophase

Two diploid (2*n*) nuclei form. After cytoplasmic division, two diploid body cells.

Meiosis II

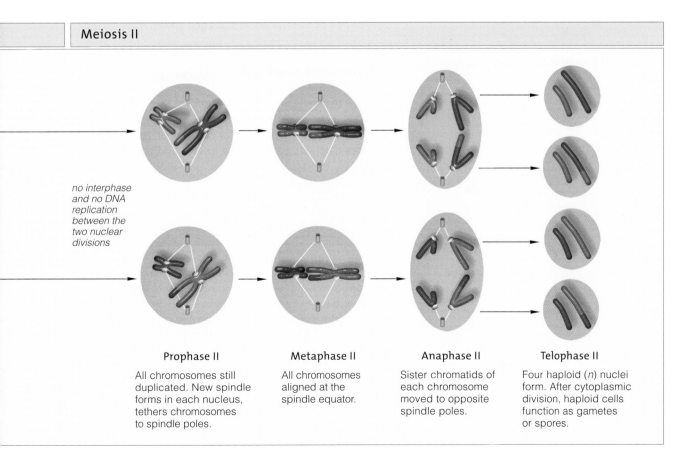

no interphase and no DNA replication between the two nuclear divisions

Prophase II

All chromosomes still duplicated. New spindle forms in each nucleus, tethers chromosomes to spindle poles.

Metaphase II

All chromosomes aligned at the spindle equator.

Anaphase II

Sister chromatids of each chromosome moved to opposite spindle poles.

Telophase II

Four haploid (*n*) nuclei form. After cytoplasmic division, haploid cells function as gametes or spores.

Summary

Section 10.1 Life cycles of eukaryotic species often have asexual and sexual phases.

Asexual reproduction by way of mitosis yields a clone, or offspring that are genetically the same as one parent. Compared with sexual modes, it is easier, requires less energy, and gives rise to huge populations in far less time.

Sexual reproduction involves two parents that engage in meiosis, gamete formation, and fertilization. It leads to novel allele combinations in offspring. Compared to asexual reproduction, the expressed range of variation offers a far greater capacity for rapid, adaptive response to novel changes in abiotic and biotic conditions.

Alleles are slightly different molecular forms of the same gene that specify different versions of the same gene product. Meiosis and fertilization mix up the alleles (and forms of traits) in each generation of offspring.

Section 10.2 Meiosis, a nuclear division process, precedes gamete formation. It divides the chromosome number characteristic of a species by half, so that fusion of two gametes at fertilization restores the chromosome number (Figure 10.12).

Offspring of most sexual reproducers inherit pairs of chromosomes, one from a maternal and one from a paternal parent. Except in individuals that have inherited

nonidentical sex chromosomes (e.g., X with Y), the pairs are homologous (alike); each pair has the same length, shape, and mostly the same gene sequence. All pairs interact at meiosis. Meiosis parcels out one chromosome of each type for forthcoming gametes.

Section 10.3 All chromosomes in a reproductive cell are duplicated in interphase, prior to meiosis. Meiosis sorts out duplicated chromosomes twice, in two divisions (meiosis I and II) that are not separated by interphase.

In meiosis I, the first nuclear division, homologous chromosomes are partitioned into two clusters, both with one of each type of chromosome.

Prophase I. Chromosomes condense into threadlike form, and each pair of homologues typically undergoes crossing over. Microtubules start forming a bipolar spindle. One of two pairs of centrioles, if present, is moved to the opposite side of the nucleus. The nuclear envelope breaks up, so microtubules growing from both spindle poles can penetrate the nuclear region and tether the chromosomes.

Metaphase I. A tug-of-war between microtubules from both poles has positioned all pairs of the tethered homologous chromosomes at the spindle equator.

Anaphase I. Microtubules pull each chromosome away from its homologue, to opposite spindle poles. Other microtubules that overlap at the spindle equator ratchet past each other to push the poles farther apart. There are now two parcels of duplicated chromosomes, one near each spindle pole.

Telophase I. Two haploid nuclei form around the parcels. Cytoplasmic division typically follows.

In meiosis II, the second nuclear division, the sister chromatids of each chromosome are pulled away from each other and partitioned into two clusters. This occurs in both haploid nuclei that formed in meiosis I. By the end of telophase II, there are four nuclei, each with a haploid chromosome number.

When the cytoplasm divides, there are four haploid cells. One or all may serve as gametes or, in plants, as spores that will give rise to gamete-producing bodies.

Biology ⒮ Now
Explore what happens during each stage of meiosis with the animation on BiologyNow.

Section 10.4 Novel combinations of alleles and of maternal and paternal chromosomes arise through events in prophase I and metaphase I.

*Non*sister chromatids of homologous chromosomes undergo crossing over during prophase I. They break and exchange segments, so that each ends up with allelic combinations that were not present in either parent.

Maternal and paternal chromosomes get tethered randomly to one spindle pole or the other. Thus they are positioned at random when they are aligned at the spindle equator at metaphase I, so alleles of either one may end up in a new nucleus, then in a gamete.

Biology ⒮ Now
Study how crossing over and metaphase I alignments affect allele combinations with the animation on BiologyNow.

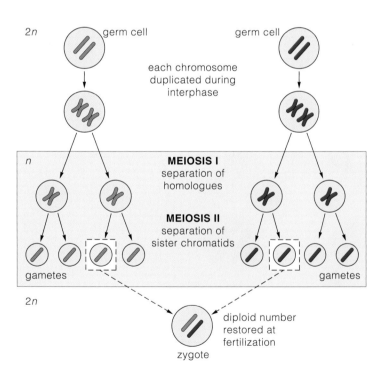

Figure 10.12 Summary of changes in chromosome number at different stages of sexual reproduction, using two diploid (*2n*) germ cells as the example. During two nuclear divisions, meiosis reduces the chromosome number by half (*n*). The union of haploid nuclei of two gametes at fertilization restores the diploid number.

Section 10.5 Life cycles of plants and animals have sexual phases. Sporophytes are a multicelled plant body that produces sexual spores. Such plant spores give rise to gametophytes, in which haploid gametes form.

In most animals, germ cells in reproductive organs give rise to sperm or eggs. Fusion of a sperm and egg nucleus at fertilization results in a zygote, the first cell of a new individual.

Biology⊘Now

Learn how gametes form with the animation on BiologyNow.

Section 10.6 Like mitosis, meiosis uses a bipolar spindle to move and sort duplicated chromosomes. But meiosis occurs only in sex cells and does not produce clones of the parent; it reduces the parental chromosome number by half. Crossing over and random alignments of different mixes of maternal and paternal chromosomes for distribution to gametes occur only in meiosis. These events, and the chance of any two gametes meeting at fertilization, contribute to enormous variation in traits among offspring.

Figure 10.13 Bdelloid rotifer.

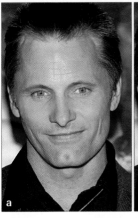

Figure 10.14 Viggo Mortensen (**a**) with and (**b**) without a chin dimple.

11. Match each term with its description.

_____ chromosome number	a. different molecular forms of the same gene
_____ alleles	b. none between meiosis I, II
_____ metaphase I	c. all chromosomes aligned at spindle equator
_____ interphase	d. sum total of all chromosomes in cells of a given type

Additional questions are available on Biology⊘Now™

Self-Quiz

Answers in Appendix II

1. Meiosis and cytoplasmic division function in _____ .
   a. asexual reproduction of single-celled eukaryotes
   b. growth, tissue repair, often asexual reproduction
   c. sexual reproduction
   d. both b and c

2. A duplicated chromosome has _____ chromatid(s).
   a. one   b. two   c. three   d. four

3. A somatic cell having two of each type of chromosome has a(n) _____ chromosome number.
   a. diploid   b. haploid   c. tetraploid   d. abnormal

4. Sexual reproduction requires _____ .
   a. meiosis   c. spore formation
   b. fertilization   d. a and b

5. Generally, a pair of homologous chromosomes _____ .
   a. carry the same genes   c. interact at meiosis
   b. are the same length, shape d. all of the above

6. Meiosis _____ the parental chromosome number.
   a. doubles   b. halves   c. maintains   d. corrupts

7. Meiosis ends with the formation of _____ .
   a. two cells   c. eight cells
   b. two nuclei   d. four nuclei

8. The cell in the diagram below is in anaphase I rather than anaphase II. I know this because _____ .

9. Sister chromatids of each duplicated chromosome separate during _____ .
   a. prophase I   c. anaphase I
   b. prophase II   d. anaphase II

10. Sexual reproducers bestow variation in traits on offspring by _____ .
   a. crossing over   c. fertilization
   b. metaphase I   d. both a and b
      random   e. All of the above
      orientations   are factors.

Critical Thinking

1. Why can you predict that meiosis will give rise to genetic variation between parent cells and daughter cells in fewer cell cycles than mitosis?

2. The bdelloid rotifer lineage started at least 40 million years ago (Figure 10.13). About 360 known species of these tiny animals live in many aquatic habitats worldwide. All are female. Do some research to identify conditions in the physical and biological environments to which they might be reproductively adapted.

3. Actor Viggo Mortensen inherited a gene that makes his chin dimple. Figure 10.14*b* shows what he might have looked like if he inherited a different form of that gene. What is the name for alternative forms of the same gene?

4. Assume you can measure the amount of DNA in the nucleus of a primary oocyte, and then in the nucleus of a primary spermatocyte. Each gives you a mass *m*. What mass of DNA would you expect to find in the nucleus of each mature gamete (egg and sperm) that forms after meiosis? What mass of DNA will be (1) in the nucleus of a zygote that forms at fertilization and (2) in that zygote's nucleus after the first DNA duplication?

5. The diploid chromosome number for the somatic cells of several eukaryotic species are listed at right. Write down the number of chromosomes that normally end up in gametes of each species. Then write what the number would be after three generations if meiosis did not occur before gamete formation.

Fruit fly, *Drosophila melanogaster*	8
Garden pea, *Pisum sativum*	14
Corn, *Zea mays*	20
Frog, *Rana pipiens*	26
Earthworm, *Lumbricus terrestris*	36
Human, *Homo sapiens*	46
Chimpanzee, *Pan troglodytes*	48
Amoeba, *Amoeba*	50
Horsetail, *Equisetum*	216

In Pursuit of a Better Rose

Researchers at Texas A&M and Clemson universities are breathing new life into *rose breeding*. People have been practicing this form of artificial selection for thousands of years. Starting with small, simple, five-petaled wild roses, they patiently cross-bred plants and in time were rewarded with a profusion of petals, fabulous fragrances, exquisite colors, and other compelling traits. Today, rose fanciers in thirty-six countries all over the world claim membership in the World Federation of Rose Societies. In any given year, people from all walks of life buy billions of dollars' worth of rosebuds and blooms. On Valentine's Day in the United States alone, 110 million cut roses are offered as symbols of love and romance. Roses are now big business.

Fossils in Colorado tell us that roses have been around for at least 40 million years. When rose breeding started, the ancestral stock had a diploid chromosome number—two sets of seven chromosomes. A great variety of cultivars now have four, seven, fourteen, even twenty-one sets of chromosomes! Within those chromosomes are genes that specify the size, number, and shape of petals and thorns, genes that deal with floral scents and colors, and genes that dictate whether plants bloom once or all year long. Other genes influence resistance to diseases and pests.

Unlike many wild roses, most of the cultivated varieties are susceptible to black spot, powdery mildew, and other diseases (Figure 11.1). The fungus that causes black spot is notably active in rainy, humid regions; the one that causes powdery mildew thrives in greenhouses. Fungicides work against pathogenic fungi, but they are costly, and many kinds also kill beneficial microorganisms.

Possibly a safer approach would be to cross-breed a wild plant known to have disease resistance with a plant known to be susceptible. However, traditional breeding practices are hit-or-miss, and they are tedious. Breeders have to wait for plants to form seeds, then plant the seeds, then observe whether any or all plants of the new generation, and the next, and the one after that show disease resistance.

Enter the new researchers. They are working to make genetic maps for all seven of the rose chromosomes. Just as road maps pinpoint cities along a highway, genetic maps can pinpoint where genes that influence specific traits are located along the length of chromosomes. By pinpointing a gene that influences a desired trait, breeders will be able to speed up their artificial selection practices.

For example, remember those radioisotopes described in Chapter 2? Researchers use them to make a DNA probe, a bit of radioactive DNA. They use it to test offspring from a cross for a specific DNA region—say, one near a gene that affects disease resistance. If the probe does not bind to the DNA of offspring, a breeder can assume the new plant has

Figure 11.1 One representative of a long history of artificial selection. Like most of the modern cultivars and unlike many wild roses, this one is vulnerable to black spot, a disease that results in the telltale destruction of leaves. Researchers are working to develop faster, more efficient ways to breed roses that have disease resistance and other desired traits.

not inherited the resistance gene and can try a new cross. Such *marker-assisted selection* is useful when several genes control a trait, as for disease resistance. A plant lineage that inherits all the genes may be the least vulnerable to attack.

In time, the maps being pieced together at Texas A&M, Clemson, and elsewhere will become consolidated into a permanent genetic map for roses. Its information will be retrieved for breeding programs. It also will be put to use for actual transfers of desirable genes into roses by way of biotechnology and genetic engineering. But these are cutting-edge topics that will not make much sense without in-depth knowledge of the structure and function of DNA, genes, and their protein products. We reserve them for later chapters in this unit. For now, start with something you already know about—the chromosomes and alleles introduced in the preceding chapter. This will be enough for you to follow the classical breeding practices that gave us our first glimpses of the principles of inheritance.

How Would You Vote?

The federal government helps support some agricultural extension programs that offer homeowners advice on gardens and ornamental plants. Do you consider this to be an appropriate use of government resources? See BiologyNow for details, then vote online.

Key Concepts

WHERE MODERN GENETICS STARTED

Gregor Mendel gathered the first experimental evidence of the genetic basis of inheritance: Each gene has a specific location on a chromosome. Organisms that have a diploid chromosome number have *pairs* of genes, at equivalent locations on pairs of homologous chromosomes. Alleles that are nonidentical may affect a trait differently. One allele is often dominant, in that its effect on a trait masks the effect of a recessive allele paired with it.   Section 11.1

INSIGHTS FROM MONOHYBRID EXPERIMENTS

Some experiments yielded evidence of gene segregation: When one chromosome is separated from its homologous partner during meiosis, their pairs of alleles also separate and end up in different gametes.   Section 11.2

INSIGHTS FROM DIHYBRID EXPERIMENTS

Other experiments yielded evidence of independent assortment: During meiosis, each pair of homologous chromosomes is sorted out for distribution into one gamete or another independently of how all of the other pairs of homologous chromosomes are assorted.   Section 11.3

VARIATIONS IN GENE EXPRESSION

Not all traits have clearly dominant or recessive forms. One allele of a pair may be fully or partially dominant over its partner or codominant with it. Two or more gene pairs often influence the same trait, and some single genes influence many traits. The environment introduces more variation in gene expression.   Sections 11.4–11.7

Links to Earlier Concepts

Before starting this chapter, review the definitions of genes, alleles, and diploid versus haploid chromosome numbers (Sections 10.1 and 10.2). As you read, you may wish to refer to the earlier introduction to natural selection (1.4) and to the visual road map for the stages of meiosis (10.3). You will be considering experimental evidence of two major topics that were introduced earlier—the effects that crossing over and metaphase I alignments have on inheritance (10.4).

11.1 Mendel, Pea Plants, and Inheritance Patterns

LINKS TO
SECTIONS
1.4, 10.1

We turn now to recurring inheritance patterns among humans and other sexually reproducing species. You already know meiosis halves the parental chromosome number, which is restored at fertilization. Here the story picks up with some observable outcomes of these events.

More than a century ago, people wondered about the basis of inheritance. Most had an idea that two parents contribute hereditary material to offspring, but few even suspected that it is organized in units, or genes.

Figure 11.2 Gregor Mendel, the founder of modern genetics.

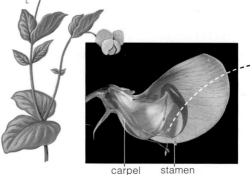

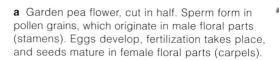

carpel   stamen

a Garden pea flower, cut in half. Sperm form in pollen grains, which originate in male floral parts (stamens). Eggs develop, fertilization takes place, and seeds mature in female floral parts (carpels).

b Pollen from a plant that breeds true for purple flowers is brushed onto a floral bud of a plant that breeds true for white flowers. The white flower had its stamens snipped off. This is one way to assure cross-fertilization of plants.

c Later, seeds develop inside pods of the cross-fertilized plant. An embryo within each seed develops into a mature pea plant.

d Each new plant's flower color is indirect but observable evidence that hereditary material has been transmitted from the parent plants.

Figure 11.3 *Animated!* Garden pea plant (*Pisum sativum*), which can self-fertilize or cross-fertilize. Experimenters can control the transfer of its hereditary material from one flower to another.

Rather, according to the prevailing view, hereditary material was fluid, with the fluids from both parents blending at fertilization like milk into coffee.

The idea of "blending inheritance" failed to explain the obvious. For example, many children who differ in eye color or hair color have the same two parents. If parental fluids blended, then the eye or hair color of children should be a blend of the parental colors. If neither parent had freckles, freckled children would never pop up. A white mare bred with a black stallion should consistently give birth to gray offspring, but as horse breeders knew, this was not always the case. Blending inheritance could scarcely explain much of the obvious variation in traits that people could see with their own eyes.

Even Charles Darwin accepted the blending notion until he and his cousin conducted experiments that disproved it. According to Darwin's theory of natural selection, individuals of a population show variation in traits. Over the generations, variations that help an individual survive and reproduce show up among more and more offspring, and less helpful variations become less frequent and might even disappear. Thus blending inheritance *seemed* to support the theory of natural selection. As it turned out, the idea of discrete units of information—genes—explain it better.

Even before Darwin presented his theory, someone was gathering evidence that eventually would help support it. A monk, Gregor Mendel (Figure 11.2), had already guessed that sperm and eggs carry distinct units of information about heritable traits. After he analyzed specific traits of pea plants, one generation after another, he found indirect but *observable* evidence of how parents transmit genes to offspring.

MENDEL'S EXPERIMENTAL APPROACH

Mendel spent most of his adult life in Brno, a city near Vienna that is now part of the Czech Republic. Yet he was not a man of narrow interests who accidentally stumbled onto dazzling principles.

Mendel's monastery was close to European capitals that were centers of scientific inquiry. Having been raised on a farm, he was keenly aware of agricultural principles and their applications. He kept abreast of literature on breeding experiments. He belonged to an agricultural society and won awards for developing improved varieties of vegetables and fruits. Shortly after entering the monastery, Mendel took courses in mathematics, physics, and botany at the University of Vienna. Few scholars of his time showed interest in both plant breeding *and* mathematics.

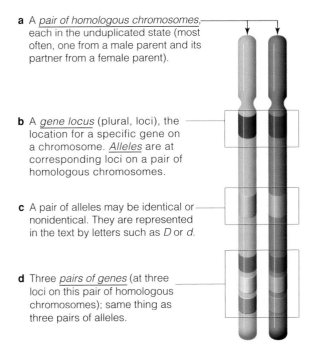

a A *pair of homologous chromosomes,* each in the unduplicated state (most often, one from a male parent and its partner from a female parent).

b A *gene locus* (plural, loci), the location for a specific gene on a chromosome. *Alleles* are at corresponding loci on a pair of homologous chromosomes.

c A pair of alleles may be identical or nonidentical. They are represented in the text by letters such as *D* or *d*.

d Three *pairs of genes* (at three loci on this pair of homologous chromosomes); same thing as three pairs of alleles.

Figure 11.4 *Animated!* A few genetic terms. Garden peas and other species with a diploid chromosome number have pairs of genes, on pairs of homologous chromosomes. Most genes come in slightly different molecular forms called alleles. Different alleles specify different versions of the same trait. An allele at any given location on a chromosome may or may not be identical to its partner on the homologous chromosome.

Shortly after his university training, Mendel started to study *Pisum sativum,* the garden pea plant (Figure 11.3). This plant is self-fertilizing. Its flowers produce both male and female gametes—call them sperm and eggs—that can come together and give rise to a new plant. One lineage of pea plants can "breed true" for certain traits. This means that successive generations will be like parents in one or more traits, as when all offspring grown from seeds of self-fertilized, white-flowered parent plants also have white flowers.

Pea plants also cross-fertilize when plant breeders transfer pollen from one plant to the flower of another plant. As Mendel knew, breeders open a floral bud of a plant that bred true for white flowers or some other trait and snip out its stamens. (Pollen grains, in which sperm develop, start forming in stamens.) The buds can be brushed with pollen from a plant that bred true for a *different* version of the trait—say, purple flowers.

As Mendel hypothesized, such clearly observable differences might help him track a given trait through many generations. If there were patterns to the trait's inheritance, *then those patterns might tell him something about heredity itself.*

TERMS USED IN MODERN GENETICS

In Mendel's time, no one knew about genes, meiosis, or chromosomes. As we follow his thinking, we will clarify the picture by substituting some modern terms used in inheritance studies, as stated here and in Figure 11.4:

1. **Genes** are units of information on heritable traits, which parents transmit to offspring. Each gene has a specific location (locus) in chromosomal DNA.

2. Cells with a **diploid** chromosome number (2*n*) have pairs of genes, on pairs of homologous chromosomes.

3. **Mutation** alters a gene's molecular structure and its message about a trait. It may cause a trait to change, as when a gene for flower color specifies yellow and a mutant form of the gene specifies white. All molecular forms of the same gene are known as **alleles**.

4. When offspring inherit a pair of *identical* alleles for a trait generation after generation, they typically are a true-breeding lineage. Offspring of a cross between two individuals that breed true for different forms of a trait are **hybrids**; each one has inherited *nonidentical* alleles for the trait.

5. A pair of identical alleles on a pair of homologous chromosomes is a *homozygous* condition. A pairing of nonidentical alleles is a *heterozygous* condition.

6. An allele is *dominant* when its effect on a trait masks the effect of any *recessive* allele paired with it. Capital letters signify dominant alleles, and lowercase letters signify recessive ones. *A* and *a* are examples.

7. Pulling this all together, a **homozygous dominant** individual has a pair of dominant alleles (*AA*) for the trait under study. A **homozygous recessive** individual has a pair of recessive alleles (*aa*), and a **heterozygous** individual has a pair of nonidentical alleles (*Aa*).

8. Two terms help keep the distinction clear between genes and the traits they specify. *Genotype* refers to the particular alleles that an individual carries. *Phenotype* refers to an individual's observable traits.

9. P stands for true-breeding parents, F_1 for the first-generation offspring, and F_2 for the second-generation offspring of self-fertilized or intercrossed F_1 individuals.

Mendel hypothesized that tracking clearly observable differences in forms of a given trait might reveal patterns of inheritance. He recognized patterns of dominance and recessiveness in certain traits, which later were connected with pairs of alleles on pairs of homologous chromosomes.

11.2 Mendel's Theory of Segregation

Mendel used monohybrid experiments to test a hypothesis: Pea plants inherit two "units" of information (genes) for a trait, one from each parent.

In **monohybrid experiments**, two homozygous parents differ in a trait that is governed by alleles of one gene. They are crossed to produce F_1 offspring that are all heterozygous ($AA \times aa \longrightarrow Aa$). Next, depending on the species, F_1 individuals are allowed to self-fertilize or mate in order to produce an F_2 generation.

MONOHYBRID EXPERIMENT PREDICTIONS

Mendel tracked seven traits for two generations. In one set of experiments, he crossed plants that bred true for purple *or* white flowers. All F_1 offspring had purple flowers, but in the next generation, some F_2 offspring had white flowers! So what was going on? Pea plants have pairs of homologous chromosomes. Assume one plant is homozygous dominant (AA) and another is homozygous recessive (aa) at the locus that governs flower color. Following meiosis, each sperm or egg that forms has only one of these alleles (Figure 11.5). Therefore, when a sperm fertilizes an egg, only one outcome is possible: $A + a \longrightarrow Aa$.

With his background in mathematics, Mendel knew about sampling error (Figure 1.12). He crossed seventy plants. He also counted and recorded the number of dominant and recessive forms of traits in thousands of offspring. On average, three of every four F_2 plants were dominant, and one was recessive (Figure 11.6).

The ratio hinted that fertilization is a chance event having a number of possible outcomes. Mendel knew about probability—*which applies to chance events and thus could help him predict possible outcomes of genetic crosses*. **Probability** means this: The chance that each outcome of an event will occur is proportional to the number of ways in which the outcome can be reached.

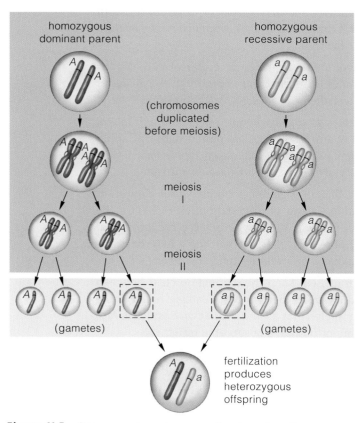

Figure 11.5 One gene of a pair segregating from the other gene in a monohybrid cross. Two parents that breed true for two versions of a trait produce only heterozygous offspring.

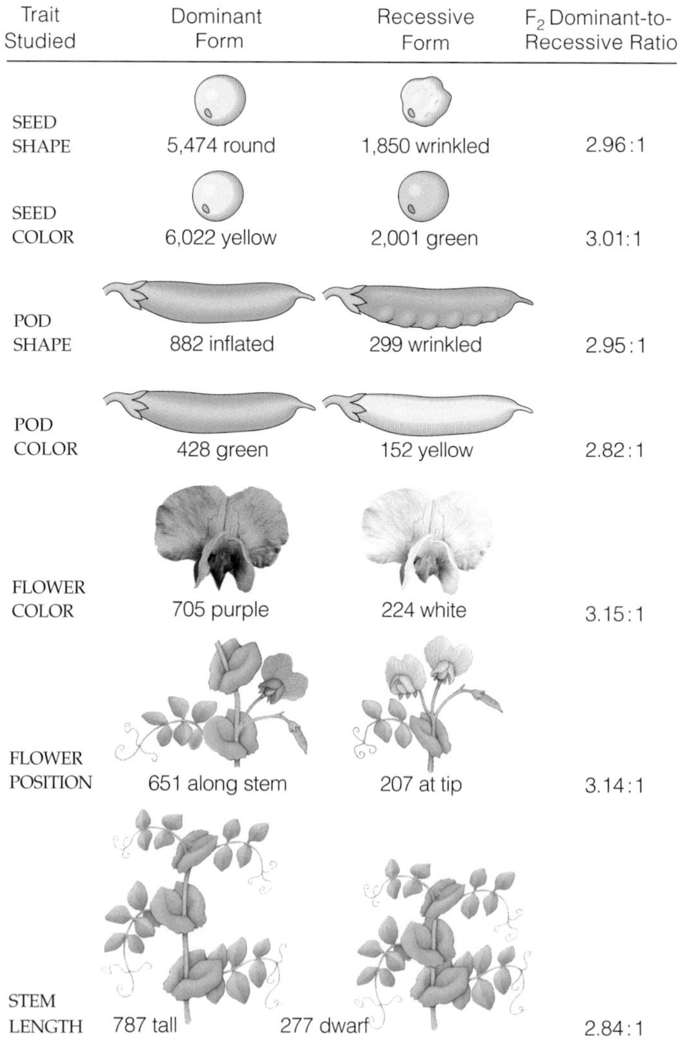

Figure 11.6 *Right*, from some of Mendel's monohybrid experiments with pea plants, counts of F_2 offspring having dominant or recessive hereditary "units" (alleles). On average, the 3:1 phenotypic ratio held for traits.

Trait Studied	Dominant Form	Recessive Form	F_2 Dominant-to-Recessive Ratio
SEED SHAPE	5,474 round	1,850 wrinkled	2.96 : 1
SEED COLOR	6,022 yellow	2,001 green	3.01 : 1
POD SHAPE	882 inflated	299 wrinkled	2.95 : 1
POD COLOR	428 green	152 yellow	2.82 : 1
FLOWER COLOR	705 purple	224 white	3.15 : 1
FLOWER POSITION	651 along stem	207 at tip	3.14 : 1
STEM LENGTH	787 tall	277 dwarf	2.84 : 1

A **Punnett-square method**, explained and applied in Figure 11.7, shows the possibilities. If half of a plant's sperm or eggs are *a* and half are *A*, then we can expect four outcomes with each fertilization:

POSSIBLE EVENT	PROBABLE OUTCOME
sperm *A* meets egg *A*	1/4 *AA* offspring
sperm *A* meets egg *a*	1/4 *Aa*
sperm *a* meets egg *A*	1/4 *Aa*
sperm *a* meets egg *a*	1/4 *aa*

Each F_2 plant has 3 chances in 4 of inheriting at least one dominant *A* allele (purple flowers). It has 1 chance in 4 of inheriting two recessive *a* alleles (white flowers). That is a probable phenotypic ratio of 3:1.

Mendel's observed ratios were not *exactly* 3:1. Yet he put aside the deviations. To understand why, flip a coin several times. As we all know, a coin is as likely to end up heads as tails. But often it ends up heads, or tails, several times in a row. If you flip the coin only a few times, the observed ratio might differ greatly from the predicted ratio of 1:1. Flip it many times, and you are more likely to approach the predicted ratio.

That is why Mendel used rules of probability and counted so many offspring. He minimized sampling error deviations in the observed results.

TESTCROSSES

Testcrosses supported Mendel's prediction. In such experimental tests, an organism shows dominance for a specified trait but its genotype may be unknown, so it is crossed with a homozygous recessive individual. The test results may reveal whether it is homozygous dominant or heterozygous.

For example, Mendel crossed F_1 purple-flowered plants with true-breeding white-flowered plants. If all were homozygous dominant, then F_2 offspring would all be purple flowered. If heterozygous, only about half would be. As it happened, about half of the testcross offspring had purple flowers (*Aa*) and half had white (*aa*). To predict outcomes of this testcross, construct a Punnett square.

The results from Mendel's monohybrid experiments became the basis of a theory of **segregation**, which we state here in modern terms:

> MENDEL'S THEORY OF SEGREGATION *Diploid cells have pairs of genes, on pairs of homologous chromosomes. The two genes of each pair are separated from each other during meiosis, so they end up in different gametes.*

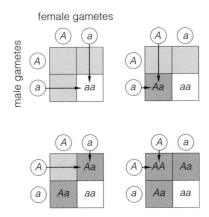

female gametes

male gametes

a Step-by-step construction of a Punnett square. Circles signify gametes. *A* and *a* signify a dominant and recessive allele, respectively. Possible genotypes among offspring are written in the squares.

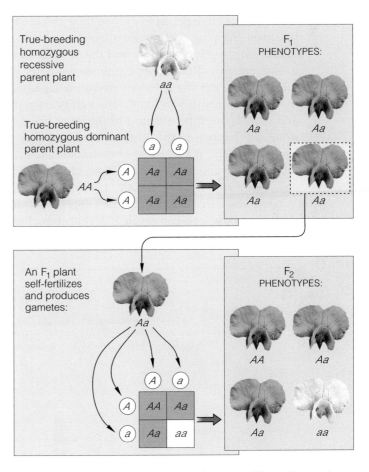

b Cross between two plants that breed true for different forms of a trait, followed by a monohybrid cross between their F_1 offspring.

Figure 11.7 *Animated!* (**a**) Punnett-square method of predicting probable outcomes of genetic crosses. (**b**) Results from one of Mendel's monohybrid experiments. On average, the ratio of dominant-to-recessive that showed up among second-generation (F_2) plants was 3:1.

11.3 Mendel's Theory of Independent Assortment

Mendel used dihybrid experiments to explain how two pairs of genes are sorted into gametes.

Dihybrid experiments start with a cross between true-breeding homozygous parents that differ in two traits governed by alleles of two genes. The F$_1$ offspring are all heterozygous for the alleles of both genes.

Let's duplicate one of Mendel's dihybrid crosses for flower color (alleles *A* or *a*) and for height (*B* or *b*):

True-breeding parents: *AABB* X *aabb*

Gametes: *AB AB ab ab*

F$_1$ hybrid offspring: *AaBb*

As Mendel would have predicted, F$_1$ offspring from this cross are all purple-flowered and tall (*AaBb*).

How will genes that control these traits assort in the F$_1$ plants? It depends in part on their chromosome locations. Suppose that the *Aa* alleles are on one pair of homologous chromosomes and the *Bb* alleles are on a different pair. Remember, chromosome pairs align midway between the spindle poles at metaphase I of meiosis (Figures 10.5 and 11.8). The pair bearing the *A* and *a* alleles will be tethered to opposite poles. The same will happen to the other chromosome pair that bears the *B* and *b* alleles. After meiosis, there can be four possible combinations of alleles in the sperm or eggs that form: 1/4 *AB*, 1/4 *Ab*, 1/4 *aB*, and 1/4 *ab*.

Given the alternative metaphase I alignments, many allelic combinations can result at fertilization. Simple

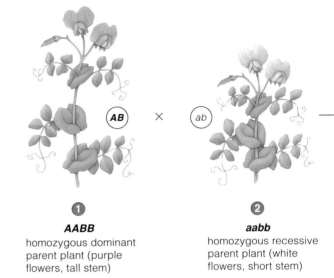

①
AABB
homozygous dominant parent plant (purple flowers, tall stem)

②
aabb
homozygous recessive parent plant (white flowers, short stem)

Figure 11.9 *Animated!* Results from one of Mendel's dihybrid experiments with the garden pea plant. The parent plants were true-breeding for different versions of two traits: flower color and plant height. *A* and *a* signify the dominant and recessive alleles for flower color. *B* and *b* signify dominant and recessive alleles for height. The Punnett square on the facing page shows all of the allelic combinations possible in the F$_2$ generation.

Adding up the corresponding F$_2$ phenotypes, we get:

- ☐ 9/16 or 9 purple-flowered, tall
- ☐ 3/16 or 3 purple-flowered, dwarf
- ☐ 3/16 or 3 white-flowered, tall
- ☐ 1/16 or 1 white-flowered, dwarf

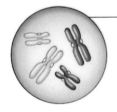

Nucleus of a diploid (2*n*) reproductive cell with two pairs of homologous chromosomes

Figure 11.8 An example of independent assortment at meiosis. Either chromosome of a pair may get tethered to either spindle pole. When just two pairs are tracked, two different metaphase I alignments are possible.

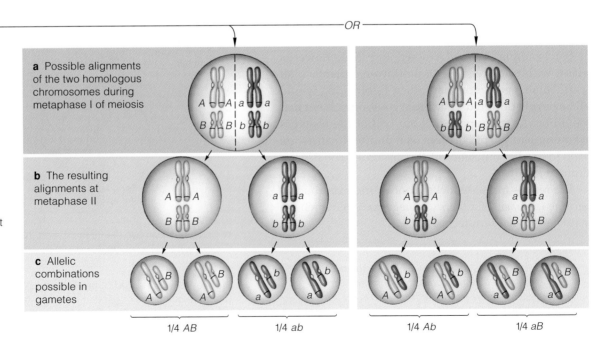

a Possible alignments of the two homologous chromosomes during metaphase I of meiosis

b The resulting alignments at metaphase II

c Allelic combinations possible in gametes

1/4 *AB*   1/4 *ab*   1/4 *Ab*   1/4 *aB*

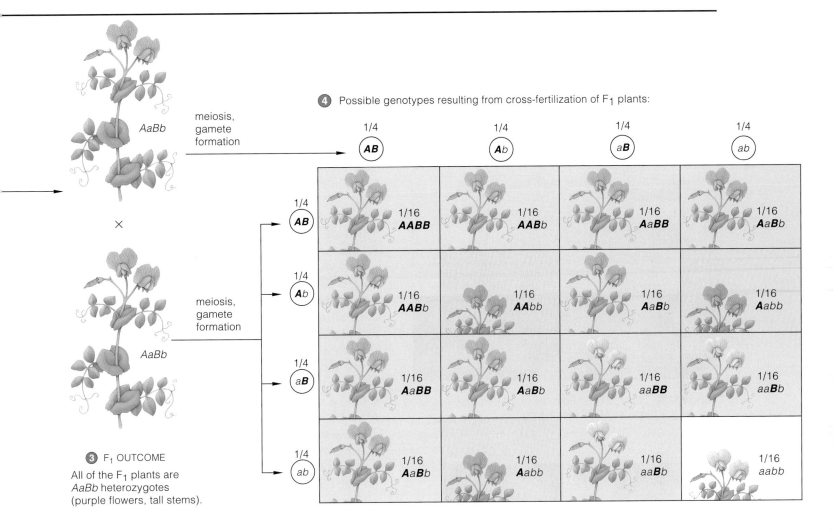

4 Possible genotypes resulting from cross-fertilization of F₁ plants:

AaBb

meiosis, gamete formation

×

meiosis, gamete formation

AaBb

3 F₁ OUTCOME

All of the F₁ plants are *AaBb* heterozygotes (purple flowers, tall stems).

	1/4 **AB**	1/4 **A**b	1/4 **aB**	1/4 ab
1/4 **AB**	1/16 **AABB**	1/16 **AAB**b	1/16 **A**a**BB**	1/16 **A**a**B**b
1/4 **A**b	1/16 **AAB**b	1/16 **AA**bb	1/16 **A**a**B**b	1/16 **A**abb
1/4 **aB**	1/16 **A**a**BB**	1/16 **A**a**B**b	1/16 aa**BB**	1/16 aa**B**b
1/4 ab	1/16 **A**a**B**b	1/16 **A**abb	1/16 aa**B**b	1/16 aabb

multiplication (four sperm types × four egg types) tells us that sixteen combinations of gametes are possible among F₂ offspring of a dihybrid cross (Figure 11.9).

Adding all possible phenotypes gives us a ratio of 9:3:3:1. We can expect to see 9/16 tall purple-flowered, 3/16 dwarf purple-flowered, 3/16 tall white-flowered, and 1/16 dwarf white-flowered F₂ plants. The results from the dihybrid experiment that Mendel reported were close to this ratio.

Mendel analyzed the numerical results from such experiments, but he did not know that seven pairs of homologous chromosomes carry a pea plant's "units" of inheritance. He could only hypothesize that two units for flower color were sorted out into gametes independently of the two units for height.

In time, his hypothesis became known as the theory of **independent assortment**. In modern terms, after meiosis ends, the genes on each pair of homologous chromosomes are sorted into gametes independently of how genes on other pairs of homologues are sorted out. Independent assortment and segregation give rise to genetic variation. In a monohybrid cross for one gene pair, three genotypes are possible: *AA*, *Aa*,

and *aa*. We represent this as 3^n, where *n* is the number of gene pairs. The more pairs, the more combinations are possible. If, say, the parents differ in twenty gene pairs, the number approaches 3.5 billion!

In 1866 Mendel published his work. Apparently his article was read by few and understood by no one. In 1871 he became monastery abbot, and his pioneering experiments ended. He died in 1884, never to know that his experiments would be the starting point for modern genetics. Mendel's theory of segregation still stands for most genes in most organisms: the units of hereditary material (genes) do retain their identity all through meiosis. However, his theory of independent assortment requires qualification, because the alleles of gene pairs do not *always* assort independently into gametes, as Section 11.5 explains.

MENDEL'S THEORY OF INDEPENDENT ASSORTMENT *As meiosis ends, genes on pairs of homologous chromosomes have been sorted out for distribution into one gamete or another, independently of gene pairs on other chromosomes.*

11.4 More Patterns Than Mendel Thought

LINKS TO
SECTIONS
4.6, 6.6

Mendel happened to focus on traits that have clearly dominant and recessive forms. However, expression of genes for some traits is not as straightforward.

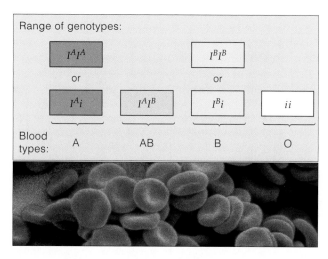

Figure 11.10 *Animated!* Possible allelic combinations that are the basis for ABO blood typing.

Cross two of the F₁ plants, and the F₂ offspring will show three phenotypes in a 1:2:1 ratio:

Figure 11.11 Incomplete dominance in heterozygous (*pink*) snapdragons, in which an allele that affects red pigment is paired with a "white" allele.

CODOMINANCE IN ABO BLOOD TYPES

In *codominance*, a pair of nonidentical alleles affecting two phenotypes are both expressed at the same time in heterozygotes. For example, red blood cells have a type of glycolipid at the plasma membrane that helps give them their unique identity. The glycolipid comes in slightly different forms. An analytical method, *ABO blood typing*, reveals which form a person has.

An enzyme dictates the glycolipid's final structure. Three alleles for this enzyme are present in all human populations. Two, I^A and I^B, are codominant when paired. (These superscripts represent two dominant alleles for the gene.) The third allele, *i*, is recessive when paired with I^A or I^B. The occurrence of three or more alleles for a single gene locus among individuals of a population is called a **multiple allele system**.

Each of these glycolipid molecules was assembled in the endomembrane system (Section 4.6). First, an oligosaccharide chain was attached to a lipid, then a series of sugars was attached to the chain. But alleles I^A and I^B specify different forms of the enzyme that attaches the last sugar. The two attach *different* sugars, which gives the glycolipid a different identity: A or B.

If you have $I^A I^A$ or $I^A i$, your blood is type A. With $I^B I^B$ or $I^B i$, it is type B. With codominant alleles $I^A I^B$, it is AB; you have both versions of the sugar-attaching enzyme. If you are (*ii*), the glycolipid molecules never did get a final sugar on the side chain, so your blood type is not A or B. It is O. Figure 11.10 is a simple way to think about these combinations.

INCOMPLETE DOMINANCE

In *incomplete* dominance, one allele of a pair is not fully dominant over its partner, so the heterozygote's phenotype is *somewhere between* the two homozygotes. Cross true-breeding red and white snapdragons and their F₁ offspring will be pink-flowered. Cross two F₁ plants and you can expect to see red, white, and *pink* flowers in a particular ratio (Figure 11.11). Why? Red snapdragons have two alleles that let them make a lot of molecules of a red pigment. White snapdragons have two mutant alleles and are pigment-free. Pink snapdragons have a "red" allele and a "white" allele; these genotypes have not "blended." Heterozygotes make enough pigment to color flowers pink, not red.

Two interacting gene pairs also can give rise to a phenotype that neither produces by itself. In chickens, interactions among alleles at the *R* and *P* gene loci specify walnut, rose, pea, and single combs, as shown in Figure 11.12.

EPISTASIS

Traits also arise through **epistasis**: interactions among products of two or more gene pairs. Two alleles might mask expression of another gene's alleles, and some expected phenotypes might not appear at all.

As an example, several gene pairs govern whether a Labrador retriever has black, yellow, or brown fur (Figure 11.13). Its coat color depends on how enzymes and other products of alleles at more than one gene locus make a dark pigment, melanin, and deposit it in tissues. Allele *B* (black) is dominant to *b* (brown). At a different locus, allele *E* promotes melanin deposition but two recessive alleles (*ee*) reduce it. In this case, fur appears yellow regardless of alleles at the *B* locus.

SINGLE GENES WITH A WIDE REACH

Alleles at one locus on a chromosome may affect two or more traits in good or bad ways, an outcome called **pleiotropy**. Many genetic disorders, including cystic fibrosis, sickle-cell anemia, and Marfan syndrome, are examples. *Marfan syndrome* arises from an autosomal dominant mutation of the gene for fibrillin, a protein in the most abundant, widespread vertebrate tissues —connective tissues. Thin, loose or crosslinked strands of fibrillin passively recoil after being stretched, as by the beating heart.

Altered fibrillin weakens the connective tissues in 1 of 10,000 men and women and puts the heart, blood vessels, skin, lungs, and eyes at risk. One mutation disrupts the synthesis of fibrillin 1, its secretion from cells, and its tissue deposition. It alters the structure and function of smooth muscle cells inside the wall of the aorta, a big vessel carrying blood out of the heart. Immune cells infiltrate and multiply inside the wall's lining. Calcium deposits accumulate and inflame the wall. Elastic fibers split into fragments. The aorta wall, thinned and weakened, can rupture abruptly during strenuous exercise. Until recent advances in medicine, Marfan syndrome killed most affected people before their fifties. Flo Hyman was one of them (Figure 11.14).

An allele at a given gene locus may be fully dominant, incompletely dominant, or codominant with its partner on a homologous chromosome.

Some gene products may interact with each other and influence the same trait through epistasis.

A single gene's product may have pleiotropic effects, or positive or negative impact on two or more traits.

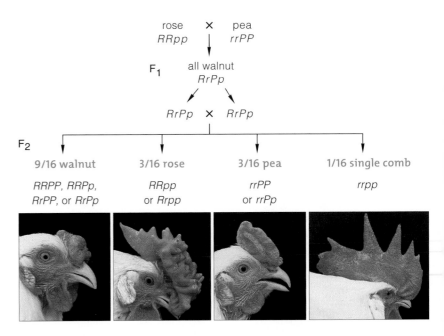

Figure 11.12 Polygenic inheritance in chickens. Interactions among alleles at two gene loci have variable effects on the comb on a chicken's head. The first cross is between a Wyandotte (rose comb) and a Brahma (pea comb).

a BLACK LABRADOR **b** YELLOW LABRADOR **c** CHOCOLATE LABRADOR

Figure 11.13 Coat color among Labrador retrievers. The trait arises through epistatic interactions among alleles of two genes.

Figure 11.14 Flo Hyman, left, captain of the United States volleyball team that won an Olympic silver medal in 1984. Two years later, at a game in Japan, she slid to the floor and died. A dime-sized weak spot in the wall of her aorta had burst. We know at least two affected college basketball stars also died abruptly as a result of Marfan syndrome.

11.5 Impact of Crossing Over on Inheritance

LINKS TO
SECTIONS
3.2, 10.3, 10.4

Crossing over between homologous chromosomes is one of the main pattern-busting events in inheritance.

We now know there are many genes on each type of autosome and sex chromosome. All the genes on one chromosome are called a **linkage group**. For instance, the fruit fly (*Drosophila melanogaster*) has four linkage groups, corresponding to its four pairs of homologous chromosomes. Indian corn (*Zea mays*) has ten linkage groups, corresponding to its ten pairs, and so on.

If genes on the same chromosome stayed together through meiosis, then there would be no surprising mixes of parental traits. You could expect parental phenotypes among, say, the F$_2$ offspring of dihybrid experiments to show up in a predictable ratio. As early experiments with fruit flies showed, however, that ratio was often predictably different for linked genes. In one dihybrid experiment, 17 percent of the F$_2$ offspring inherited a new combination of alleles that did not occur in either of their parents.

Many genes on the same chromosome do not stay linked through meiosis, but some stay together more often than others. Why? They are closer together on the chromosome, and so they are separated less often by crossing over. *The probability that crossing over will disrupt the linkage between any two genes is proportional to the distance between the two genes.*

If genes *A* and *B* are twice as far apart as genes *C* and *D* on a chromosome, then we can expect crossing over to disrupt the linkage between genes *A* and *B* more frequently than between the other two genes:

Two genes are very closely linked when the distance between them is small. Their combinations of alleles nearly always end up in the same gamete. Linkage is more vulnerable to crossing over when the distance between two gene loci is greater (Figure 11.15). When two loci are far apart, crossing over is so frequent that the genes assort independently into gametes.

Human gene linkages were identified by tracking DNA inheritance in families over the generations. One thing is clear from such studies: Crossovers are not rare. For most eukaryotes, meiosis cannot even be completed properly until at least one crossover occurs between each pair of homologous chromosomes.

> *All of the genes at different locations along the length of a chromosome belong to the same linkage group.*
>
> *Crossing over between homologous chromosomes disrupts gene linkages and results in nonparental combinations of alleles in chromosomes.*
>
> *The farther apart two genes are on a chromosome, the greater will be the frequency of crossing over and genetic recombination between them.*

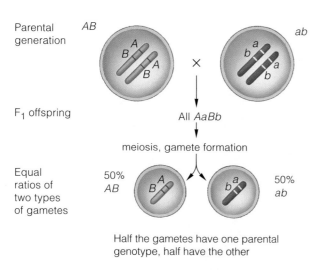

a Full linkage between two genes; no crossing over. Genes very close together along the length of the same chromosome typically stay together during gamete formation.

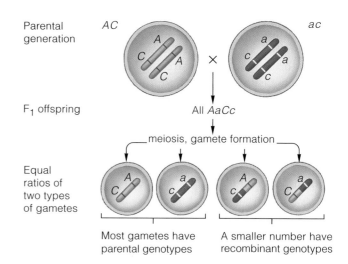

b Incomplete linkage; crossing over affected the outcome. Any two genes that are far apart along the length of a chromosome are more vulnerable to crossing over.

**Figure 11.15 *Animated!* Examples of outcomes of crossing over between two gene loci:
(a) full linkage and (b) incomplete linkage.**

11.6 Genes and the Environment

The environment often contributes to variable gene expression among a population's individuals.

Possibly you have noticed a Himalayan rabbit's coat color. Like a Siamese cat, this mammal has dark hair in some parts of its body and lighter hair in others. The Himalayan rabbit is homozygous for the c^h allele of the gene specifying tyrosinase. Tyrosinase is one of the enzymes involved in melanin production. The c^h allele specifies a heat-sensitive form of this enzyme. This form is active only when the temperature around body cells is below 33°C, or 91°F.

When cells that give rise to this rabbit's hair grow under warmer conditions, they cannot make melanin, so hairs appear light. This happens in body regions that are massive enough to conserve a fair amount of metabolic heat. The ears and other slender extremities tend to lose metabolic heat faster, so they are cooler. Figure 11.16 shows one experiment that demonstrated how the environmental temperature can influence the production of melanin.

One classic experiment identified environmental effects on yarrow plants. These plants can grow from cuttings, so they are a useful experimental organism. Why? Cuttings from the same plant all have the same genotype, so experimenters can discount genes as a basis for differences that show up among them.

In this study, cuttings (clones) from each of several yarrow plants were grown at three elevations. The researchers periodically observed the growth of the plants in their habitats. They found that cuttings from the same parent plants grew differently at different altitudes. For example, cuttings from one plant grew tall at the lowest and the highest elevation, but a third cutting remained short at mid-elevation (Figure 11.17). Even though these plants were genetically identical, their phenotypes differed in different environments.

Similarly, plant a hydrangea in a garden and it may have pink flowers instead of the expected blue ones. Soil acidity affects the function of gene products that color hydrangea flowers.

What about humans? One of our genes codes for a transporter protein that moves serotonin across the plasma membrane of brain cells. This gene product has several effects, one of which is to counter anxiety and depression when traumatic events challenge us. For a long time, researchers have known that some people handle stress without getting too upset, while others spiral into a deep and lasting depression.

Mutation of the gene for the serotonin transporter compromises responses to stress. It is as if some of us are bicycling through life without an emotional helmet.

Figure 11.16 *Animated!* Observable effect of an environmental factor that alters gene expression. A Himalayan rabbit normally has black hair only on its long ears, nose, tail, and leg regions farthest from the body mass. In one experiment, a patch of a rabbit's white coat was removed and an icepack was placed over the hairless patch. Where the colder temperature had been maintained, the hairs that grew back were black.

Himalayan rabbits are homozygous for an allele that encodes a mutant version of tyrosinase, an enzyme required to make melanin. As described in the text, this allele encodes a heat-sensitive form of the enzyme, which functions only when air temperature is below about 33°C.

a Mature cutting at high elevation (3,060 meters above sea level)

b Mature cutting at mid-elevation (1,400 meters above sea level)

c Mature cutting at low elevation (30 meters above sea level)

Figure 11.17 Experiment demonstrating the impact of environmental conditions of three different habitats on phenotype in yarrow (*Achillea millefolium*). Cuttings from the same parent plant were grown in the same kind of soil at three different elevations.

Only when we take a fall does the phenotypic effect—depression—appear. Other genes also affect emotional states, but mutation of this particular gene reduces our capacity to snap out of it when bad things happen.

Variation in traits arises not only from gene mutations and interactions, but also in response to variations in environmental conditions that each individual faces.

11.7 Complex Variations in Traits

For most populations or species, individuals show rich variation for many of the same traits. Sometimes the phenotypes cannot be predicted, and most of the time they are part of a continuous range of variation.

REGARDING THE UNEXPECTED PHENOTYPE

Think back on Mendel's dihybrid crosses. Nearly all of the traits that he tracked occurred in predictable ratios because the two genes happened to be on different chromosomes or far apart on the same chromosome. They tended to segregate cleanly. Track two or more different pairs of genes—as Mendel did—and you might observe phenotypes that you would not have predicted at all. And not all of the variation is a result of tight linkage or crossing over.

As one example, *camptodactyly*, a rare abnormality, affects the shape and movement of fingers. Some of the people who carry a mutant allele for this heritable trait have immobile, bent fingers on both hands. Others have immobile, bent fingers on the left or right hand only. Fingers of still other people who have the mutant allele are not affected in any obvious way at all.

What causes such odd variation? Remember, most organic compounds are synthesized by a sequence of metabolic steps. *Different enzymes, each a gene product, control different steps.* One gene may have mutated in a number of ways. A gene product might be blocking some pathway or making it run nonstop or not long enough. Perhaps poor nutrition or some other variable factor in the individual's environment is influencing the activity of one of the pathway's enzymes. Such variable factors can introduce big or small variations even in otherwise expected phenotypes.

CONTINUOUS VARIATION IN POPULATIONS

Another point: Individuals of populations generally show a range of small differences in most traits. This feature of natural populations is known as **continuous variation**. It arises through **polygenic inheritance**, or the inheritance of multiple genes that affect the same trait. The distribution of all forms of a trait becomes more and more continuous when greater numbers of genes and environmental factors are involved.

Look in a mirror at your eye color. The colored part is the iris, a doughnut-shaped, pigmented structure just under the cornea (Figure 11.18). The color results from several gene products. Some products help make and distribute different kinds and amounts of melanins, which are similar to the light-absorbing pigment that affects coat color in mammals. Almost black irises have

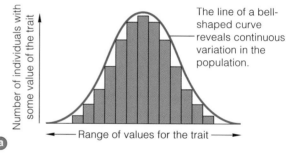

The line of a bell-shaped curve reveals continuous variation in the population.

a ← Range of values for the trait →

Number of individuals with some value of the trait

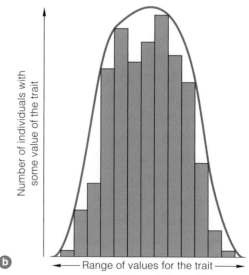

b ← Range of values for the trait →

Number of individuals with some value of the trait

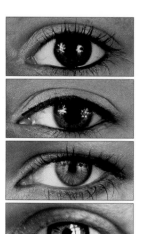
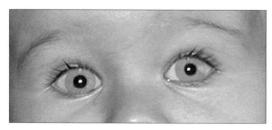

Figure 11.18 Sampling of the range of continuous variation in human eye color. Products of different gene pairs interact in making and distributing the pigment melanin, which helps color the iris. Small color differences arise from different combinations of alleles. The frequency distribution for the eye-color trait is continuous over a far larger range than this, from black to light blue.

Figure 11.19 *Animated!* Continuous variation. (**a**) A bar graph can reveal continuous variation in a population. The proportion of individuals in each category is plotted against the range of measured phenotypes. (**b**) The curved line above this particular set of bars is a real-life example of a bell-shaped curve that emerged for the population in Figure 11.20. It reflects continuous variation in body height, one of the traits that help characterize human populations.

dense melanin deposits, which can absorb most of the incoming light. Deposits are not as extensive in brown eyes, so some unabsorbed light is reflected out. Light brown or hazel eyes have even less melanin.

Green, gray, or blue eyes have lesser amounts of the pigments. Many or most of the blue wavelengths of light that enter the eyeball are simply reflected out.

How can you describe the continuous variation of some trait in a group? Divide the range of phenotypes for a trait—say, height—into measurable categories, such as numbers of inches. Next, do a count of how many individuals fall into each category; this will give you the relative frequencies of phenotypes across the range of measurable values. Finally, plot out the data as a bar chart, such as the one in Figure 11.19a.

In this figure, the shortest bars represent categories having the fewest individuals. The tallest bar signifies the category that has the most individuals. In this case, a graph line skirting the top of all of the bars will be a bell-shaped curve. Such **bell curves** are typical of

any trait showing continuous variation. Figure 11.19b is a bell curve based on real-life measurements at the University of Florida (Figure 11.20).

And so we conclude this chapter, which introduces heritable and environmental factors that give rise to great variation in traits. What is the take-home lesson? Simply this: An individual's phenotype is an outcome of complex interactions among its genes, enzymes and other gene products, and the environment. Chapter 18 will consider some of the evolutionary consequences.

Enzymes and other gene products control each step of most metabolic pathways. Mutations, interactions among genes, and environmental conditions typically affect one or more steps in ways that contribute to variation in phenotypes.

Individuals of populations or species show continuous variation—a range of small differences. Usually, the more genes and environmental factors that influence a trait, the more continuous the distribution of phenotypes.

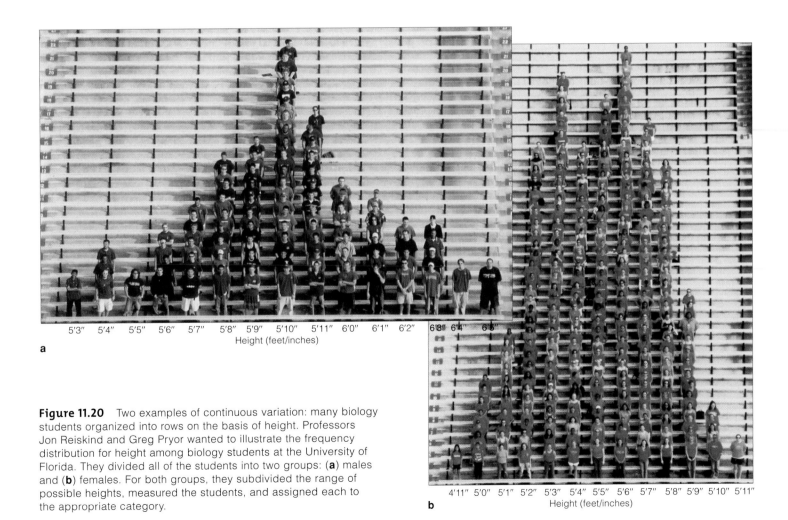

5'3" 5'4" 5'5" 5'6" 5'7" 5'8" 5'9" 5'10" 5'11" 6'0" 6'1" 6'2" 6'3" 6'4" 6'5"

Height (feet/inches)

a

4'11" 5'0" 5'1" 5'2" 5'3" 5'4" 5'5" 5'6" 5'7" 5'8" 5'9" 5'10" 5'11"

Height (feet/inches)

b

Figure 11.20 Two examples of continuous variation: many biology students organized into rows on the basis of height. Professors Jon Reiskind and Greg Pryor wanted to illustrate the frequency distribution for height among biology students at the University of Florida. They divided all of the students into two groups: (**a**) males and (**b**) females. For both groups, they subdivided the range of possible heights, measured the students, and assigned each to the appropriate category.

Summary

Section 11.1 Genes are heritable units of information about traits. Each gene has its own locus, or location, along the length of a particular chromosome. Different molecular forms of the same gene are known as alleles.

By experimenting with garden pea plants, Mendel was the first to gather evidence of patterns by which genes are transmitted from parents to offspring.

Offspring of a cross between two individuals that breed true for different forms of a trait are hybrids; each inherited nonidentical alleles for a trait being studied.

An individual with two dominant alleles for a trait (*AA*) is homozygous dominant. A homozygous recessive has two recessive alleles (*aa*). A heterozygote has two nonidentical alleles (*Aa*) for a trait. A dominant allele may mask the effect of a recessive allele partnered with it on the homologous chromosome.

Genotype refers to the particular alleles at any or all gene locations on an individual's chromosomes. *Phenotype* refers to an individual's observable traits.

Biology⊗Now
Learn how Mendel crossed garden pea plants and the definitions of important genetic terms on BiologyNow.

Section 11.2 A cross between parents of different genotypes yields hybrid offspring. For monohybrid experiments, two parents that bred true for different forms of a trait produce F_1 heterozygotes that are identical for one pair of genes. Mendel's monohybrid experiments gave indirect evidence that some forms of a gene may be dominant over recessive forms.

All F_1 offspring of a parental cross *AA* x *aa* were *Aa*. Crosses between F_1 monohybrids resulted in these allelic combinations among the F_2 offspring:

	A	a
A	*AA*	*Aa*
a	*Aa*	*aa*

AA (dominant)
Aa (dominant)   the expected
Aa (dominant)   phenotypic
aa (recessive) ratio of 3:1

Mendel's monohybrid experiment results led to a theory of segregation: Diploid organisms have pairs of genes, on pairs of homologous chromosomes. Genes of each pair segregate from each other at meiosis, so each gamete formed gets one or the other gene.

Biology⊗Now
Carry out monohybrid experiments with the interaction on BiologyNow.

Section 11.3 Dihybrid experiments start with a cross between true-breeding heterozygous parents that differ for alleles of two genes (*AABB* x *aabb*). All F_1 offspring are heterozygous for both genes (*AaBb*). In Mendel's dihybrid experiments, phenotypes of the F_2 offspring of F_1 hybrids were close to a 9:3:3:1 ratio:

9 dominant for both traits
3 dominant for *A*, recessive for *b*
3 dominant for *B*, recessive for *a*
1 recessive for both traits

His results support a theory of independent assortment: Before gamete formation, meiosis assorts gene pairs of homologous chromosomes independently of how gene pairs of all the other chromosomes are sorted. Random alignment of all pairs of homologous chromosomes at metaphase I is the basis of this outcome.

Biology⊗Now
Observe the results of a dihybrid cross with the interaction on BiologyNow.

Section 11.4 Inheritance patterns are not always straightforward.

Some alleles are not fully dominant over their partner allele on the homologous chromosomes, and both are expressed at the same time. The phenotype that results from this allelic combination is somewhere between the two homozygous conditions.

Some alleles are codominant and are expressed at the same time in heterozygotes. An example occurs in the multiple allele system underlying ABO blood typing.

Also, products of one or more genes commonly interact in ways that influence the same trait, and a single gene may have effects on two or more traits.

Biology⊗Now
Explore patterns of non-Mendelian inheritance with the interactions on BiologyNow.

Section 11.5 A linkage group consists of all genes along the length of one chromosome. Crossing over between pairs of homologous chromosomes disrupts expected inheritance patterns by breaking linkages. Its outcome is nonparental combinations of alleles in gametes. The farther apart two genes are on a chromosome, the greater will be the frequency of crossing over and genetic recombination between them.

Section 11.6 Environmental factors also can alter how genes are expressed in individuals of a population. An example is a difference in temperature that affects the activity of a heat-sensitive form of an enzyme—a gene product—that helps produce a coat color pigment.

Biology⊗Now
See how the environment can affect phenotype with animation on BiologyNow.

Section 11.7 Gene interactions and environmental factors influence many enzymes differently among individuals, and many phenotypes result. They also contribute to small, incremental differences—a range of continuous variation—in a population.

Biology⊗Now
Plot the continuous distribution of height for a class with the interaction on BiologyNow.

Self-Quiz *Answers in Appendix II*

1. Alleles are _____ .
   a. different molecular forms of a gene
   b. different phenotypes
   c. self-fertilizing, true-breeding homozygotes

2. A heterozygote has a _____ for a trait being studied.
   a. pair of identical alleles
   b. pair of nonidentical alleles
   c. haploid condition, in genetic terms

3. The observable traits of an organism are its _____ .
   a. phenotype   c. genotype
   b. sociobiology   d. pedigree

4. Second-generation offspring of a cross between parents who are homozygous for different alleles are the _____ .
   a. F_1 generation   c. hybrid generation
   b. F_2 generation   d. none of the above

5. F_1 offspring of the cross $AA \times aa$ are _____ .
   a. all AA c. all Aa
   b. all aa d. 1/2 AA and 1/2 aa

6. Refer to Question 5. Assuming complete dominance, the F_2 generation will show a phenotypic ratio of _____ .
   a. 3:1   b. 9:1   c. 1:2:1   d. 9:3:3:1

7. Crosses between two dihybrid F_1 pea plants, which are offspring from a parental cross $AABB \times aabb$, result in F_2 phenotypic ratios close to _____ .
   a. 1:2:1   b. 3:1   c. 1:1:1:1   d. 9:3:3:1

8. The probability of a crossover occurring between two genes on the same chromosome is _____ .
   a. unrelated to the distance between them
   b. increased if they are close together
   c. increased if they are far apart

9. Two genes that are close together on the same chromosome are _____ .
   a. linked   c. homologous   e. all of the
   b. identical alleles d. autosomes   above

10. Match each example with the most suitable description.
   ____ dihybrid experiment   a. bb
   ____ monohybrid experiment   b. $AABB \times aabb$
   ____ homozygous condition   c. Aa
   ____ heterozygous condition   d. $Aa \times Aa$

Additional questions are available on **Biology Now™**

Genetics Problems   *Answers in Appendix III*

1. A gene encodes the second enzyme in a melanin-synthesizing pathway. An individual who is homozygous for a recessive mutant allele of this gene cannot produce or deposit melanin in body tissues. *Albinism*, the absence of melanin, is the result.

   Humans and a number of other organisms can have this phenotype. Figure 11.21 shows two examples. In the following situations, what are the possible genotypes of the father, the mother, and their children?

   a. Both parents have normal phenotypes; some of their children are albino and others are unaffected.

   b. Both parents are albino and have albino children.

   c. The woman is unaffected, the man is albino, and they have one albino child and three unaffected children.

2. As rose breeders know, several alleles influence specific traits, such as long, symmetrical, urn-shaped buds, double flowers, glossy leaves, and resistance to mildew (Figure 11.22). Alleles of a single gene govern whether a plant will

Figure 11.21 Two albino organisms. By not posing his subjects as objects of ridicule, the photographer of human albinos is attempting to counter the notion that there is something inherently unbeautiful about them.

dominant   dominant

recessive   recessive

Figure 11.22 (**a**) Climbing rose and (**b**) shrub rose. (**c**) Globe-shaped buds versus (**d**) urn-shaped buds.

be a climber (dominant) or shrubby (recessive). When a true-breeding climber is crossed with a shrubby plant, all F_1 offspring are climbers. If an F_1 plant is crossed with a shrubby plant, about 50 percent of the offspring will be shrubby and 50 percent will be climbers. Using symbols A and a to represent the dominant and recessive alleles, make a Punnett-square diagram of the expected genotypic and phenotypic outcomes in the F_1 offspring and the offspring of the cross between an F_1 plant and a shrubby plant.

Figure 11.23 The Manx, a breed of cat that has no tail.

3. One gene has alleles *A* and *a*. Another has alleles *B* and *b*. For each of the following genotypes, what type(s) of gametes will form, assuming independent assortment during meiosis occurs?

 a. *AABB*   c. *Aabb*

 b. *AaBB*   d. *AaBb*

4. Refer to Problem 3. What will be the genotypes of offspring from the following matings? Indicate the frequencies of each genotype among them.

 a. *AABB* × *aaBB*   c. *AaBb* × *aabb*

 b. *AaBB* × *AABb*   d. *AaBb* × *AaBb*

5. Return to Problem 3. Assume you now study a third gene having alleles *C* and *c*. For each genotype listed, what type(s) of gametes will be produced, assuming that independent assortment occurs?

 a. *AABBCC*   c. *AaBBCc*

 b. *AaBBcc*   d. *AaBbCc*

6. Certain alleles are so essential for normal development that an individual who is homozygous recessive for a mutant form cannot survive. Such recessive, *lethal alleles* can be perpetuated in the population by heterozygotes.

 Consider the allele *Manx* (M^L) in cats. Homozygous cats ($M^L M^L$) die when they are still embryos inside the mother cat. In heterozygotes ($M^L M$), the spine develops abnormally. The cats end up with no tail (Figure 11.23).

 Two $M^L M$ cats mate. What is the probability that any one of their *surviving* kittens will be heterozygous?

7. In one experiment, Mendel crossed a pea plant that bred true for green pods with one that bred true for yellow pods. All the F$_1$ plants had green pods. Which form of the trait (green or yellow pods) is recessive? Explain how you arrived at your conclusion.

8. Mendel crossed a pea plant that produced plump and rounded seeds with a pea plant that produced wrinkled seeds. In the F$_1$ generation, all seeds were round. Mendel planted the F$_1$ seeds, which grew into plants that, when self-fertilized, produced 5,474 round seeds and 1,850 wrinkled seeds in the F$_2$ generation. The alleles that govern seed shape are designated *R* and *r*.

 a. What are the genotypes of the parents?

 b. What are the possible outcomes of a cross between a homozygous round-seeded plant and a wrinkle-seeded plant?

9. Mendel crossed a true-breeding tall, purple-flowered pea plant with a true-breeding dwarf, white-flowered plant. All F$_1$ plants were tall and had purple flowers. If an F$_1$ plant self-fertilizes, then what is the probability that a randomly selected F$_2$ offspring will be heterozygous for the genes specifying height and flower color?

10. Suppose you identify a new gene in mice. One of its alleles specifies white fur. A second allele specifies brown fur. You want to determine whether the relationship between the two alleles is one of simple dominance or incomplete dominance. What sorts of genetic crosses would give you the answer? What types of observations would you require to form conclusions?

11. In sweet pea plants, an allele for purple flowers (*P*) is dominant to an allele for red flowers (*p*). An allele for long pollen grains (*L*) is dominant to an allele for round pollen grains (*l*). Bateson and Punnett crossed a plant having purple flowers/long pollen grains with one having white flowers/round pollen grains. All F$_1$ offspring had purple flowers and long pollen grains. In the F$_2$ generation, the researchers observed the following phenotypes:

 296 purple flowers/long pollen grains

 19 purple flowers/round pollen grains

 27 red flowers/long pollen grains

 85 red flowers/round pollen grains

 What is the best explanation for these results?

12. A dominant allele *W* confers black fur on guinea pigs. A guinea pig that is homozygous recessive (*ww*) has white fur. Fred would like to know whether his pet black-furred guinea pig is homozygous (*WW*) or heterozygous (*Ww*). How might he determine his pet's genotype?

13. Red-flowering snapdragons are homozygous for allele R^1. White-flowering snapdragons are homozygous for a different allele (R^2). Heterozygous plants ($R^1 R^2$) bear pink flowers. What phenotypes should appear among first-generation offspring of the crosses listed? What are the expected proportions for each phenotype?

 a. $R^1 R^1$ × $R^1 R^2$ c. $R^1 R^2$ × $R^1 R^2$

 b. $R^1 R^1$ × $R^2 R^2$ d. $R^1 R^2$ × $R^2 R^2$

(In cases of incomplete dominance, alleles are usually designated by superscript numerals, as shown here, not by the uppercase letters for dominance and lowercase letters for recessiveness.)

For each cross, list which of these F$_1$ phenotypes show up as well as the proportion of each:

 a. _____red   _____ pink   _____ white

 b. _____red   _____ pink   _____ white

 c. _____red   _____ pink   _____ white

 d. _____red   _____ pink   _____ white

14. Two pairs of genes affect comb type in chickens (Figure 11.12), and they assort independently. When both are homozygous for recessive alleles, a chicken has a single comb. But a dominant allele of one gene, *P*, gives rise to a pea comb, and a dominant allele of the other gene (*R*) gives rise to a rose comb. An *epistatic* interaction occurs when a chicken has at least one of both dominant alleles, $P\_\_ R \_\_$, which gives rise to a walnut comb.

   Predict the ratios resulting from a cross between two walnut-combed chickens that are heterozygous for both genes (*PpRr*) and list them below:

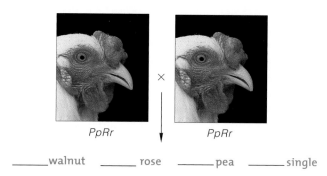

PpRr   ×   *PpRr*

_____walnut _____ rose _____pea _____ single

15. As Section 3.6 explains, a single mutant allele gives rise to an abnormal form of hemoglobin (Hb^S instead of Hb^A). Homozygotes (Hb^SHb^S) develop the genetic disease sickle-cell anemia. Heterozygotes (Hb^AHb^S) show few obvious symptoms.

   A couple who are both heterozygous for the Hb^A allele plan to have children. For *each* of the pregnancies, state the probability that this couple will have a child who is:
   a. homozygous for the Hb^S allele
   b. homozygous for the Hb^A allele
   c. heterozygous Hb^AHb^S

16. Watermelons (*Citrullus*) are important crops around the world (Figure 11.24). A single gene determines the density of green pigment that colors the rind, with solid light green (*g*) recessive to solid dark green (*G*). When a true-breeding plant having a dark-green rind is crossed with a plant having a light-green rind, what fraction of the dark-green F$_2$ offspring is expected to be heterozygous for this trait?

17. The rind of a watermelon that is homozygous for recessive allele *e* bursts, or splits explosively, when cut. Genotype *EE* results in a "nonexplosive" rind that is better for shipping watermelons to market. The rind of a watermelon that is homozygous for the recessive allele *f* has a furrowed surface. A furrowed rind has less market appeal than a smooth rind, which results from expression of dominant allele *F*.

   For one testcross, a dihybrid plant that produces melons with a smooth, nonexplosive rind is crossed with a plant that produces melons with a furrowed, explosive rind. Make a Punnett square of the following results:

   118 smooth, nonexplosive
   112 smooth, explosive
   109 furrowed, nonexplosive
   121 furrowed, explosive

What is the smooth rind/furrowed rind ratio among the testcross offspring? What is the ratio of nonexplosive rind/explosive rind? Are the two gene loci assorting independently of each other?

18. Two pairs of genes determine kernel color in wheat plants. Alleles of one pair show incomplete dominance over the other pair. The product of allele A^1 at one locus produces enough pigment to add a dose of red color to the kernels, but that of allele A^2 does not. The product of allele B^1 at the second locus also adds a dose of red color to the kernels, but that of allele B^2 does not.

   The chart shown below lists the numbers of different wheat kernel colors observed during a recent harvest, together with their corresponding genotypes. Using the information in this table, draw a graph showing the percentage of kernels in the wheat population that inherited each of the five kernel colors.

   Explain why the kernel color in wheat plants shows a varied phenotypic distribution.

Figure 11.24 A sampling of the variation in the rind characteristics of watermelon (*Citrullus*).

Genotype	Phenotype	Number Displaying the Trait	Percentage of Population
$A^1A^1B^1B^1$	Dark red	181	
$A^1A^1B^1B^2$ or $A^1A^2B^1B^1$	Red	360	
$A^1A^2B^1B^2$ or $A^1A^1B^2B^2$ or $A^2A^2B^1B^1$	Salmon	922	
$A^1A^2B^2B^2$ or $A^2A^2B^1B^2$	Pink	358	
$A^2A^2B^2B^2$	White	179	
Totals		2,000	

12 CHROMOSOMES AND HUMAN INHERITANCE

Strange Genes, Richly Tortured Minds

"This man is brilliant." That was the extent of a letter of recommendation from Richard Duffin, a mathematics professor at Carnegie Mellon University. Duffin wrote the line in 1948 on behalf of John Forbes Nash, Jr. (Figure 12.1). Nash was twenty years old at the time and applying for admission to Princeton University's graduate school.

Over the next decade, Nash made his reputation as one of America's foremost mathematicians. He was socially awkward, but so are many highly gifted people. Nash showed no symptoms of paranoid schizophrenia, a mental disorder that eventually debilitated him.

Full-blown symptoms emerged in his thirtieth year. Nash had to abandon his position at the Massachusetts Institute of Technology. Two decades passed before he was able to return to his pioneering work in mathematics.

Of every hundred people worldwide, one is affected by *schizophrenia*. This neurobiological disorder (NBD) is characterized by delusions, hallucinations, disorganized speech, and abnormal social behavior. As researchers know, exceptional creativity often accompanies schizophrenia. It also accompanies other NBDs, including autism, chronic depression, and bipolar disorder, which manifests itself as jarring swings in mood and social behavior.

Compared to the general population, highly intelligent individuals are *less* likely to develop NBDs—unless they also happen to be outside-the-box creative thinkers. Disturbingly creative writers alone are eighteen times more suicidal, ten times more likely to be depressed, and twenty times more likely to have bipolar disorder. Virginia Woolf's suicide after a prolonged mental breakdown is a tragic example.

We now have evidence that even emotionally healthy people who show creative brilliance have more personality traits in common with the mentally impaired than they do with individuals closer to the norm. For instance, they, too, are hypersensitive to environmental stimuli. Some may be on a razor's edge between mental stability and instability. Those who do go on to develop NBDs become part of a crowd that includes Socrates, Newton, Beethoven, Darwin, Lincoln, Poe, Dickens, Tolstoy, van Gogh, Freud, Churchill, Einstein, Picasso, Woolf, Hemingway, and Nash.

We have not yet identified all of the interactions among genes and the environment that might tip such individuals one way or the other. But we do know about several mutant genes that predispose them to develop NBDs.

Creatively gifted people, as well as those affected by NBDs, often turn up in the same family tree—which points

Figure 12.1 John Forbes Nash, Jr., a prodigy who solved problems that had baffled some of the greatest minds in mathematics. His early work in economic game theory won him a Nobel Prize. He is shown here at a premier of *A Beautiful Mind*, an award-winning film based on his battle with schizophrenia. His neural disorder places him in the ranks of other highly creative, distinguished, yet troubled individuals, including Abraham Lincoln, Virginia Woolf, and Pablo Picasso.

to a genetic basis for their special traits. Also, those affected by bipolar disorder and schizophrenia show altered gene expression in certain brain regions. Cells make too many or too few of the enzymes of electron transfer phosphorylation. Remember, this stage of aerobic respiration yields the bulk of the body's ATP. Does its disruption alter brain cells in ways that boost creativity but also invite illness? Perhaps.

With this intriguing connection, we invite you to reflect on how far you have come in this unit of the book. You first surveyed mitotic and meiotic cell divisions. You looked at how chromosomes and genes become shuffled during meiosis and then during fertilization. You also became acquainted with Gregor Mendel's discovery of major patterns of inheritance. This knowledge is your portal to the chromosomal basis of human inheritance.

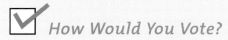
Watch the video online!

How Would You Vote?

Diagnostic tests for predisposition to neurobiological disorders will soon be available. Individuals might use knowledge of their susceptibility to modify choices in life-styles. Insurance companies and employers might also use that information to exclude predisposed but otherwise healthy individuals. Would you support legislation governing these tests? See BiologyNow for details, then vote online.

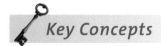
Key Concepts

AUTOSOMES AND SEX CHROMOSOMES

Sexually reproducing species have pairs of autosomes, which are chromosomes that are the same in length, shape, and which genes they carry. Nearly all animals also have a pair of sex chromosomes.

Karyotyping, a diagnostic tool, helps reveal changes in the structure or number of an individual's chromosomes. Section 12.1, 12.2

AUTOSOMAL INHERITANCE

Many alleles on autosomes are expressed in Mendelian patterns of simple dominance and recessiveness. Sections 12.3, 12.4

SEX-LINKED INHERITANCE

The pairing of sex chromosomes in human females (XX) differs from the pairing in males (XY). One of the genes on the Y chromosome dictates gender. Many alleles on the X chromosome are expressed in Mendelian patterns of simple dominance and recessiveness. Sections 12.5–12.7

CHANGES IN CHROMOSOME STRUCTURE

On rare occasions, a chromosome may undergo permanent change in its structure, as when a segment of it is deleted, duplicated, inverted, or translocated. Section 12.8

CHANGES IN CHROMOSOME NUMBER

Also on rare occasions, the parental number of autosomes or sex chromosomes changes. In humans, the change usually results in problems. Section 12.9

HUMAN GENETIC ANALYSIS AND OPTIONS

Various analytical and diagnostic procedures often reveal genetic disorders. Risks and benefits are associated with what individuals as well as society at large do with the information. Sections 12.10, 12.11

Links to Earlier Concepts

You will be drawing on your knowledge of chromosome structure (Sections 9.1, 9.3), meiosis (10.3, 10.4), and gamete formation (10.5). Be sure you understand dominance, recessiveness, and the homozygous and heterozygous conditions (11.1). Remember, environmental factors influence gene expression (11.6). Colchicine (4.10) will turn up again. So will glycolysis (8.2), this time in the context of a genetic disorder. You also will consider whether the hemoglobin family evolved after changes in chromosome structure (3.6).

12.1 Human Chromosomes

LINKS TO
SECTIONS
4.10, 9.1, 9.5, 10.3

You already know quite a bit about chromosomes and their roles in inheritance. Let's now focus on human autosomes and sex chromosomes.

Like nearly all animals, humans normally are male or female. Also like many species, they have a diploid chromosome number (2*n*), meaning that body cells have pairs of homologous chromosomes. Remember, all but one of the pairs are alike in their length, shape, and gene sequence. One member of the last pairing is a unique sex chromosome that is present in males or females, but not in both.

For instance, a diploid cell in a human female has two X chromosomes (XX). A diploid cell in a human male has one X and one Y chromosome (XY). This is a common inheritance pattern among mammals, fruit flies, and many other animals. It is not the only one, however. Among butterflies, moths, birds, and certain fishes, the males have two identical sex chromosomes and females do not.

Human X and Y chromosomes differ physically and in which genes they carry. Recall, from Section 10.3, that each pair of homologous chromosomes synapses (zippers together tightly) in prophase I of meiosis. An X chromosome and Y chromosome synapse in a small region along their length, but that is enough to allow the two to interact as homologues during meiosis.

Human X and Y chromosomes fall into the general category of **sex chromosomes**. As you will see later, when sex chromosomes are inherited in certain combinations, they dictate the gender of the new individual—that is, whether it will become a male or a female.

All of the other chromosomes in our body cells are the same in both sexes. We categorize them as **autosomes**.

The duplicated human chromosome shown in Figure 12.2 has a targeted band (*yellow*), an artistic way of introducing a key point: Molecular biology increased the power of diagnostic tools that were already in use to analyze chromosomes —as with fluorescent dyes that can label DNA regions linked to genetic disorders. In the next section, you will read about two of the diagnostic procedures.

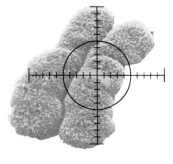

Figure 12.2 Long before the spectacular discoveries of molecular biology, researchers started identifying regions on chromosomes that probably held the genes responsible for certain genetic disorders.

Autosomes are pairs of chromosomes that are the same in males and females of a species. One other pairing, of sex chromosomes, differs between males and females.

12.2 What Is Karyotyping?

With karyotyping, a diagnostic tool, images are constructed to analyze the structure and number of chromosomes in an individual's cells.

How do we know about an individual's autosomes and sex chromosomes? *Karyotyping* is one of the earliest diagnostic tools. A typical **karyotype** is a preparation of an individual's metaphase chromosomes, sorted out by length, shape, centromere location, and other defining features. Gross abnormalities in chromosome structure or an altered chromosome number can be pinpointed by comparing the individual's karyotype against a standard karyotype for the species.

Making a Karyotype Human chromosomes are in their most condensed form and easiest to identify when a cell is at metaphase of mitosis (Sections 9.1 and 9.3). Technicians do not count on finding dividing cells in the body. They culture cells and induce mitosis artificially. They place a sample of cells, usually from blood, into a solution that stimulates growth and mitotic cell division. They add colchicine to the sample to arrest the cell cycle at metaphase. Colchicine, remember, is a poison that blocks spindle formation by preventing microtubules from forming (Section 4.10).

As Figure 12.3 explains, the cell culture is centrifuged to isolate all the metaphase cells. A hypotonic solution makes the cells swell, by way of osmosis, and move away from each other. The chromosomes inside them move away from each other, also. Then the cells are mounted on slides, fixed, and stained for microscopy.

Once the chromosomes are brought into focus, they are photographed. The photograph is cut with scissors or with a computer's cut-and-paste tools to separate the chromosomes. Then the chromosomes are lined up by size and shape, as in Figure 12.3*f*.

Spectral Karyotypes *Spectral karyotyping*, a more recent diagnostic tool, uses a range of colored fluorescent dyes that bind to specific parts of chromosomes. Analysis of the resulting rainbow-hued karyotype often reveals abnormalities that would not otherwise be discernible.

Figure 12.4 shows a spectral karyotype. The Philadelphia chromosome in this karyotype, named after the city where someone discovered it, was the first chromosome to be specifically correlated with cancer—one of the leukemias. The Philadelphia chromosome was already known to be longer than human chromosome 9, which is its normal counterpart. But spectral karyotyping identified the extra length as a piece of chromosome 22.

By chance, both chromosomes broke inside a stem cell in bone marrow. Such cells give rise to blood cells. Enzymes reattached the pieces—but on the wrong chromosomes. You can identify the translocated parts in the Figure 12.4 karyotype. We will be returning to this type of change in the structure of chromosomes in Section 12.8.

Figure 12.3 *Animated!* Karyotyping, in which an image of metaphase chromosomes is cut apart. Individual chromosomes are aligned by their centromeres and arranged according to size, shape, and length.

(**a**) A sample of cells from an individual is put in a medium that stimulates cell growth and mitotic division. Colchicine is added to arrest the cell cycle at metaphase. (**b**) The culture is subjected to *centrifugation*, which works because cells have greater mass and density than the solution bathing them. A centrifuge's spinning force moves the cells farthest from the center of rotation, so they collect at the base of the centrifuge tubes.

(**c**) The culture medium is removed; a hypotonic solution is added. The cells swell, and chromosomes move apart. (**d**) The cells are mounted on a microscope slide and stained to make the chromosomes show up.

(**e**) A photograph of one cell's chromosomes is cut up and organized, as in the human karyotype in (**f**), which shows 22 pairs of autosomes and 1 pair of sex chromosomes—XX *or* XY. Scissors or computer tools do the cuts.

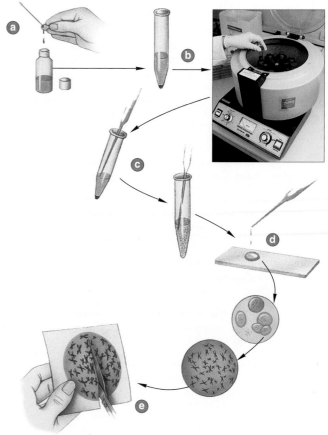

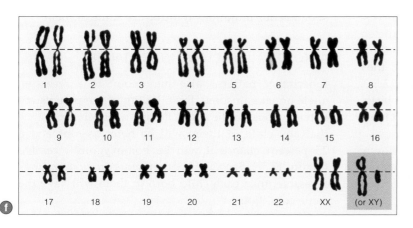

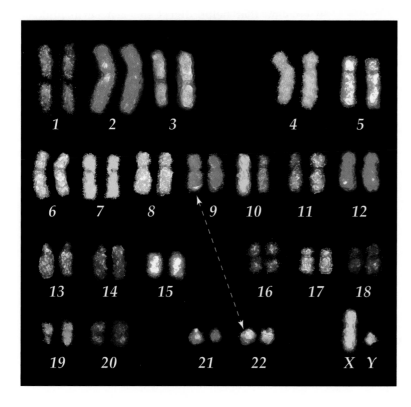

Figure 12.4 Image of a killer—the Philadelphia chromosome, as revealed by the artificial colors of spectral karyotyping. Its normal counterpart is human chromosome 9.

This chromosome exchanged a segment of itself with the nonhomologous chromosome 22. The broken end of chromosome 9 contained a gene that affects mitotic cell division. This gene fused with a DNA sequence in chromosome 22 that controls expression of another gene.

The fused gene is transcribed far more than it should be, and the cell cycle spins out of control (Section 9.5). The phenotypic outcome is *chronic myelogenous leukemia* (CML)—a rare form of leukemia in which the body produces far too many white blood cells. Uncontrolled divisions give rise to masses of malignant cells in bone tissues, where stem cells that give rise to white blood cells originate.

12.3 Examples of Autosomal Inheritance Patterns

LINKS TO
SECTIONS
8.2, 8.6

Most human traits arise from complex gene interactions, but many can be traced to autosomal dominant or recessive alleles that are inherited in simple Mendelian patterns. Some of these alleles cause genetic disorders.

AUTOSOMAL DOMINANT INHERITANCE

Figure 12.5*a* shows a typical inheritance pattern for an autosomal dominant allele. If one of the parents is heterozygous and the other homozygous, any child of theirs has a 50 percent chance of being heterozygous. The trait usually appears every generation. Why? The allele is expressed even in heterozygotes.

One autosomal condition, *achondroplasia*, affects 1 in 10,000 or so people. While they were still embryos, the cartilage model on which a skeleton is constructed did not form properly. Adults have abnormally short arms and legs relative to other body parts and they are only about four feet, four inches tall (Figure 12.5*a*). Most homozygotes die before or not long after birth. The allele does not affect the capacity of the survivors to grow and reproduce.

In *Huntington's disease*, the nervous system slowly deteriorates, and involuntary muscle action increases.

Symptoms often do not start until past age thirty, and those affected die during their forties or fifties. Many unknowingly transmit the mutant allele to children before then. The mutation causing the disorder alters a protein required for normal brain cell development. It is one of the *expansion* mutations, in which three nucleotides are repeated in series along the length of DNA. Hundreds of thousands of repeats occur within and between genes on human chromosomes, but this one (CAG) disrupts a gene product's function.

A few dominant alleles that cause severe problems persist in populations because expression of the allele may not interfere with reproduction, or affected people reproduce before the symptoms become severe. Also, spontaneous mutations reintroduce some of them.

AUTOSOMAL RECESSIVE INHERITANCE

Inheritance patterns also may point to a recessive allele on an autosome. First, if both of the parents are heterozygous for the allele, there is a 50 percent chance that any child of theirs will be heterozygous and a 25 percent chance it will be homozygous recessive (Figure 12.5*b*). Second, if both parents are homozygous recessive, then each child born to them will have the same condition.

Galactosemia is a heritable metabolic disorder that affects about 1 in every 100,000 newborns. This case of autosomal recessive inheritance involves alleles for an enzyme that helps digest the lactose in milk or milk products. The body normally converts lactose to glucose and galactose, then three enzymes convert the galactose to glucose–1–phosphate (Figure 12.6). This intermediate can enter glycolysis or be converted to glycogen (Sections 8.2 and 8.6). But galactosemics do not have functional copies for one of these three enzymes; they are homozygous recessive for a mutant

Figure 12.5 *Animated!* (**a**) Example of autosomal dominant inheritance. One dominant allele (*red*) is fully expressed in carriers. Achondroplasia, an autosomal dominant disorder, affects the three males shown above. At center, Verne Troyer (or Mini Me in the Mike Myers spy movies), stands two feet, eight inches tall.

(**b**) An autosomal recessive pattern. In this example, both of the parents are heterozygous carriers of the recessive allele (*red*).

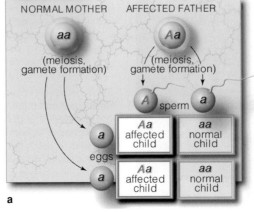

a

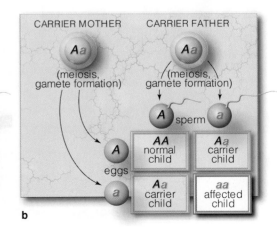

b

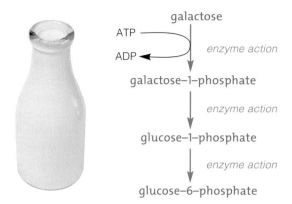

galactose

ATP

ADP ← *enzyme action*

galactose–1–phosphate

↓ *enzyme action*

glucose–1–phosphate

↓ *enzyme action*

glucose–6–phosphate

Figure 12.6 How galactose is normally converted to a form that can enter the breakdown reactions of glycolysis. A mutation that affects the second enzyme in the conversion pathway gives rise to galactosemia.

allele that encodes it. Galactose–1–phosphate builds up to toxic levels in their body. High levels of this intermediate can be detected in urine. The excess leads to malnutrition, diarrhea, vomiting, and damage to the eyes, liver, and brain.

When they do not receive treatment, galactosemics typically die young. When they are quickly placed on a diet that excludes all dairy products, the symptoms may not be as severe.

WHAT ABOUT NEUROBIOLOGICAL DISORDERS?

Those human neurobiological disorders introduced at the start of the chapter do not follow simple patterns of Mendelian inheritance. In most cases, a lone gene does not give rise to depression, schizophrenia, or bipolar disorder. Still, it is useful to search for mutations that make some people more vulnerable, as long as we recognize that many genes and environmental factors contribute in individually small ways to the outcome.

For example, researchers who conducted extensive family studies and twin studies have predicted that mutant alleles in specific regions of autosomes 1, 3, 5, 6, 8, 11 through 15, 18, and 22 increase the chance of developing schizophrenia. Similarly, several mutant alleles have been reportedly linked to bipolar disorder and depression.

Some traits can be traced to dominant or recessive alleles on autosomes because they are inherited in simple Mendelian patterns. Certain alleles on these chromosomes give rise to genetic abnormalities and genetic disorders.

Sometimes textbook examples of the human condition seem a bit abstract, so take a moment to think about two boys who were too young to be old.

Imagine being ten years old with a mind trapped in a body that is getting a bit more shriveled, more frail—*old*—every day. You are barely tall enough to peer over the top of the kitchen counter. You weigh less than thirty-five pounds. Already you are bald and have a crinkled nose. Possibly you have a few more years to live. Would you, like Mickey Hays and Fransie Geringer, still be able to laugh?

On average, of every 8 million newborn humans, one will grow old far too soon. On one of its autosomes, that rare individual carries a mutant allele that gives rise to *Hutchinson–Gilford progeria syndrome*. While that new individual was still an embryo in its mother, billions of DNA replications and mitotic cell divisions distributed the information encoded in that gene to each newly formed body cell. Its legacy will be an accelerated rate of aging and a sharply reduced life span.

The mutation grossly disrupts gene interactions that are essential for growth and development. Observable symptoms start before age two. Skin that should be plump and resilient starts to thin. Skeletal muscles weaken. Limb bones that should lengthen and grow stronger soften. Premature baldness is inevitable (Figure 12.7). There are no documented cases of progeria running in families, so spontaneous mutation must be the cause. In one recent study, researchers examined twenty affected children. All the children carried a mutant gene that specifies lamin A, a structural protein that helps organize the nucleus.

Most progeriacs can expect to die in their early teens as a result of strokes or heart attacks. These final insults are brought on by a hardening of the wall of arteries, a condition typical of advanced age. Fransie was seventeen when he died. Before Mickey died at age twenty, he was the oldest living progeriac.

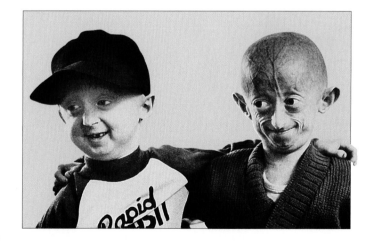

Figure 12.7 Two boys who met at a gathering of progeriacs at Disneyland, California, when they were not yet ten years old.

12.5 | Sex Determination in Humans

Expression of one of the genes on the Y chromosome—that is what it takes to become a human male.

Every normal egg produced by a human female has one X chromosome. Half of the sperm cells formed in a male carry an X chromosome, and half carry a Y. If an X-bearing sperm fertilizes an X-bearing egg, then the resulting zygote will develop into a female. If the sperm carries a Y chromosome, it will develop into a male (Figure 12.8*a*).

With only 255 genes, the human Y chromosome might seem relatively puny. But one of them is the *SRY* gene—which happens to be the master gene for male sex determination. Its expression in XY embryos triggers the formation of testes, or male gonads, as shown in Figure 12.8*b*. What do these primary male reproductive organs do? For one thing, some of their cells make testosterone, a sex hormone that controls the emergence of male secondary sexual traits.

An XX embryo has no Y chromosome, no *SRY* gene, and much less testosterone. Therefore, primary female reproductive organs—ovaries—form instead. Ovaries make estrogens and other sex hormones that govern the development of female secondary sexual traits.

The human X chromosome carries 1,141 genes. Like other chromosomes, it includes some genes associated with sexual traits, such as the distribution of body fat and hair. But most of its genes deal with *nonsexual* traits, such as blood-clotting functions. Such genes can be expressed in males as well as in females. Males, remember, also inherit one X chromosome.

> *Expression of the SRY gene on the human Y chromosome triggers testosterone synthesis, which makes a developing embryo become a male. In the absence of the Y chromosome (and the SRY gene), a developing embryo becomes a female.*

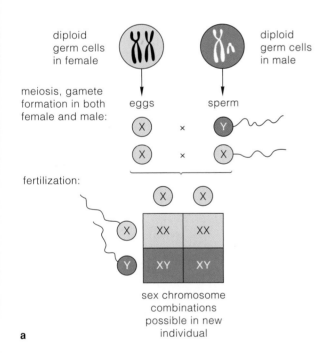

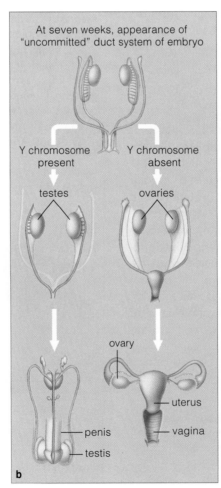

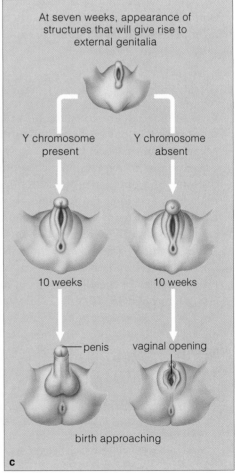

Figure 12.8 *Animated!* (**a**) Punnett-square diagram showing the sex determination pattern in humans.

(**b**) Early on, a human embryo is neither male nor female. Then tiny ducts and other structures that can develop into male *or* female reproductive organs start forming. In an XX embryo, ovaries form *in the absence of the Y chromosome and its SRY gene*. In an XY embryo, the gene product triggers the formation of testes. A hormone secreted from testes calls for development of male traits. (**c**) External reproductive organs in human embryos.

After Mendel passed away in 1884, his paper on pea plants gathered dust in a hundred libraries. Then microscopists discovered chromosomes, and interest in the cellular basis of inheritance was rekindled. In 1900 researchers came across Mendel's paper while checking literature on genetic crosses. Their results confirmed what Mendel had already found out. Later, other researchers went on to discover something Mendel did not know about—genes on sex chromosomes.

By the early 1900s, researchers suspected that each gene has a specific location on a chromosome. Thomas Morgan and his coworkers confirmed it through their hybridization experiments with mutant forms of a fruit fly, *Drosophila melanogaster*. For instance, they found evidence that this fly's X chromosome has a gene for eye color and another gene for body color. They asked: Were the two genes linked on the same chromosome? That is, do they stay together during meiosis and end up in the same gamete?

Thomas Morgan, an embryologist, already had come across a relationship between sex determination and some *nonsexual* traits. For instance, human males and females both have blood-clotting factors. And yet, males are far more likely to develop hemophilia, which is a blood-clotting disorder. This sex-linked outcome was probably related to recessive forms of genes. But it was not like anything that Mendel identified in the results from his experimental crosses of pea plants. For pea plants, it made no difference which parent carried a recessive allele.

Morgan decided to study eye color and other nonsexual traits in *D. melanogaster*. This type of fruit fly has since become a favorite experimental organism. It can live in small bottles on nothing more expensive than a bit of agar, cornmeal, molasses, and yeast. A female lays hundreds of eggs in a few days, and her offspring reproduce in less than two weeks. Morgan knew that in a single year, he could use experimental tests to track observable traits for nearly thirty generations of thousands of fruit flies.

At first, all of the fruit flies in Morgan's bottles were wild type for eye color; they had brick-red eyes (Figure 12.9). Then Morgan got lucky. A gene that controls eye color mutated, and a white-eyed male appeared in a bottle, and Morgan quickly conducted **reciprocal crosses**. In the second of such paired crosses, a trait of each sex is reversed compared to the original cross to determine the role of parental sex on inheritance. In this case:

First cross: white-eyed male   ×   red-eyed female
Second cross: red-eyed male   ×   white-eyed female

White-eyed males mated with homozygous red-eyed females. All of the F_1 offspring had red eyes, and after the F_1 individuals had mated with each other, some of their F_2 offspring were males with white eyes (Figure 12.9c).

Then Morgan allowed true-breeding red-eyed males to mate with white-eyed females. *Half* of the F_1 offspring of this cross turned out to be red-eyed females, and *half* were white-eyed males. Later, the phenotypes of F_2 offspring were 1/4 red-eyed females, 1/4 white-eyed females, 1/4 red-eyed males, and 1/4 white-eyed males. The results did not fit with the straightforward inheritance patterns of pea plants. Could Mendel's theories explain them?

They could if the locus of an eye-color gene were on a sex chromosome. But which one? Because females (XX) could be white-eyed, the recessive allele had to be on one of their X chromosomes. What if white-eyed males (XY) had the recessive allele on their X chromosome and their Y chromosome had no corresponding eye-color allele? In that case, they would have white eyes. They would have no dominant allele to mask the effect of the recessive one, as the Punnett-square diagram in Figure 12.9c shows.

And so Morgan's idea of an X-linked gene dovetailed with Mendel's concept of segregation. By proposing that a specific gene is located on an X chromosome but not on the Y chromosome, Morgan explained his reciprocal crosses. His experimental results matched predicted outcomes.

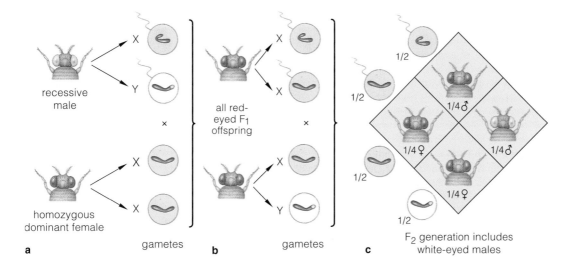

a recessive male × homozygous dominant female — gametes X, Y and X, X

b all red-eyed F_1 offspring — gametes X, X and X, Y

c F_2 generation includes white-eyed males — 1/4♂, 1/4♀, 1/4♂, 1/4♀

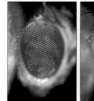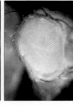

Figure 12.9 One of the experiments that pointed to sex-linked genes in *Drosophila melanogaster*. In this fruit fly, a wild-type allele specifies red eyes, and a mutant allele for the same locus specifies white eyes. *Wild-type* refers to a gene's most common form (either in nature or in standardized, laboratory-bred strains of a species) compared to less common, *mutant* alleles.

12.7 Examples of X-Linked Inheritance Patterns

LINK TO
SECTION
4.10

Alleles on an X chromosome give rise to phenotypes that also reflect simple Mendelian patterns of inheritance. Many of the recessive ones cause problems.

A recessive allele on an X chromosome often leaves certain clues when it causes a genetic disorder. First, more males than females are affected. Heterozygous females still have a dominant allele on their other X chromosome that masks the recessive allele's effects.

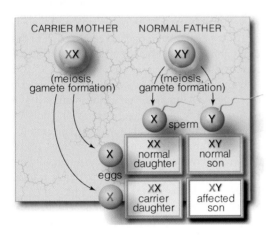

Figure 12.10 *Animated!* One pattern for X-linked recessive inheritance. In this case, the mother carries a recessive allele on one of her X chromosomes (*red*).

Males are not protected, because they inherit only one X chromosome along with one Y chromosome. Figure 12.10 reinforces this point. Second, a heterozygous female must be the bridge between an affected male and an affected grandson; an affected father cannot pass on the recessive allele to his son.

HEMOPHILIA A

Hemophilia A, a type of blood-clotting disorder, is one of the classic cases of X-linked recessive inheritance. Most of us have a functional clotting mechanism that quickly puts a stop to bleeding from minor injuries, in the manner explained in Section 38.9. The mechanism involves the synthesis of proteins that are products of genes on the X chromosome. Bleeding is prolonged in males who carry a mutant form of one of these X-linked genes. The affected males bruise easily, and the internal bleeding can cause problems in their muscles and joints.

This disorder affects 1 in 7,000 males, on average, but new mutations may account for a third of them. In heterozygous females, clotting time is close to normal. The disorder's frequency was relatively high among royal families of Europe and Russia in the nineteenth century. Figure 12.11 is a classic example of a pedigree for hemophilia A.

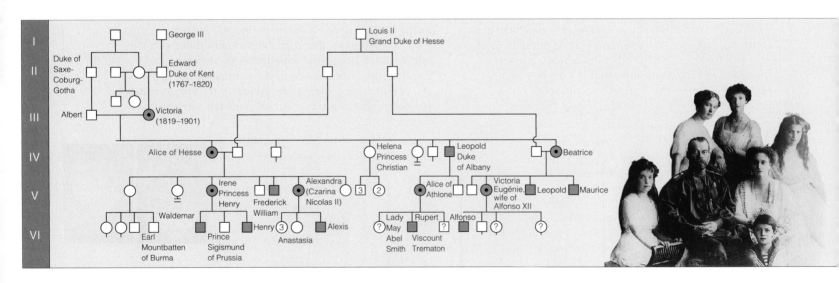

Figure 12.11 A classic case of X-linked recessive inheritance. This is a partial pedigree, or a chart of genetic connections, among descendants of Queen Victoria of England. It focuses on carriers and affected males who inherited the X-linked allele for hemophilia A (*white* circles and squares). At one time, the recessive allele was present in eighteen of Victoria's sixty-nine descendants, who sometimes intermarried. Of the Russian royal family members shown, the mother was a carrier. Through her obsession with the vulnerability of Crown Prince Alexis, a hemophiliac, she was sucked into political intrigue that helped trigger the Russian Revolution of 1917.

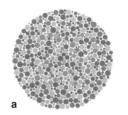
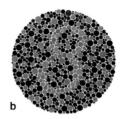

Figure 12.12 *Left,* what red–green color blindness means, using ripe red cherries on a green-leafed tree as an example. In this case, the perception of blues and yellows is normal, but the affected individual has difficulty distinguishing red from green.

Above, two of many Ishihara plates, which are standardized tests for different forms of color blindness. (**a**) You may have one form of red–green color blindness if you see the numeral "7" instead of "29" in this circle. (**b**) You may have another form if you see a "3" instead of an "8."

RED–GREEN COLOR BLINDNESS

The pattern of X-linked recessive inheritance shows up among individuals who have some degree of *color blindness*. The term refers to a range of conditions in which an individual cannot distinguish among some or all of the colors in the spectrum of visible light. Mutant gene products alter the structure and function of photoreceptors (light-sensitive receptors) in eyes.

Normally, humans can sense the differences among 150 colors. A person who is *red–green* color blind sees fewer than 25 because some or all of the receptors that respond to red and green wavelengths are weakened or absent. Other people confuse red and green colors. Still others see shades of gray instead of green but see blues and yellows quite well. Figure 12.12*a* represents this condition. Figure 12.12*b* is part of a standardized set of tests for color blindness.

Color blindness is more common in men, who are about twelve times more likely than women to develop the condition. Heterozygous women show symptoms as well. Can you explain why?

DUCHENNE MUSCULAR DYSTROPHY

Duchenne muscular dystrophy (DMD) is one of a group of X-linked recessive disorders characterized by rapid degeneration of muscles, starting early in life. About 1 in every 3,500 boys is affected.

The recessive allele encodes dystrophin. This is the protein that structurally supports fused-together cells in muscle fibers. It anchors much of the cell cortex to the plasma membrane (Section 4.10). In cases where

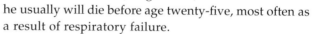

dystrophin is abnormal or absent, the cell cortex weakens and muscle cells die. The debris left behind in tissues triggers inflammation that becomes chronic.

Most individuals are diagnosed between ages three and seven. The progression of the disorder cannot be stopped. When the affected boy is about twelve years old, he will start using a wheelchair. His heart muscles will start to break down. Even with the best managed care, he usually will die before age twenty-five, most often as a result of respiratory failure.

Recently, researchers mapped all of the genes on the X chromosome. They discovered two things. First, only 5 percent of all of the genes we have reside on this sex chromosome. Second, the mutant alleles that cause or contribute to many known genetic disorders can occur at locations along this chromosome. More than 300 such connections have been identified.

Diverse recessive alleles on the human X chromosome are implicated in more than 300 genetic disorders.

A heterozygous female may not show symptoms if she has a dominant allele on her other X chromosome, which masks the effect of the recessive allele. Males (XY) cannot transmit any X-linked allele to their sons.

12.8 Heritable Changes in Chromosome Structure

LINK TO
SECTION
10.3

On rare occasions, a chromosome's structure changes. Many of the alterations have severe or lethal outcomes.

MAIN CATEGORIES OF STRUCTURAL CHANGE

One or more changes in the physical structure of a chromosome may give rise to a genetic disorder or abnormality. Such changes are rare, but they do occur spontaneously in nature. Some also can be induced by exposure to certain chemicals or irradiation. Either way, the alteration may be detected by microscopic examination and karyotype analysis of cells during mitosis or meiosis. Four kinds of structural changes are chromosomal duplications, deletions, inversions, and translocations.

Duplication Even normal chromosomes have DNA sequences that are repeated two or more times. These are called **duplications**:

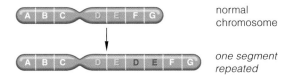

normal chromosome

one segment repeated

Duplications can occur through unequal crossovers at prophase I. Homologous chromosomes align side by side, but their DNA sequences misalign at some point along their length. The probability of this happening is greater in regions where DNA has long repeats of the same series of nucleotides. A stretch of DNA gets deleted from one chromosome and is spliced into the partner chromosome. Some duplications cause neural problems and physical abnormalities. As you will see, others apparently were important in the evolution of primates that were ancestral to humans.

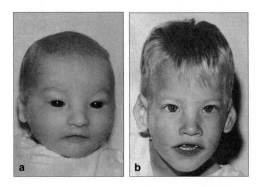

Figure 12.13 (**a**) A male infant who developed cri-du-chat syndrome. His ears are low on the side of the head relative to the eyes. (**b**) Same boy, four years later. By this age, affected humans stop making the mewing sounds typical of the syndrome.

Deletion A **deletion** is the loss of some portion of a chromosome, as by unequal crossovers, inversions, or chemical attacks:

segment C deleted

In mammals, most deletions cause serious disorders or death. Missing or broken genes disrupt the body's growth, development, and metabolism. For instance, a tiny deletion from human chromosome 5 results in an abnormally shaped larynx and mental impairment. Crying infants sound like cats meowing (Figure 12.13). Hence the name of the disorder, *cri-du-chat* (cat-cry in French). Inversions and deletions often occur together as an outcome of unequal recombination events.

Inversion With an **inversion**, part of the sequence of DNA within the chromosome becomes oriented in the reverse direction, with no molecular loss:

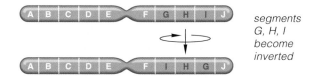

segments G, H, I become inverted

An inversion is not a problem for a carrier if it does not alter a crucial gene region. It can cause problems in meiosis. Chromosomes may mispair, and deletions may occur that can reduce the viability of gametes. Some individuals do not even know that they have an inverted chromosome region until a genetic disorder or abnormality surfaces in one or more children.

Translocation In Section 12.2, you came across a case of a broken part of one chromosome becoming attached to another chromosome. This type of change in chromosome structure is known as a **translocation**. Most translocations are reciprocal, in that both of the two chromosomes exchange broken parts:

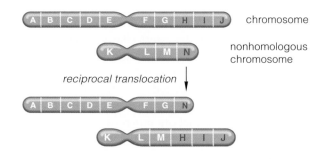

chromosome

nonhomologous chromosome

reciprocal translocation

Translocations often cause reduced fertility, because affected chromosomes have difficulty segregating in meiosis. Severe problems are rare, but they do arise. They include some sarcomas, lymphomas, myelomas, and leukemias.

DOES CHROMOSOME STRUCTURE EVOLVE?

Alterations in the structure of chromosomes generally are not good and may be selected against. Even so, many alterations with neutral effects have been built into the DNA of all species over evolutionary time.

Duplicates of genes could have bestowed adaptive advantages on descendants of their original bearers. Two or more copies of some gene means that one is free to mutate while the other continues to carry out its normal function. The slightly modified products of mutant genes can behave in slightly different or novel ways, some of which are beneficial.

Some duplications have proved adaptive. Reflect on the four globin chains of hemoglobin (Section 3.6). In humans and other primates, several genes for these polypeptide chains are strikingly similar. Apparently they evolved through duplications, mutations, and transpositions. They have slightly different molecular structures and slightly different capacities to bind and transport oxygen under a range of cellular conditions.

Alterations in chromosome structure might have contributed to the differences among closely related organisms, such as apes and humans. Eighteen of the twenty-three pairs of human chromosomes are almost identical with chimpanzee and gorilla chromosomes. The other five differ only at inverted and translocated regions. Figure 12.14 shows the striking similarities between some gibbon and human chromosomes that may have arisen by duplications and translocations.

To give one more example, human body cells have twenty-three pairs of chromosomes, but those of a chimpanzee, gorilla, or orangutan have twenty-four. Compare the banding patterns of these chromosomes. During human evolution, two chromosomes in an early ancestor fused, end to end, to form chromosome 2. In the fused region, researchers have discovered remnants of a telomere—the signature DNA sequence that caps the *ends* of all chromosomes (Figure 12.15).

A segment of a chromosome may be duplicated, deleted, inverted, or moved to a new location. Such changes can be harmful or lethal. Others have been conserved over time; they confer advantages or have had neutral effects.

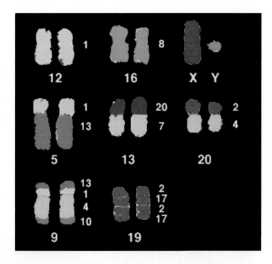

Figure 12.14 Spectral karyotype of duplicated chromosomes of the gibbon, one of the apes. The colors identify regions of gibbon chromosomes that are structurally identical with human chromosomes.

Top row: Chromosomes 12, 16, X, and Y are structurally the same in both primates. *Second row:* Translocations are present in gibbon chromosomes 5, 13, and 20, and they correspond to regions of human chromosomes 1, 13, 20, 7, 2, and 4.

Third row: Gibbon chromosome 9 corresponds to several human chromosome regions. In addition, duplications in gibbon chromosome 19 are present in human chromosomes 2 and 17.

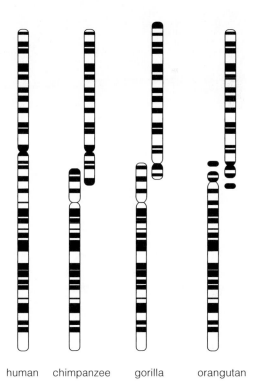

human   chimpanzee   gorilla   orangutan

Figure 12.15 Banding patterns of human chromosome 2 (*left*), compared with the patterns on two of the chromosomes in cells of the chimpanzee, gorilla, and orangutan. Such bands appear because different chromosome regions preferentially take up different kinds of stains. Their response to a given stain depends on their base composition and packing organization.

12.9 Heritable Changes in the Chromosome Number

LINKS TO
SECTIONS
9.3, 10.3

Occasionally, abnormal events occur before or during cell division, and gametes and new individuals end up with the wrong chromosome number. Consequences range from minor to lethal changes in form and function.

In **aneuploidy**, cells usually have one extra or one less chromosome. Autosomal aneuploidy is usually fatal for humans and is linked to most miscarriages. Aneuploidy typically arises through **nondisjunction**, whereby one or more pairs of chromosomes do not separate as they should during mitosis or meiosis. Figure 12.16 shows an example. In **polyploidy**, cells have three or more of each type of chromosome. Half of all species of flowering plants, some insects, fishes, and other animals are polyploid.

Such changes affect the chromosome number at fertilization. Suppose a normal gamete fuses with an $n+1$ gamete, with one extra chromosome. The new individual will be trisomic ($2n+1$), with three of one type of chromosome and two of every other type. Or what if an $n-1$ gamete and a normal n gamete fuse? In this case, the new individual will be monosomic, or $2n-1$. Mitotic divisions perpetuate such mistakes when an embryo is growing in size and developing.

AUTOSOMAL CHANGE AND DOWN SYNDROME

A few trisomics are born alive, but only trisomy 21 individuals reach adulthood. A newborn with three chromosomes 21 will develop *Down syndrome*. This autosomal disorder is the most frequent type of altered chromosome number in humans; it occurs once in every 800 to 1,000 births. It affects more than 350,000 people in the United States. Figure 12.16*a* shows a karyotype for a trisomic 21 female. About 95 percent of all cases arise through nondisjunction at meiosis. Affected individuals have upward-slanting eyes, a fold of skin that starts at the inner corner of each eye, a deep crease across each palm and foot sole, one (not two) horizontal furrows on their fifth fingers, and somewhat flattened facial features.

Not all of the outward symptoms develop in every individual. That said, trisomic 21 individuals do have moderate to severe mental impairment and heart problems. Also, their skeleton develops abnormally, so older children have shortened body parts, loose joints, and misaligned bones in hips, fingers, and toes. Their muscles and reflexes are weak. Motor skills, including speech, develop very slowly. With medical care, individuals can live for fifty-five years, on average.

The incidence of nondisjunction rises with increasing age of potential mothers (Figure 12.17). Nondisjunction might occur in the father, although less often. Trisomy 21 is just one of hundreds of conditions that can be detected through prenatal diagnosis (Section 12.11). With early training and medical intervention, individuals still can take part in normal activities. As a group, trisomics 21 tend to be cheerful and sociable.

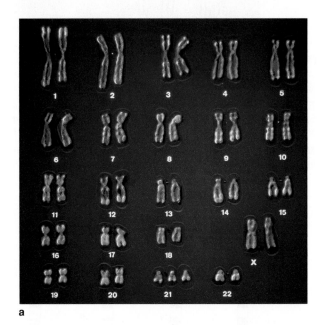

a

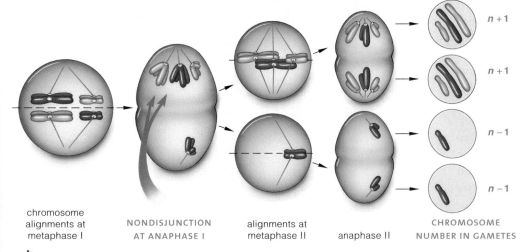

chromosome
alignments at
metaphase I

b

NONDISJUNCTION
AT ANAPHASE I

alignments at
metaphase II

anaphase II

CHROMOSOME
NUMBER IN GAMETES

$n+1$

$n+1$

$n-1$

$n-1$

Figure 12.16 (**a**) A case of nondisjunction. This karyotype reveals the trisomic 21 condition of a human female. (**b**) One example of how nondisjunction arises. Of the two pairs of homologous chromosomes shown here, one fails to separate during anaphase I of meiosis. The chromosome number is altered in the gametes that form after meiosis.

CHANGE IN THE SEX CHROMOSOME NUMBER

Nondisjunction also causes most of the alterations in the number of X and Y chromosomes. The frequency of such changes is 1 in 400 live births. Usually, they lead to difficulties in learning and motor skills, such as speech, although problems can be so subtle that the underlying cause is not even diagnosed.

Female Sex Chromosome Abnormalities *Turner syndrome* individuals have an X chromosome and no corresponding X or Y chromosome (XO). About 1 in 2,500 to 10,000 newborn girls are XO. Nondisjunction originating with the father accounts for 75 percent of the cases. Yet cases are few, compared with other sex chromosome abnormalities. At least 98 percent of XO embryos may spontaneously abort early in pregnancy.

Despite the near lethality, XO survivors are not as disadvantaged as other aneuploids. On average, they are well proportioned, as shown here, but only four feet, eight inches tall. Most cannot make enough sex hormones; they do not have functional ovaries. The condition affects the development of secondary sexual traits, such as breast development. A few eggs form in the ovaries but degenerate by the time the girls are two years old.

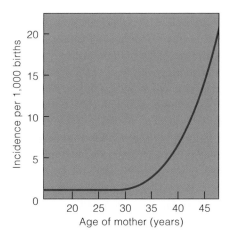

A few females inherit three to five X chromosomes. An *XXX syndrome* occurs in about 1 of 1,000 live births. Adults are fertile. Except for slight learning difficulties, most fall in the normal range of social behavior.

Male Sex Chromosome Abnormalities About 1 of every 500 males has an XXY karyotype, with an extra chromosome inherited from the mother. Two-thirds of the cases are an outcome of nondisjunction at meiosis. Among the remainder, failure of the Y chromosome to separate at mitosis gave rise to a mosaic karyotype (XY in some cells and XXY in other cells).

The resulting *Klinefelter syndrome* develops during puberty. XXY males tend to be overweight and tall. The testes and the prostate gland usually are smaller than average. Many XXY males are within the normal range of intelligence, although some have short-term memory loss and learning disabilities. They make less testosterone and more estrogen than normal males, with feminizing effects. Sperm counts are low. Hair is sparse, the voice is pitched high, and the breasts are enlarged somewhat. When affected individuals enter

Figure 12.17 Relationship between the frequency of Down syndrome and mother's age at childbirth. The data are from a study of 1,119 affected children. The risk of having a trisomic 21 baby rises with the mother's age. This may seem odd, because about 80 percent of trisomic 21 individuals are born to women not yet thirty-five years old. But these women are in the age categories with the highest fertility rates, and they simply have more babies.

puberty, they can receive testosterone injections that can reverse the feminized traits.

About 1 in 500 to 1,000 males has one X and two Y chromosomes, an *XYY condition*. They tend to be taller than average, with mild mental impairment, but most fall in the normal phenotypic range. They were once thought to be genetically predisposed to a life of crime. This misguided view was based on a sampling error (too few cases of narrowly chosen groups, such as prison inmates) and were biased (the same researchers gathered karyotypes *and* personal histories). Fanning the stereotype was a report that a mass murderer of young nurses was XYY. He wasn't.

In 1976 a Danish geneticist reported results from his study of 4,139 tall males, all twenty-six years old, who had registered at their draft board. Besides their data from physical examinations and intelligence tests, the draft records offered clues to social and economic status, education, and criminal convictions, if there were any. Twelve of the males studied were XYY, which meant the "control group" had more than 4,000 males. The only finding was that mentally impaired, tall males who engage in criminal deeds are just more likely to get caught—irrespective of karyotype.

The majority of XXY, XXX, and XYY children may not even be diagnosed. Some are dismissed unfairly as being underachievers.

Nondisjunction in germ cells, gametes, or early embryonic cells changes the number of autosomes or the number of sex chromosomes. The change affects development and the resulting phenotypes.

Nondisjunction at meiosis causes most sex chromosome abnormalities, which typically lead to subtle difficulties with learning, and speech and other motor skills.

12.10 Human Genetic Analysis

Some organisms, including pea plants and fruit flies, are ideal for genetic analysis. They do not have very many chromosomes. They can grow and reproduce fast in small spaces, under controlled conditions. It does not take long to track a trait through many generations. Humans, however, are another story.

Unlike the flies in laboratory bottles, we humans live under variable conditions in diverse environments, and we live as long as the geneticists who study us. Most of us select our own mates and reproduce if and when we want to. Most families are not large, which means that there are not enough offspring available for researchers to make easy inferences.

Geneticists often gather information from several generations to increase the numbers for analysis. If a trait follows a simple Mendelian inheritance pattern, geneticists can be more confident about predicting the probability of its showing up again. The pattern also can be a clue to the past (Figure 12.18).

Such information is often displayed in **pedigrees**, or charts of genetic connections among individuals. Standardized methods, definitions, and symbols that

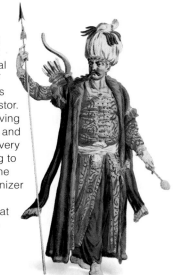

Figure 12.18 An intriguing pattern of inheritance. Eight percent of the men in Central Asia carry nearly identical Y chromosomes, which implies descent from a shared ancestor. If so, then 16 million males living between northeastern China and Afghanistan—close to 1 of every 200 men alive today—belong to a lineage that started with the warrior and notorious womanizer Genghis Khan. In time, his offspring ruled an empire that stretched from China all the way to Vienna.

represent different kinds of individuals are used to construct these charts. You already came across one in Section 12.7. Figures 12.19 and 12.20 are two more.

Those who analyze pedigrees rely on knowledge of probability and patterns of Mendelian inheritance that may yield clues to a trait. Such researchers have traced many genetic abnormalities and disorders to a dominant or recessive allele and often to its location on an autosome or a sex chromosome. Table 12.1 is a list of the ones used as examples in this book.

As individuals and as members of society, what do we do with the information? The next section gets into options. When considering them, keep in mind some important distinctions. First, a genetic *abnormality* is only a rare or uncommon version of a trait, as when a person is born with six digits on each hand or foot instead of the usual five. Whether you view such an abnormality as disfiguring or merely interesting is subjective only; there is nothing inherently life-threatening about it. By contrast, a genetic *disorder* is a heritable condition that sooner or later gives rise to mild to severe medical problems. Each

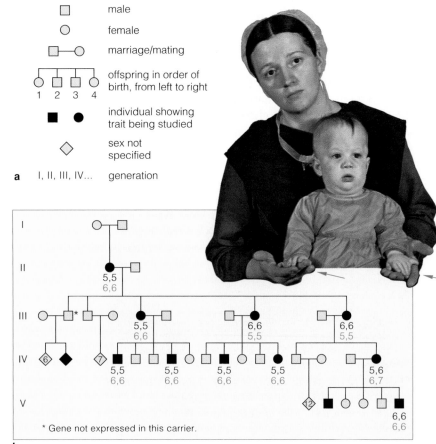

a

☐	male	
○	female	
☐—○	marriage/mating	

offspring in order of birth, from left to right
1 2 3 4

■ ● individual showing trait being studied

◇ sex not specified

I, II, III, IV... generation

b

* Gene not expressed in this carrier.

II 5,5 / 6,6

III 5,5 / 6,6   6,6 / 5,5   6,6 / 5,5

IV ◇6   ◇7   5,5 / 6,6   5,5 / 6,6   5,5 / 6,6   5,5 / 6,6   5,6 / 6,7

V ◇12   6,6 / 6,6

Figure 12.19 *Animated!* (**a**) Standardized symbols used in pedigrees. (**b**) A pedigree for *polydactyly*, characterized by extra fingers, toes, or both. *Black* numerals signify the number of fingers on each hand; *blue* numerals signify the number of toes on each foot. This condition recurs as one symptom of Ellis–van Creveld syndrome.

Disorder or Abnormality	Main Symptoms	Disorder or Abnormality	Main Symptoms
Autosomal recessive inheritance		**X-linked recessive inheritance**	
Albinism	Absence of pigmentation	Androgen insensitivity syndrome	XY individual but having some female traits; sterility
Blue offspring	Bright blue skin coloration	Red–green color blindness	Inability to distinguish among some or all shades of red and green
Cystic fibrosis	Abnormal glandular secretions leading to tissue, organ damage		
Ellis–van Creveld syndrome	Extra fingers, toes, short limbs	Fragile X syndrome	Mental impairment
Fanconi anemia	Physical abnormalities, bone marrow failure	Hemophilia	Impaired blood-clotting ability
		Muscular dystrophies	Progressive loss of muscle function
Galactosemia	Brain, liver, eye damage	X-linked anhidrotic dysplasia	Mosaic skin (patches with or without sweat glands); other effects
Phenylketonuria (PKU)	Mental impairment		
Sickle-cell anemia	Adverse pleiotropic effects on organs throughout body		
		Changes in chromosome structure	
		Chronic myelogenous leukemia (CML)	Overproduction of white blood cells in bone marrow; organ malfunctions
Autosomal dominant inheritance		Cri-du-chat syndrome	Mental impairment; abnormally shaped larynx
Achondroplasia	One form of dwarfism		
Camptodactyly	Rigid, bent fingers		
Familial hypercholesterolemia	High cholesterol levels in blood; eventually clogged arteries	**Changes in chromosome number**	
		Down syndrome	Mental impairment; heart defects
Huntington's disease	Nervous system degenerates progressively, irreversibly	Turner syndrome	Sterility; abnormal ovaries, abnormal sexual traits
Marfan syndrome	Abnormal or no connective tissue	Klinefelter syndrome	Sterility; mild mental impairment
Polydactyly	Extra fingers, toes, or both	XXX syndrome	Minimal abnormalities
Progeria	Drastic premature aging	XYY condition	Mild mental impairment or no effect
Neurofibromatosis	Tumors of nervous system, skin		

genetic disorder is characterized by a specific set of symptoms—a **syndrome**.

One more point to keep in mind: Alleles that give rise to severe genetic disorders are generally rare in populations, because they put their bearers at risk. Why don't they disappear entirely? Rare mutations introduce new ones. In addition, in heterozygotes, a normal allele masks harmful effects that may result from expression of the mutant recessive allele. This means that heterozygotes can transmit harmful alleles to their offspring. The next section addresses how we may address the consequences.

Pedigree analysis may reveal simple patterns of Mendelian inheritance. From such patterns, specialists can infer the probability that offspring will inherit certain alleles.

A genetic abnormality is a rare or less common version of a heritable trait. A genetic disorder is a heritable condition that results in mild to severe medical problems.

Figure 12.20 Pedigree for Huntington's disease, a progressive degeneration of the nervous system. Researcher Nancy Wexler and her team constructed this extended family tree for nearly 10,000 Venezuelans. Their analysis of unaffected and affected individuals revealed that a dominant allele on human chromosome 4 is the culprit. Wexler has a special interest in the disease; it runs in her family.

12.11 Prospects in Human Genetics

With the arrival of their newborn, parents typically ask, "Is our baby all right?" Quite naturally, they want their baby to be free of genetic disorders, and most babies are. But what are the options when something goes wrong?

Many prospective parents have difficulty coming to terms with the possibility that a child of theirs might develop a severe genetic disorder. What are their options?

Phenotypic Treatments   Surgery, prescription drugs, hormone replacement therapy, and often dietary controls can minimize and in some cases eliminate the symptoms of many genetic disorders.

For instance, strict dietary controls work in cases of *phenylketonuria*, or PKU. Individuals affected by this genetic disorder are homozygous for a recessive allele on an autosome. They cannot make a functional form of an enzyme that catalyzes the conversion of the amino acid phenylalanine to tyrosine. Because the conversion is blocked, phenylalanine accumulates and is diverted into other metabolic pathways. The outcome is an impairment of brain function.

Affected people who restrict phenylalanine intake can lead essentially normal lives. They must avoid soft drinks and other products that are sweetened with aspartame, a compound that contains phenylalanine.

Genetic Screening   The idea behind genetic screening is to detect alleles that cause genetic disorders, provide information on reproductive risks, and help families who are already affected. Often, carriers or affected individuals are detected early enough to start countermeasures for minimizing the damage before symptoms develop.

A few large-scale screening programs are operational. Besides helping individuals, the information they generate is being used to estimate the prevalence and distribution of harmful alleles in populations. In the United States, for instance, most hospitals routinely screen newborns for PKU, so we now see fewer individuals with symptoms of the disorder.

There are social risks that must be considered. How would you feel if you were labeled as someone with "bad" alleles? Would the knowledge invite chronic anxiety? Would potential employers or insurance companies turn you down? How would you interact with an affected child that you brought into the world if you had known about the risk in advance? No easy answers here.

Prenatal Diagnosis   Doctors and clinicians commonly use methods of *prenatal diagnosis* to determine the sex of embryos or fetuses and to screen for more than 100 known genetic problems. *Prenatal* means before birth. *Embryo* is a term that applies until eight weeks after fertilization, after which the term *fetus* is appropriate.

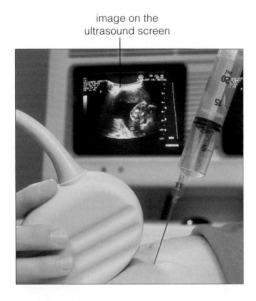

image on the ultrasound screen

Figure 12.21 *Animated!*   Amniocentesis, a prenatal diagnostic tool. A pregnant woman's doctor holds an ultrasound emitter against her abdomen while drawing a sample of amniotic fluid into a syringe. He monitors the path of the needle with an ultrasound screen, in the background. Then he directs the needle into the amniotic sac that holds the developing fetus and withdraws twenty milliliters or so of amniotic fluid. The fluid contains fetal cells and wastes that can be analyzed for genetic disorders.

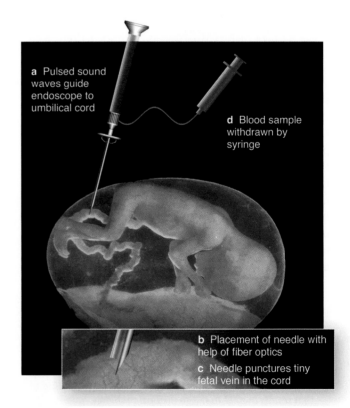

a Pulsed sound waves guide endoscope to umbilical cord

d Blood sample withdrawn by syringe

b Placement of needle with help of fiber optics

c Needle punctures tiny fetal vein in the cord

Figure 12.22   Fetoscopy for prenatal diagnosis.

Suppose a forty-five-year-old woman is pregnant and worries about Down syndrome. Between eight and twelve weeks after conception, she might opt for *amniocentesis* (Figure 12.21). By this diagnostic procedure, a clinician uses a syringe to withdraw a small sample of fluid from the amniotic cavity. The "cavity" is a fluid-filled sac, bounded by a membrane—the amnion—that encloses the fetus. The fetus normally sheds some cells into the fluid. Cells suspended in the fluid sample can be analyzed for many genetic disorders, including Down syndrome, cystic fibrosis, and sickle-cell anemia.

Chorionic villi sampling (CVS) is a similar diagnostic procedure. A clinician withdraws a few cells from the chorion, a membrane that surrounds the amnion and helps form the placenta. Unlike amniocentesis, however, CVS can be requested to find out information as early as eight weeks into pregnancy.

It is now possible to see a live, developing fetus with the aid of an endoscope, a fiber-optic device. In *fetoscopy*, sound waves are pulsed across the mother's uterus. Images of parts of the fetus, umbilical cord, or placenta show up on a computer screen that is connected to the endoscope (Figure 12.22). A sample of fetal blood is often drawn at the same time. This procedure can be used to diagnose many blood cell disorders, such as sickle-cell anemia and hemophilia.

There are risks to a fetus associated with all three procedures, including punctures or infections. Also, if the amnion does not reseal itself quickly, too much fluid can leak out of the amniotic cavity and endanger the fetus. Amniocentesis increases the risk of miscarriage by 1 to 2 percent. CVS may disrupt the placenta's development, which can cause missing or underdeveloped fingers and toes in 0.3 percent of newborns. Fetoscopy raises the risk of a miscarriage by 2 to 10 percent.

Genetic Counseling Parents-to-be commonly ask genetic counselors to compare the risks associated with diagnostic procedures against the likelihood that their future child will be affected by a severe genetic disorder. At the time of counseling, they also should discuss the small overall risk (3 percent) that complications can affect *any* child during the birth process. They should talk about how old they are. The older either prospective parent is, the greater the risk may be.

As a case in point, suppose a first child or a close relative has a severe disorder. Genetic counselors come up with a program of diagnosis of parental genotypes, pedigrees, and genetic testing for known disorders. Using this information, counselors can predict risks for disorders in future children. They should remind prospective parents that the same risk usually applies to each pregnancy.

Regarding Abortion What happens after prenatal diagnosis reveals a severe problem? Do prospective parents opt for an induced abortion? An *abortion* is an expulsion

Figure 12.23 Eight-cell and multicelled stages of human development.

of a pre-term embryo or fetus from the uterus. We can only say here that individuals must weigh awareness of the severity of the genetic disorder against their ethical and religious beliefs. Worse, today they must play out their personal tragedy on a larger stage that is dominated by a nationwide battle between highly vocal "pro-life" and "pro-choice" factions. We return to this volatile topic in Section 44.15, after explaining the stages of human embryonic development.

Preimplantation Diagnosis This procedure relies on *in vitro fertilization*. Sperm and eggs from prospective parents are mixed in a sterile culture medium. One or more eggs may get fertilized. If this happens, mitotic cell divisions can turn an egg into a ball of eight cells within forty-eight hours (Figure 12.23).

According to one view, the tiny, free-floating ball is a pre-pregnancy stage. Like all of the unfertilized eggs that a woman's body discards monthly during her reproductive years, it has not attached to the uterus. All of its cells have the same genes. However, its cells are not yet committed to being specialized one way or another. Doctors carefully remove one of these undifferentiated cells and analyze its genes. If it has no detectable genetic defects, the ball is inserted into the uterus. The withdrawn cell will not be missed. Many of the resulting "test-tube babies" are born in good health.

Some couples who are at risk of passing on the alleles for cystic fibrosis, muscular dystrophy, or some other genetic disorder have opted for this procedure.

Summary

Section 12.1 Of twenty-three pairs of homologous chromosomes in human body cells, one is a pairing of sex chromosomes. The other chromosomes are called autosomes; in both sexes, they are the same length and shape, have the same centromere location, and carry the same genes along their length.

Section 12.2 In karyotyping, a diagnostic tool, an individual's metaphase chromosomes are prepared for microscopy, photographed, and arranged in sequence in a chart on the basis of their defining features.

Biology⊜Now
Learn how to create a karyotype with the animation on BiologyNow.

Sections 12.3, 12.4 Some dominant and recessive alleles on autosomes are inherited in simple Mendelian patterns that can be predictably connected with specific phenotypes. Some mutant forms of these alleles give rise to genetic abnormalities or genetic disorders.

Biology⊜Now
Investigate autosomal inheritance with the interaction on BiologyNow.

Sections 12.5, 12.6 Human females have identical sex chromosomes (XX) and males have nonidentical ones (XY). The *SRY* gene on the Y chromosome is the basis of sex determination. Its expression starts the synthesis of testosterone, a hormone that causes a human embryo to develop into a male. If an embryo has no Y chromosome (no *SRY* gene), it develops into a female.

Experiments with fruit flies yielded the first evidence that specific genes that give rise to nonsexual traits are located on the X chromosome.

Biology⊜Now
See how gender is determined in humans with the interaction on BiologyNow.

Section 12.7 Certain dominant and recessive alleles on the X chromosome are inherited in simple patterns. A number of alleles on the X chromosome contribute to more than 300 known genetic disorders. Males cannot transmit a recessive X-linked allele to their sons; an affected female must be the bridge of inheritance.

Biology⊜Now
Investigate X-linked inheritance with the interaction on BiologyNow.

Section 12.8 On rare occasions, a chromosome's physical structure undergoes abnormal alterations. Part of it may be duplicated, deleted, inverted, or moved to a new location (translocated) in the same chromosome or a different one.

Most alterations are harmful or lethal. Even so, many have accumulated in the chromosomes of all species over evolutionary time. Either they had neutral effects or they later proved to be useful. Many duplications, inversions, and translocations are built into primate chromosomes. They are strikingly similar among human, chimpanzee, gorilla, orangutan, and gibbon chromosomes, which is strong evidence of divergences from a common ancestor.

Section 12.9 The parental chromosome number can change permanently. Most often, this is an outcome of nondisjunction: the failure of one or more pairs of duplicated chromosomes to separate from each other, most often during meiosis.

Aneuploids have inherited one extra or one less chromosome than their parents. In the human population, trisomy 21, the most well-known form of aneuploidy, results in Down syndrome. Most human autosomal aneuploids die before birth.

Polyploids inherited three or more of each type of chromosome from their parents. About half of all flowering plants and some insects, fishes, and other animals are polyploid.

Changes in the number of sex chromosomes usually cause problems with learning and motor skills. Problems can be so subtle that the underlying cause may not be diagnosed, as among XXY, XXX, and XYY children.

Sections 12.10, 12.11 Traditionally, geneticists have constructed pedigrees, or charts of genetic connections among individuals, to estimate the probability that offspring will inherit a trait of interest. Phenotypic treatments, genetic screening, genetic counseling, prenatal diagnosis, and preimplantation diagnosis are options available for potential parents who are at risk of transmitting a harmful allele to offspring.

Biology⊜Now
Examine a human pedigree with the animation on BiologyNow.
Explore amniocentesis with the animation on BiologyNow.

Self-Quiz
Answers in Appendix II

1. The _____ of chromosomes in a cell are compared to construct karyotypes.
   a. length and shape   c. gene sequence
   b. centromere location   d. both a and b

2. The _____ determines gender in humans.
   a. X chromosome   c. *SRY* gene
   b. *Dll* gene   d. both b and c

3. If one parent is heterozygous for a dominant allele on an autosome and the other parent is homozygous, any child of theirs has a _____ chance of being heterozygous.
   a. 25 percent   c. 75 percent
   b. 50 percent   d. no chance; it will die

4. Expansion mutations occur _____ within and between genes in human chromosomes.
   a. only rarely   c. not at all
   b. frequently   d. only in multiples of ten

5. Galactosemia is a case of _____ inheritance.
   a. autosomal dominant   c. X-linked dominant
   b. autosomal recessive   d. X-linked recessive

6. Is this statement true or false: A son can inherit an X-linked recessive allele from his father.

7. Color blindness is a case of _____ inheritance.
   a. autosomal dominant   c. X-linked dominant
   b. autosomal recessive   d. X-linked recessive

8. A (An) _____ can alter chromosome structure.
   a. deletion   c. inversion   e. all of the
   b. duplication   d. translocation   above

9. Nondisjunction may occur during _____ .
   a. mitosis   c. fertilization
   b. meiosis   d. both a and b

10. Is this statement false: Body cells sometimes inherit three or more of each type of chromosome characteristic of the species, a condition called aneuploidy.

11. The karyotype for Klinefelter syndrome is _____ .
   a. XO   c. XXY
   b. XXX   d. XYY

12. A recognized set of symptoms that characterize a specific disorder is a _____ .
   a. syndrome   b. disease   c. pedigree

13. Match the chromosome terms appropriately.
   ____ polyploidy   a. number and defining
   ____ deletion   features of an individual's
   ____ nondisjunction   metaphase chromosomes
   ____ translocation   b. segment of a chromosome
   ____ karyotype   moves to a nonhomologous
   ____ aneuploidy   chromosome
                         c. extra chromosome sets
                         d. one outcome: gametes with
                            wrong chromosome number
                         e. a chromosome segment lost
                         f. change by one chromosome

Additional questions are available on **Biology ⒺNow™**

Genetics Problems   *Answers in Appendix III*

1. Human females are XX and males are XY.
   a. Does a male inherit the X from his mother or father?
   b. With respect to X-linked alleles, how many different types of gametes can a male produce?
   c. If a female is homozygous for an X-linked allele, how many types of gametes can she produce with respect to that allele?
   d. If a female is heterozygous for an X-linked allele, how many types of gametes might she produce with respect to that allele?

2. In Section 11.4, you read about a mutation that causes a serious genetic disorder, *Marfan syndrome*. A mutant allele responsible for the disorder follows a pattern of autosomal dominant inheritance. What is the chance that any child will inherit it if one parent does not carry the allele and the other is heterozygous for it?

3. Somatic cells of individuals with Down syndrome usually have an extra chromosome 21; they contain forty-seven chromosomes.
   a. At which stages of meiosis I and II could a mistake alter the chromosome number?
   b. A few individuals with Down syndrome have forty-six chromosomes, two of which are normal-appearing

Figure 12.24   A case of Klinefelter syndrome. Until his teenage years, Stefan was shy, reserved, and prone to rage for no apparent reason. Psychologists and doctors assumed he had learning disabilities that affected comprehension, auditory processing, memory, and abstract thinking. One told Stefan he was stupid and lazy, and would be lucky to graduate from high school. In time, Stefan was graduated from college with degrees in business administration and sports management. He never discussed his learning disabilities. Instead, he took pride in doing the work on his own and not being treated differently.

Stefan was twenty-five years old before laboratory tests as well as karyotyping revealed a 46XY/47XXY mosaic condition. That same year, he started a job as a software engineer. Having a full-time position helped him open doors to volunteer work with the Klinefelter syndrome network. During his volunteer work, he met his future fiancée, whose son also has the syndrome.

chromosomes 21 and a longer-than-normal chromosome 14. Speculate on how this chromosome abnormality may have arisen.

4. As you read earlier, *Duchenne muscular dystrophy* is a genetic disorder that arises through the expression of a recessive X-linked allele. Usually, symptoms start in childhood. Gradual, progressive loss of muscle function leads to death, usually by age twenty or so. Unlike color blindness, the disorder is nearly always restricted to males. Suggest why.

5. In the human population, mutation of two genes on the X chromosome causes two types of X-linked *hemophilia* (A and B). In a few cases, a woman is heterozygous for both mutant alleles (one on each of the X chromosomes). All of her sons should have either hemophilia A or B.
   However, on very rare occasions, one of these women gives birth to a son who does not have hemophilia, and his one X chromosome does not have either mutant allele. Explain how such an X chromosome could arise.

6. Does the phenotype indicated by red circles and squares in this pedigree show a Mendelian inheritance pattern that is autosomal dominant, autosomal recessive, or X-linked?

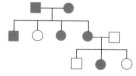

7. When it comes to acceptance of a genetic condition that is out of the ordinary, people tend to be subjective. As an example, consider the individual described in Figure 12.24. How would you have categorized him without knowing the genetic basis of his early behavior? How would you categorize him now in terms of what we as a society consider to be "ideal" phenotypes?

13 DNA STRUCTURE AND FUNCTION

Goodbye, Dolly

In 1997 in Scotland, geneticist Ian Wilmut made headlines when he did not bother with the union of sperm and eggs to produce a new lamb. He wanted to make a genetic copy —a **clone**—of a fully grown sheep. He thought he could do so by slipping the nucleus of an adult's body cell into an unfertilized egg that had its own nucleus gently sucked out beforehand. He succeeded. One egg that his team modified developed into a cloned lamb, which they named Dolly.

Dolly grew up and later gave birth to six lambs of her own (Figure 13.1). Since then, researchers all over the world have been using adult DNA to make identical copies of other adult mammals. Mice, rabbits, pigs, cattle, goats, mules, deer, horses, and cats have all been cloned.

Sheep normally do not show symptoms of old age until they are about ten years old. By age five, Dolly had become arthritic and overweight. Less than a year later, an infection in her lungs proved irreversible and she was put to sleep.

Did Dolly develop health problems simply because she was a clone? Earlier studies of her telomeres had raised suspicions. Telomeres are short segments that cap the ends of chromosomes and stabilize them. They become shorter and shorter as an animal ages. When Dolly was only two years old, telomeres in some of her cells were as short as those of a six-year-old sheep—the exact age of the adult animal that was her genetic donor.

Using adult DNA to clone mammals is challenging. Most clones die before birth or shortly afterward. It took almost seven hundred attempts to get a clone of a guar, a wild ox on the endangered species list. Less than two days after his birth, he died of complications following an infection.

The clones that do survive often have health problems. Like Dolly, many become unusually overweight as they age. Other clones are exceptionally large from birth or have some enlarged organs. Cloned mice develop lung and liver problems, and almost all die prematurely. Cloned pigs have heart problems, they limp, and one never did develop a tail or, worse still, an anus.

Physically moving the DNA from an adult's cell into an egg stripped of its own nucleus is only part of the challenge. Most genes in a mature cell are inactive. To guide the reproductive process, they have to be reprogrammed or switched on in controlled ways. Apparently, not all of the genes in all clones are being properly activated.

Some people want to put a stop to adult cloning because the risk of bringing defective mammals into the world troubles them deeply. Other people want research into reprogramming DNA to continue. For instance, they point to patients who are desperate for organ transplants. People are already cloning pigs that were genetically modified to produce organs that human donors are less likely to reject. A few people on the fringes of bioethical common sense are toying with the idea of reprogramming DNA to clone an adult human.

Is DNA amazing? It certainly is. Are we amazing in our capacity to use and abuse its potential? You bet.

Figure 13.1 Dolly and one of her lambs. Dolly was the first mammal to be formed by way of adult DNA cloning. She awakened society to the goings-on of the molecular revolution by jarring our notions of what it takes to reproduce a complex animal.

Watch the video online!

Figure 13.2 Watson, Crick, and the model for DNA that brought about a revolution in molecular biology.

With this chapter, we move past the chromosomal basis of inheritance and turn to the investigations and models that led to our current understanding of DNA (Figure 13.2). The story is more than a march through the details of how its molecular structure encodes hereditary information. *It also is revealing of how ideas are generated in science.*

On the one hand, having a shot at fame and fortune quickens the pulse of men and women in any profession, and scientists are no exception. On the other hand, science proceeds as a community effort, with individuals sharing not only what they can explain but also what they do not understand. Even when an experiment fails to produce the anticipated results, it may turn up information that others can use or lead to questions that others can answer. Unexpected results, too, might be clues to something important about the natural world.

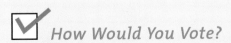 ## How Would You Vote?

Abnormal animals often form during animal cloning experiments, but cloning research may also result in new drugs and organ replacements for human patients. Should animal cloning be banned? See BiologyNow for details, then vote online.

 ## Key Concepts

DISCOVERY OF DNA'S FUNCTION

In all living cells, DNA molecules are storehouses of information that governs heritable traits. Section 13.1

THE DNA DOUBLE HELIX

DNA is a double-stranded molecule consisting of four kinds of nucleotides: adenine, thymine, guanine, and cytosine. The two strands coil together helically, like a spiral stairway.

Each nucleotide base of one strand is hydrogen-bonded to a base of the other strand. As a rule, adenine pairs with thymine, and guanine with cytosine.

The order in which one kind of base follows the next in a strand encodes heritable information. The DNA of each species has at least some unique base sequences that are not found in the DNA of any other species. Section 13.2

HOW CELLS DUPLICATE THEIR DNA

Before a cell divides, enzymes and other proteins replicate the DNA; they make a copy of it. Different kinds unwind the double helix and construct a new, complementary strand on the exposed bases of each parent strand. They do so according to the base-pairing rule for DNA.

Repair enzymes monitor mismatched base pairings and other changes in the DNA strands. Section 13.3

DNA AND THE CLONING CONTROVERSIES

What does it mean when we say that DNA holds heritable information? The answer hits home when the messages encoded in its base sequences are tapped to make an exact copy of an adult animal. Section 13.4

THE FRANKLIN FOOTNOTE

As in any profession, some were winners and some losers in the DNA chase. Some players might have taken less-than-noble shortcuts. Section 13.5

 ## Links to Earlier Concepts

This chapter builds on your understanding of hydrogen bonding (Section 2.4), condensation reactions (3.2), and the earlier overview of DNA structure and function (3.7). Your knowledge of chromosomes, mitosis, and meiosis will help you understand the nuclear transfers that are part of cloning procedures (9.3, 10.3). Keep the image of the eight-cell stage of human development in mind when you read about embryo cloning, because the cells used are no more developed than this (12.11).

13.1 The Hunt for Fame, Fortune, and DNA

Why, in the spring of 1868, was Johann Miescher collecting cells from the pus of open wounds and, later, from sperm of a fish? This physician wanted to identify the chemical composition of the nucleus. Such cells have little cytoplasm, which makes it easier to isolate the nuclear material. In time he isolated an acidic compound that contains nitrogen and phosphorus. He had discovered what came to be known many years later as **deoxyribonucleic acid**, *or* **DNA**.

EARLY AND PUZZLING CLUES

At the time Miescher made his discovery, no one knew much about the physical basis of inheritance. That is, *which substance encodes the information about reproducing parental traits in offspring?* Few researchers thought that DNA might hold the answer. For a long time, most were thinking PROTEINS! Because heritable traits are so diverse, they assumed that molecules of inheritance had to be structurally diverse, too. Proteins, they said, consist of unlimited combinations of twenty kinds of amino acids. Other molecules just seemed too simple.

Now fast-forward to 1928. An army medical officer, Frederick Griffith, wanted to develop a vaccine against the bacterium *Streptococcus pneumoniae*, a major cause of pneumonia. He did not succeed, but he isolated and cultured two strains that unexpectedly shed light on inheritance. The colonies of one strain had a rough surface appearance; colonies of the other appeared smooth. Griffith designated the strains *R* and *S*, and he used them in a series of experiments (Figure 13.3).

First, he injected mice with live *R* cells. The mice did not develop pneumonia. *The R strain was harmless.*

Second, he injected other mice with live *S* cells. The mice died. Blood samples from them teemed with live *S* cells. *The S strain was pathogenic; it caused the disease.*

Third, he killed *S* cells by exposing them to high temperature. *Mice injected with dead S cells did not die.*

Fourth, he mixed live *R* cells with heat-killed *S* cells. He injected them into mice. The mice died—*and blood samples drawn from them teemed with live S cells!*

What went on in the fourth experiment? Maybe heat-killed *S* cells in the mix were not really dead. But if that were so, then the mice injected with heat-killed *S* cells in experiment 3 would have died. Or maybe the harmless *R* cells had mutated into a killer strain. But if that were so, then the mice injected with the *R* cells only in experiment 1 would have died.

The simplest explanation was this: *Heat had killed the S cells but did not destroy their hereditary material—including the part that specified "how to cause infection."* Somehow, that material had been transferred from the dead *S* cells into living *R* cells, which put it to use.

Further tests made it clear that the transformation was permanently heritable. Even after a few hundred generations, *S* cell descendants were still infectious.

What was the hereditary material that caused the transformation? Scientists started looking in earnest, but most were still thinking PROTEINS!

Still, Griffith's results intrigued Oswald Avery, who began to transform harmless bacteria by mixing them with extracts of killed pathogenic cells. Avery asked: What part of the extracts caused the transformation? He found that adding protein-digesting enzymes to the extracts had no effect; cells were still transformed. However, adding a DNA-digesting enzyme to extracts prevented transformation. DNA was looking good.

CONFIRMATION OF DNA FUNCTION

By the 1950s, Max Delbrück, Alfred Hershey, Martha Chase, Salvador Luria, and other molecular sleuths were using viruses for experiments. These infectious particles hold information on substances required to make new virus particles. After viruses infect a host cell, their enzymes trick its metabolic machinery into synthesizing those substances. **Bacteriophages**, which only infect certain bacteria, were the viruses of choice for the early experiments.

As researchers knew, some bacteriophages consist only of DNA and a coat, probably of protein. Also, as

Figure 13.3
Animated!
Summary of results from Fred Griffith's experiments with *Streptococcus pneumoniae* and laboratory mice.

1 Mice injected with live cells of harmless strain *R*.

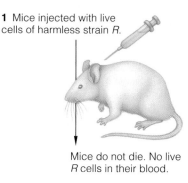

Mice do not die. No live *R* cells in their blood.

2 Mice injected with live cells of killer strain *S*.

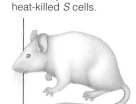

Mice die. Live *S* cells in their blood.

3 Mice injected with heat-killed *S* cells.

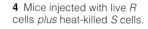

Mice do not die. No live *S* cells in their blood.

4 Mice injected with live *R* cells *plus* heat-killed *S* cells.

Mice die. Live *S* cells in their blood.

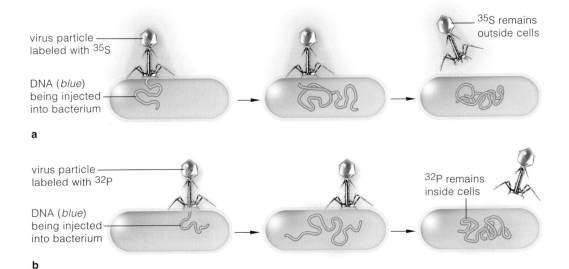

virus particle labeled with ^{35}S

DNA (*blue*) being injected into bacterium

^{35}S remains outside cells

a

virus particle labeled with ^{32}P

DNA (*blue*) being injected into bacterium

^{32}P remains inside cells

b

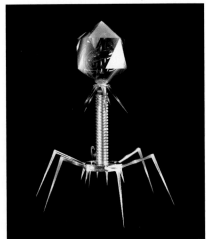

Figure 13.4 *Animated!* Example of the landmark experiments that tested whether genetic material resides in bacteriophage DNA, proteins, or both. Alfred Hershey and Martha Chase knew that sulfur (S) but not phosphorus (P) is a component of bacteriophage proteins. They also knew that phosphorus but not sulfur is a component of DNA.

(**a**) In one experiment, bacteria were grown in a culture medium with a tracer, the radioisotope ^{35}S. The cells used the ^{35}S when they built proteins. Bacteriophages infected the labeled cells, which started to make viral proteins. So the proteins of new virus particles became labeled with the ^{35}S. The labeled virus particles infected a new batch of unlabeled cells. The mixture was whirred in a kitchen blender. Whirring dislodged the viral coats from infected cells. Chemical analysis revealed the presence of labeled protein in the solution but only traces of it inside the cells.

(**b**) In another experiment, bacteriophages infected cells that had taken up the radioisotope ^{32}P. Later, the cells used ^{32}P when they built viral DNA. This labeled the DNA and new virus particles. The labeled viruses were used to infect bacteria in solution, then were dislodged from them. Most labeled viral DNA stayed in the cells—evidence that DNA is the genetic material of this virus.

c *Above*, model for a bacteriophage. *Below*, micrograph of virus particles injecting their DNA into an *E. coli* cell.

micrographs revealed, the coat remains on the *outer surface* of infected cells. Did viruses inject hereditary material only into cells? If so, then was the material protein, DNA, or both? Figure 13.4 outlines just two of many experiments that pointed to DNA.

Then Linus Pauling did something no one had done before. With his training in biochemistry, a talent for model building, and a dose of intuition, he deduced the structure of a protein—collagen. His discovery was electrifying. If someone could pry open the secrets of proteins, then why not DNA? And if DNA's structural details were deduced, would those details hold clues to how DNA functions in inheritance? *Someone could go down in history as having discovered the secret of life!*

ENTER WATSON AND CRICK

Scientists started to scramble after the prize. Among them were Francis H. Crick, a Cambridge University, researcher, and James Watson, a postdoctoral fellow recently arrived from Indiana University. They spent

hours arguing over everything they had read about DNA's size, shape, and bonding requirements. They fiddled with cardboard cutouts, and they badgered chemists to help them identify possible bonds they might have overlooked. They built models from thin bits of metal connected with wire "bonds."

In 1953, Watson and Crick built a model that fit all the pertinent biochemical rules and all the clues they had gleaned from other sources. They had discovered the structure of DNA. The molecule has breathtaking simplicity, and it helped Crick answer another riddle —*how life can show unity at the molecular level and still give rise to such spectacular diversity at the level of whole organisms.* Turn now to high points in the community effort that gave us these insights.

DNA functions as the cell's treasurehouse of inheritance. The cumulative efforts of many scientists, building on one another's work, resulted in the discovery of that function.

13.2 The Discovery of DNA's Structure

LINKS TO
SECTIONS
3.2, 3.7

Long before the bacteriophage studies were under way, biochemists knew that DNA contains only four kinds of nucleotides that are the building blocks of nucleic acids. But how were the nucleotides arranged in DNA?

DNA'S BUILDING BLOCKS

Recall, from Section 3.7, that a **nucleotide** in DNA has a five-carbon sugar (deoxyribose), a phosphate group, and one of the following nitrogen-containing bases:

adenine	guanine	thymine	cytosine
A	G	T	C

T and **C** are pyrimidines, with a backbone of carbon and nitrogen that forms a single ring. **A** and **G** are purines—larger, bulkier molecules having two rings. Overall, the four types of nucleotides have the same bonding pattern, as Figure 13.5 indicates.

By 1949, the biochemist Erwin Chargaff had shared with the scientific community two insights about the proportions of nucleotides in DNA. First, the amount of adenine relative to guanine differs among species. Second, the amounts of thymine and adenine in DNA are identical, and so are the amounts of cytosine and guanine. We may show this as **A=T** and **G=C**.

These symmetrical proportions had to mean something. As biochemists already knew, the nucleotides in DNA are joined to one another by way of condensation reactions that form long chains (Section 3.2). But how were the four kinds arranged in a chain, and in what order?

The first convincing clue to the actual arrangement emerged from Maurice Wilkins's research laboratory at Cambridge, England. Researcher Rosalind Franklin made exceptional **x-ray diffraction images** of DNA. Such images form after a beam of x-rays is directed at a molecule, which scatters the x-rays in a pattern that can be captured on film. The pattern consists only of dots and streaks; in itself, it is not the structure of the molecule. However, researchers can use it to calculate the positions of the molecule's atoms.

Before Franklin, researchers had been working with dehydrated DNA molecules. Franklin was the first to put DNA into a "wet" form—which is the form that occurs in cells—and make an exceptionally clear image of it. With that image, she painstakingly calculated that the DNA molecule is long and thin, and that it has a 2-nanometer diameter. She also found repeats of some molecular configuration every 0.34 nanometer along its length, and another repeat every 3.4 nanometers. These were crucial clues, but her part in the discovery process was downplayed until recently (Section 13.5).

What did the repeating variation in DNA mean? Could DNA be coiled along its length, like a circular stairway? Certainly Pauling thought so. After all, he had calculated that collagen is helically coiled. Like

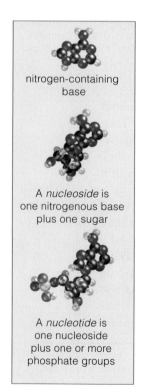

nitrogen-containing base

A *nucleoside* is one nitrogenous base plus one sugar

A *nucleotide* is one nucleoside plus one or more phosphate groups

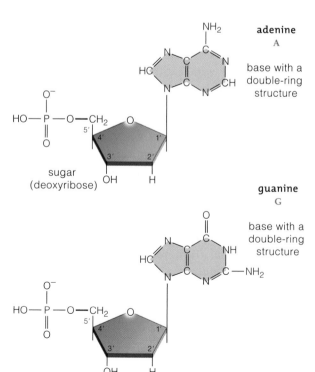

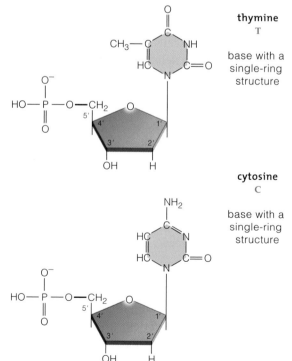

Figure 13.5 Four kinds of nucleotides in the DNA molecule.

many others—including Wilkins, Watson, and Crick—he was thinking "helix." As Watson later wrote, "We thought, why not try it on DNA? We were worried that *Pauling* would say, why not try it on DNA? Certainly he was a clever man. He was a hero of mine. But we beat him at his own game. I still can't figure out why."

Pauling, it turned out, made a big chemical mistake. His model had all the negatively charged phosphate groups facing the interior of the DNA helix instead of facing outward. If they were that close together, they would repel each other too much to remain stable.

PATTERNS OF BASE PAIRING

Franklin filed away her image of wet DNA, but it still came to the attention of Watson and Crick. From all the clues that had accumulated, they perceived that DNA must consist of two strands of nucleotides, held together at their bases by hydrogen bonds (Figure 13.6). Such bonds form when the two strands run in opposing directions and twist to form a double helix. Only two kinds of base pairings typically form along the molecule's length: **A—T** and **G—C**.

This bonding pattern accommodates variation in the order of bases. For instance, a stretch of DNA from a rose, a human, or any other organism might be:

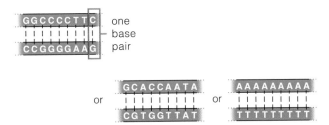

All DNA molecules show the same bonding pattern. Many stretches of base sequences are the same in all of them. But some are unique for each species and even vary among individuals of a species! *The constancy in DNA's bonding pattern is the basis for life's unity—and variation in base sequences is the basis for life's diversity.*

Intriguingly, computer simulations show that if you want to pack a string into the least space, coil it into a helix. Was this space-saving advantage a factor in the molecular origin of the DNA double helix? Maybe.

The pattern of base pairing between the two strands in DNA is constant for all species—A with T, and G with C. However, each species has a number of unique sequences of base pairs along the length of its DNA molecules.

Figure 13.6 *Animated!*
Composite of different ways to represent the DNA double helix. Two sugar–phosphate backbones run in *opposing* directions. Think of the sugar units (deoxyribose) of one strand as being upside down.

By comparing the numerals used to identify each carbon atom of DNA's backbone (1′, 2′, 3′, and so on), you see that one strand runs in the 5′→3′ direction and the other runs in the 3′→5′ direction.

2-nanometer diameter overall

0.34-nanometer distance between each pair of bases

3.4-nanometer length of each full twist of the double helix

In all respects shown here, the Watson–Crick model for DNA structure is consistent with the known biochemical and x-ray diffraction data.

The pattern of base pairing (A with T, and G with C) is consistent with the known composition of DNA (A = T, and G = C).

13.3 DNA Replication and Repair

LINKS TO
SECTIONS
9.2, 9.5, 12.8

The discovery of DNA structure was a turning point in the studies of inheritance. Crick saw at once how a parent cell could make copies of that molecule for its daughter cells.

HOW DNA IS DUPLICATED

Semiconservative Replication Until Watson and Crick presented their model, no one could explain **DNA replication**, or how the molecule of inheritance is duplicated before a cell divides. Such replication takes place in interphase of the cell cycle.

Enzymes easily break the hydrogen bonds between the two nucleotide strands of a DNA molecule. When enzymes and other proteins act on the molecule, one strand unwinds from the other and exposes stretches of its nucleotide bases. Cells contain stockpiles of free nucleotides that can pair with the exposed bases.

Each parent strand stays intact, and a companion strand is assembled on each one according to the base-pairing rules **A** to **T**, and **G** to **C**. As soon as a stretch of a new, partner strand forms on a stretch of a parent strand, the two twist together in a double helix. Each parent strand is conserved during replication, so that half of every double-stranded DNA molecule is "old" and half is "new." Figures 13.7 and 13.8 offer a look into this process, called *semiconservative* replication.

Replication Enzymes DNA replication uses a team of molecular workers; Table 13.1 lists important ones. Signaling molecules that are part of the controls over the cell cycle activate replication enzymes. Starting at certain regions along the length of the DNA double helix, enzymes called **helicases** unzip the hydrogen bonds, which are individually weak and easy to break (Section 2.4). The two strands of the double helix now unwind from each other in both directions from the unzipped sites. The two parent strands are prevented from winding back together because small proteins temporarily bind with them.

Next, **DNA polymerases** catalyze the formation of two brand-new strands of DNA from free nucleotides. These enzymes also catalyze the hydrogen bonding of each new strand to the unwound region of one of the two parent DNA strands. But they can assemble new strands only in the 5′ → 3′ direction.

Check out Figure 13.5 to see what this directional strand assembly means. A free nucleotide has a tail of three phosphate groups dangling from the 5′ carbon of their sugar component. DNA polymerase splits off two. The energy released drives the attachment of the last phosphate to an —OH group dangling from the 3′ carbon of a sugar—which belongs to the most recent nucleotide addition to the growing strand.

What about the parent strand that runs the other way? Nucleotides are assembled in short stretches on the parent strand, then **DNA ligases** seal the stretches

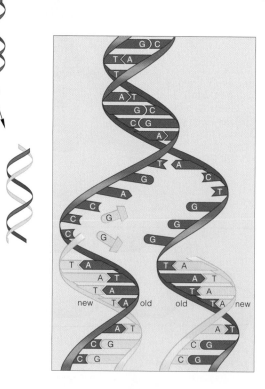

Figure 13.7 The semiconservative nature of DNA replication. The original two-stranded DNA molecule is coded *blue*. Each parent strand remains intact. One new strand (*gold*) is assembled on each of the parent strands.

Table 13.1	Three of the Enzymes With Roles in DNA Replication and Repair

Helicases

Catalyze the breaking of hydrogen bonds between base pairs in the DNA molecule, which unzips in two directions from double-stranded to single-stranded form. Protein factors work with helicases to keep the two parent strands unwound. The helicases are ATP-driven motors, similar to ATP synthases.

DNA polymerases

Catalyze the additions of free nucleotides to each new strand of deoxyribonucleases on a parent DNA template. Also proofread; some DNA polymerases can reverse direction by one base pair and correct mismatches, which occur once in every thousand or so additions.

DNA ligases

Catalyze the sealing-together of short stretches of new nucleotides, which are assembled discontinuously on one of the parent DNA strands. Also can seal strand breaks.

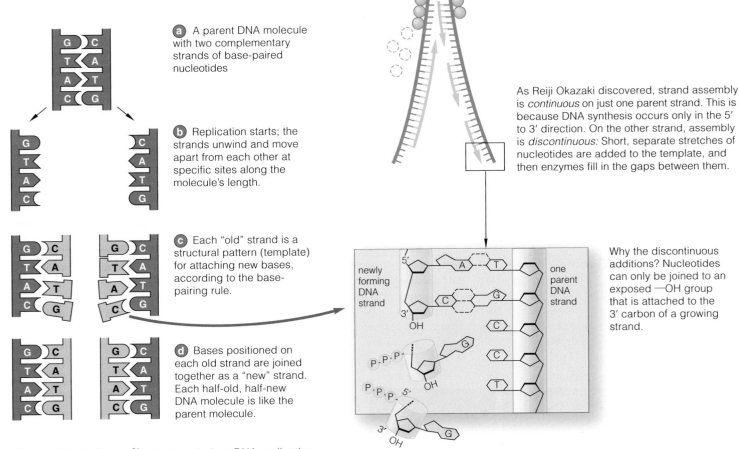

a A parent DNA molecule with two complementary strands of base-paired nucleotides

b Replication starts; the strands unwind and move apart from each other at specific sites along the molecule's length.

c Each "old" strand is a structural pattern (template) for attaching new bases, according to the base-pairing rule.

d Bases positioned on each old strand are joined together as a "new" strand. Each half-old, half-new DNA molecule is like the parent molecule.

As Reiji Okazaki discovered, strand assembly is *continuous* on just one parent strand. This is because DNA synthesis occurs only in the 5' to 3' direction. On the other strand, assembly is *discontinuous*: Short, separate stretches of nucleotides are added to the template, and then enzymes fill in the gaps between them.

Why the discontinuous additions? Nucleotides can only be joined to an exposed —OH group that is attached to the 3' carbon of a growing strand.

Figure 13.8 Animated! A closer look at DNA replication.

together into a continuous strand. A complementary strand now winds up with the parent strand that was its template. Two DNA double helixes are the result.

As you will read in Chapter 16, DNA polymerases and DNA ligases also have been put to good use as tools in recombinant DNA technology.

FIXING MISMATCHES AND BREAKS

Over the long term, changes in chromosomal DNA can give rise to variations in traits that help define a species (Section 12.8). However, in terms of a lifetime, the individual may not survive if something changes its DNA. For instance, on rare occasions, the wrong nucleotide is base-paired to a parent template (Table 13.1). Unless such mistakes are reversed, they might alter or weaken the functions of genes or the protein products. Chapter 12 has a number of examples of the kinds of genetic disorders that can follow changes.

DNA proofreading mechanisms swiftly fix most errors in replication and most of the strand breaks. For example, some DNA polymerases proofread new base pairings. They can reverse catalytic additions by one base and correct a mismatch. When they cannot,

replication is arrested, and controls over the cell cycle come into play (Sections 9.2 and 9.5).

Mismatches that slip past the proofreaders are only one type of DNA damage. One or both backbones of the double helix may break, as by ionizing radiation and some chemicals (Section 14.5). Also, a base in one strand may become covalently bonded to a base in the same strand or the partner strand.

Specialized sets of **repair enzymes** can repair some changes; they recognize and snip out a damaged site or mismatches. For instance, some glycosylases excise a mismatched base and replace it with a suitable one in a tiny burst of DNA synthesis.

DNA is replicated prior to cell division. At certain sites, helicases unzip hydrogen bonds between its two strands, so the double helix unwinds. Each strand remains intact —it is conserved—and DNA polymerases assemble a new, complementary strand on each parent strand.

DNA proofreading mechanisms and special sets of repair enzymes fix nearly all mismatched base pairs. Different kinds also seal or bypass most breaks.

13.4 Using DNA To Duplicate Existing Mammals

LINKS TO
SECTIONS
9.3, 10.2, 12.11

Here we return to a topic introduced at the start of this chapter. Researchers can now isolate DNA from an adult mammal and use it to bypass meiosis, gamete formation, and fertilization. Unlike sexual reproduction, which gives rise to mixes of traits from two parents, adult DNA cloning produces an exact genetic copy of a single adult.

"Cloning" can be a confusing word. It can apply to a method in recombinant DNA technology that makes multiple copies of DNA fragments. It also applies to natural and manipulated interventions in the steps of reproduction and development. These interventions are called embryo cloning, adult cloning, and therapeutic cloning (Table 13.2).

Embryo Cloning Embryo cloning occurs all the time in nature. For instance, soon after fertilized human eggs start dividing, a tiny ball of cells has formed. You saw one of these early developmental stages in Section 12.11. Once in every seventy-five or so pregnancies, the ball splits, and the two parts grow and develop into identical twins. A laboratory

procedure, "artificial twinning," simply simulates what goes on in nature. For instance, the balls of cells grown from fertilized cattle eggs in petri dishes are encouraged to split into identical-twin embryos. Then they are implanted in surrogate mothers, which give birth to cloned calves.

Embryo cloning has been practiced for decades. However, embryo clones inherit DNA from two parents, not one. If breeders are looking for a particularly valued trait, such as better milk production or mating vigor, they have to wait for clones to grow up to see if they display the trait.

Adult Cloning Because it takes so long to observe the outcome, some researchers were looking for an alternative to embryo cloning of cattle and other complex animals. It seemed that cloning a differentiated cell would be far more efficient, because the desired phenotype was already there, right in front of them.

You may wonder what "differentiated" means. As you already know after reading about mitosis and meiosis, all cells descended from a fertilized egg inherit the same chromosomal DNA (Sections 9.3 and 10.2). However, as the new embryo develops, different cells inside it start to select and use DNA's information in different ways. Their selections commit them to becoming liver cells,

 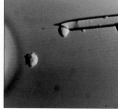 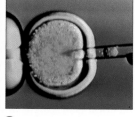

① A microneedle is about to remove the nucleus from an unfertilized sheep egg (*center*).

② The microneedle has now emptied the sheep egg of its own nucleus, which held the DNA.

③ A nucleus from a donor cell (in the microneedle) is about to be deposited in the egg that was stripped of its nucleus.

④ An electric current will stimulate the egg to enter mitotic cell division. After a few rounds of divisions, the ball of cells will be implanted in the womb of a female sheep (ewe).

the first sheep cloned from adult DNA

Figure 13.9 *Animated!* Transfer of an adult nucleus that led to Dolly.

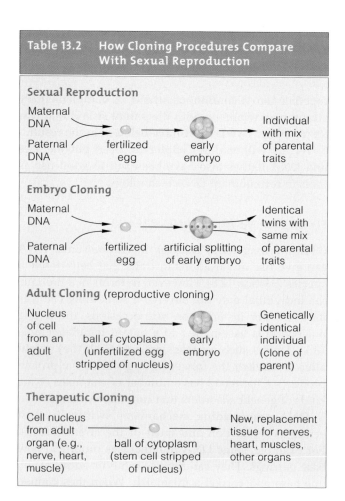

Table 13.2 How Cloning Procedures Compare With Sexual Reproduction

Sexual Reproduction

Maternal DNA / Paternal DNA → fertilized egg → early embryo → Individual with mix of parental traits

Embryo Cloning

Maternal DNA / Paternal DNA → fertilized egg → artificial splitting of early embryo → Identical twins with same mix of parental traits

Adult Cloning (reproductive cloning)

Nucleus of cell from an adult → ball of cytoplasm (unfertilized egg stripped of nucleus) → early embryo → Genetically identical individual (clone of parent)

Therapeutic Cloning

Cell nucleus from adult organ (e.g., nerve, heart, muscle) → ball of cytoplasm (stem cell stripped of nucleus) → New, replacement tissue for nerves, heart, muscles, other organs

Figure 13.10 *Left*, DNA donor Tahini, a Bengal cat. *Right*, Tabouli and Baba Ganoush, two of her clones. Eye color changes as a Bengal cat matures; in both clones, it will in time be the same as Tahini's. All three are household pets of the founder of a pet cloning company that, precisely because of its success, is at the center of a cloning controversy.

blood cells, or other specialists in structure, composition, and function in the adult. The mechanisms by which this happens are topics of later chapters.

With respect to adult cloning, a "grown-up" cell—one already differentiated—must be tricked into rewinding the developmental clock. Each of its descendants will have to start using a different subset of its DNA all over again to form a clone of its original owner.

Nuclear transfer is one way to trick a differentiated cell. A researcher with a good microscope and a very steady hand replaces the nucleus of an unfertilized egg with one from a differentiated cell of an adult animal, as in Figure 13.9. Small doses of chemicals or electric shocks may induce the cell to divide. If all goes well, then a cluster of embryonic cells forms and can be implanted inside a surrogate mother. In Dolly's case, the donor nucleus came from a cell from the lining of a sheep's udder.

A recipient egg is not always tricked into using the DNA as it is supposed to. For example, researchers studied many genes in the first cloned mice and found that 4 percent or so were being abnormally expressed. The cloning procedure had disrupted when and how mouse cells were supposed to use their genes. Even so, researchers around the world are becoming better at rewinding the developmental clock.

For instance, Genetic Savings and Clone is a company that uses adult DNA to clone beloved pet cats that are old and dying. So far, the pet owners who have requested the procedure report that their clones are healthy, lively, and uncannily like the DNA donor, not only in appearance but also in behavior. Figure 13.10 shows an example. Compare the markings on the top of the head, the face, the legs, and the tail of the donor and of the two clones.

Some people are deeply offended by the idea of spending tens of thousands of dollars to clone a cat when so many lost or orphaned cats are awaiting adoption in animal shelters. Some owners say that they have bonded deeply with their particular pet, that they would grieve mightily over its loss, and that it is, after all, their money.

The real issue, of course, is that *humans*—like cats, mice, and pigs—are mammals. Not very long ago, mammalian cloning was fraught with technical problems. It still is. As researchers get better at what they do, however, the use of adult DNA cloning to make a genetic copy of a human no longer seems in the realm of science fiction. That is why most countries recently banned the use of federal funding for any research into adult human cloning.

Therapeutic Cloning *Therapeutic* cloning also uses nuclear transfers. In this case, the idea is to transplant DNA of a somatic cell from the heart, liver, muscles, or nerves into a stem cell. A *stem cell*, recall, is one that has not yet differentiated and retains the capacity to divide (Section 12.2). The descendant cells go on to differentiate into cell types of specific tissues and organs. The process is known as *somatic cell nuclear transfer* (SCNT).

SCNT already has produced stem cells that are an exact genetic match to an individual. This is not reproduction; no sperm are used. However, descendants of the modified cells may be able to regenerate a tissue for transplant back into a patient affected by an incurable disease or a spinal cord injury. Potentially, SCNT could regenerate organs. There are long waiting lists for organ transplants, and those who receive them have to take drugs to suppress the immune system for the rest of their lives.

We will return to some of these issues later on, after you learn more about the molecular basis of life. Make your own informed decisions about them, and remember this: For better or worse, our capacity to manipulate DNA started with Watson, Crick, and so many others who shared knowledge that became the underpinnings of a brave new world.

13.5 Rosalind's Story

FOCUS ON BIOETHICS

There is a saying among researchers in any discipline—publish or perish. As soon as Watson and Crick's structural model of DNA fell into place, they immediately published a one-page paper that dazzled the world. All others who had helped fill in crucial pieces of the puzzle, including Franklin, received little or no recognition. Franklin's contribution is now receiving more attention.

Rosalind Franklin arrived at King's Laboratory in London with impressive credentials (Figure 13.11). She developed a refined x-ray diffraction method while studying the structure of coal in Paris. She took a new mathematical approach to interpreting x-ray diffraction images and, like Pauling, had built three-dimensional molecular models. Now she was asked to create and run a state-of-the-art x-ray crystallography laboratory. Her assignment was to investigate the structure of DNA.

No one bothered to tell Franklin that, just down the hall, Maurice Wilkins was already working on the puzzle. Even a graduate student assigned to assist her failed to mention it. No one bothered to tell Wilkins about Franklin's assignment; he assumed that she was a technician hired to do his x-ray crystallography work because he didn't know how to do it himself. And so a clash began. To Franklin, Wilkins seemed inexplicably prickly. To Wilkins, Franklin was appalling in her lack of deference to him.

Wilkins had a prized cache of DNA, which he gave to his "technician." Five months later, Franklin gave a talk on what she had learned so far. DNA, she said, may have two, three, or four parallel chains twisted into a helix, with phosphate groups projecting outward.

With his crystallography background, Crick would have recognized the significance of her report—*if* he had been there. (A *pair* of chains oriented in opposing directions would be the same even if flipped 180 degrees. Two pairs of chains? No. DNA's density ruled that out. But one pair of chains? Yes!) Watson was in the audience but did not know what Franklin was talking about.

Later on, Franklin produced her superb x-ray diffraction image of wet DNA fibers (Figure 13.12). The image fairly screamed *HELIX!* Franklin also worked out the length and the diameter of DNA. However, she had been working with dry fibers for a long time, and she chose not to dwell on the meaning of her new data. Wilkins did.

In 1953, without Franklin's knowledge, he let Watson see that image and reminded him of what she had reported more than a year before. When Watson and Crick did focus on her data, they had the final bit of information that they needed to build a plausible model of DNA—one with two helically twisted chains running in opposing directions.

Figure 13.11 Portrait of Rosalind Franklin arriving at Cambridge in style, from Paris.

Figure 13.12 Franklin's best x-ray diffraction image of DNA fibers.

Summary

Section 13.1 Experimental tests that used bacteria and bacteriophages offered the first solid evidence that DNA is the hereditary material in living organisms.

Biology⊛Now
Learn about experiments that revealed the function of DNA with the animation on BiologyNow.

Section 13.2 The nucleotide monomers of DNA have a five-carbon sugar (deoxyribose), a phosphate group, and one of four kinds of nitrogen-containing bases: adenine, thymine, guanine, or cytosine.

A DNA molecule consists of two nucleotide strands coiled together into a double helix. The bases of one strand hydrogen-bond with bases of the other.

Bases of the two DNA strands pair in a constant way. Adenine pairs with thymine (A–T), and guanine with cytosine (G–C). Which base follows another along a strand varies among species. The DNA of each species incorporates some number of unique sequences of base pairs that set it apart from the DNA of all other species.

Biology⊛Now
Investigate the structure of DNA with the animation on BiologyNow.

Section 13.3 In DNA replication, enzymes called helicases unwind the DNA double helix, and small proteins hold the two strands apart. DNA polymerases covalently bond free nucleotides into chains in a base sequence complementary to the parent strand that serves as its template. Two double-stranded DNA molecules result. One strand of each is old (is conserved); the other strand is new.

Strand assembly occurs only at an exposed —OH group at the 3′ end of a growing nucleotide strand. DNA ligases seal tiny gaps between short stretches of nucleotides in one of the growing strands.

Proofreading mechanisms fix most base-pairing mistakes and strand breaks. Special repair enzymes recognize and snip out damaged sites in the DNA as well as mismatches.

Biology⊛Now
See how a DNA molecule is replicated with the animation on BiologyNow.

Section 13.4 "Cloning" means copying fragments of DNA in recombinant DNA work. It also refers to three other procedures: Embryo cloning results in genetically identical twins. Adult cloning results in a genetically identical copy of an existing adult. Therapeutic cloning is a proposed method of producing stem cells that are an exact genetic match of a patient, the idea being to regenerate tissues and possibly organs.

Biology⊛Now
Observe the procedure used to create Dolly and other clones with animation on BiologyNow.

Section 13.5 Science advances as a community effort that is both cooperative and competitive. Ideally, individuals share their work and recognition for honors that come their way. As in all human endeavors, some fail to receive suitable recognition for their contribution.

Self-Quiz

Answers in Appendix II

1. Which is *not* a nucleotide base in DNA?
   a. adenine   c. uracil   e. cytosine
   b. guanine   d. thymine   f. All are in DNA.

2. What are the base-pairing rules for DNA?
   a. A–G, T–C   c. A–U, C–G
   b. A–C, T–G   d. A–T, G–C

3. One species' DNA differs from others in its _____ .
   a. sugars   c. base sequence
   b. phosphates   d. all of the above

4. When DNA replication begins, _____ .
   a. the two DNA strands unwind from each other
   b. the two DNA strands condense for base transfers
   c. two DNA molecules bond
   d. old strands move to find new strands

5. DNA replication requires _____ .
   a. free nucleotides   c. many enzymes
   b. new hydrogen bonds   d. all of the above

6. _____ is the basis for the life's diversity.
   a. constancy in DNA's   b. variation in the base
       bonding pattern   sequences in DNA

7. Adult cloning starts with _____ .
   a. an early embryo   c. artificial twinning
   b. nuclear transfers   d. both b and c

8. Match the terms appropriately.
   ____ bacteriophage   a. nitrogen-containing base
   ____ clone   bonded to a sugar and one
   ____ nucleotide   or more phosphate groups
   ____ helicase   b. breaks hydrogen bonds,
   ____ DNA ligase   starts unwinding of DNA
   ____ DNA polymerase   during replication
                      c. only DNA and protein
                      d. fills in gaps, seals breaks
                        in a DNA strand
                      e. carbon copy of dad or mom
                      f. adds nucleotides to a
                        growing DNA strand

Additional questions are available on Biology ⒺNow™

Critical Thinking

1. A pathogenic strain of *E. coli* has acquired an ability to produce a dangerous toxin that causes medical problems and fatalities. It is especially dangerous to young children who eat undercooked or raw contaminated beef. Develop hypotheses to explain how a normally harmless bacterium such as *E. coli* can become a pathogen.

2. Matthew Meselson and Franklin Stahl's experiments supported a semiconservative model of DNA replication. These researchers obtained "heavy" DNA by growing *Escherichia coli* in a medium enriched with ^{15}N, a nitrogen radioisotope. They prepared "light" DNA by growing *E. coli* in the presence of ^{14}N, the more common isotope. An available technique helped them identify which replicated

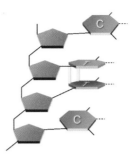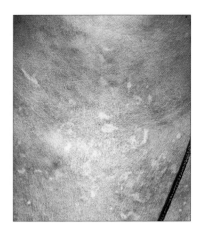

Figure 13.13 *Above*, a thymine dimer in a DNA strand. It can lead to xeroderma pigmentosum (*right*), a genetic disorder.

molecules were heavy, light, or hybrid (one heavy strand and one light). Use pencils of two colors, one for heavy strands and one for light. Assuming a DNA molecule has two heavy strands, arrange pencils to show how daughter molecules form after replication in a medium with ^{14}N. Represent four DNA molecules that form if the daughter molecules are replicated in the ^{14}N medium.

3. If you are part of a biology class, split into groups. See which one writes up the clearest, most concise description of the component parts and organization of this molecule:

4. Mutations, remember, are permanent changes in DNA base sequences—the original source of genetic variation and the raw material of evolution. Yet how can mutations accumulate, given that cells have repair systems that fix changes or breaks in DNA strands?

5. In 1999, scientists discovered a woolly mammoth that had been frozen in glacial ice for the past 20,000 years. They thawed it very carefully so they could use its DNA to clone a woolly mammoth. It turns out there was not enough preserved material to work with. They plan to try again the next time a frozen woolly mammoth comes along. Reflect on Section 13.4, then speculate on the pros and cons of cloning an extinct animal.

6. *Xeroderma pigmentosum* (XP) is an autosomal recessive disorder that is characterized by the rapid formation of skin sores that can develop into cancers (Figure 13.13). Affected individuals have no mechanism for dealing with the damage that ultraviolet (UV) light can inflict on skin cells. They must avoid all forms of radiation—including sunlight and fluorescent lights.

Affected individuals do not have functioning DNA repair mechanisms. James Cleaver discovered this when he studied what happens in cells when DNA's nitrogen-containing bases absorb UV light. A covalent bond can form between two thymine bases in the same DNA strand. The resulting thymine dimer puts a kink in the strand. Propose what some of the consequences might be during interphase, when most proteins are synthesized, and then during DNA replication.

Ricin and Your Ribosomes

In 2003, police acted on an intelligence tip and stormed a London apartment, where they found laboratory glassware and castor oil beans (Figure 14.1). They arrested a few young men and reminded the world that unconscionable people still view ricin as a bioweapon.

The castor oil plant (*Ricinus communis*) has ricin in all of its tissues, but its oil is valued as an ingredient in many plastics, paints, cosmetics, textiles, and adhesives. The oil —and ricin—is most concentrated in the seeds (beans), but the ricin is discarded when the oil is extracted.

A dose of ricin as small as a grain of salt can kill you; only plutonium and botulism toxin are more deadly. Researchers knew about ricin's lethal effects as long ago as 1888. During World War I, when deadly chlorine and mustard gases were wafting across battlefields, England and the United States investigated ricin's potential use as a weapon. Both countries shelved the research when the war ended.

Now fast-forward to 1969, at the height of the Cold War between Russia and the West. Georgi Markov, a Bulgarian writer, had defected to England. As he strolled down a busy London street, an assassin jammed the tip of a modified umbrella into one of Markov's legs. The tip held a tiny ball laced with ricin. Markov died in agony three days later.

Ricin is on stage once again. In 2004, traces were found in a United States Senate mailroom and State Department building, and in an envelope addressed to the White House. In 2005, the FBI arrested a man who had castor oil beans, substances consistent with ricin production, and an AK-47 stashed in a home in Florida.

How does ricin exert its deadly effects? *It inactivates ribosomes, the protein-building machinery of all cells.*

Ricin is a protein with two polypeptide chains. One chain helps ricin insert itself into cells. The other chain serves as an enzyme. Its catalytic action wrecks part of the ribosome where amino acids are assembled into proteins. It yanks adenine subunits out of an RNA molecule that is a crucial component of the ribosome's three-dimensional structure. Once that happens, the ribosome's shape unravels, protein synthesis stops, and cells spiral toward death. So does the individual. There is no antidote.

You can go about your business without ever knowing what a ribosome is or what it does. However, you also can recognize that protein synthesis is not a topic invented to torture biology students. It is something worth knowing about and appreciating for how it keeps us alive—and for appreciating anti-terrorism researchers who are working to keep us that way.

Watch the video online!

Figure 14.1 *Left*, castor oil plant seeds, source of the ribosome-busting ricin. *Right*, model for one of ricin's two polypeptide chains. This chain helps ricin penetrate living cells. The other one destroys the capacity for protein synthesis, and for life.

Start with what you already know about DNA, the book of protein-building information in cells. The alphabet used to write the book seems simple enough—just A, T, G, and C, for the four nucleotide bases adenine, thymine, guanine, and cytosine. But how do you get from an alphabet to a "word"— a protein? The answer starts with the order, or sequence, of the four nucleotide bases in a DNA molecule.

As you know, when a cell replicates its DNA, the two nucleotide strands of the DNA double helix unwind from each other completely. At other times, however, enzymes selectively unwind the two strands in certain regions, which exposes the base sequences of genes. Most genes encode information about specific proteins.

It takes two big steps, **transcription** and **translation**, to get from the sequence of nucleotide bases in genes to the sequence of amino acids in a protein. In eukaryotic cells, the first step occurs inside the nucleus. A newly exposed DNA base sequence functions as a structural pattern, or a template, for making a strand of ribonucleic acid (RNA) from the cell's pool of free ribonucleotides.

The RNA moves into the cytoplasm, where it becomes translated. In this second step of protein synthesis, the RNA guides the assembly of amino acids into a new polypeptide chain. These are the chains that twist and fold into the three-dimensional shapes of proteins.

In short, RNA is transcribed on DNA templates, then RNA is translated into proteins:

$$\text{DNA} \xrightarrow{\textit{transcription}} \text{RNA} \xrightarrow{\textit{translation}} \text{PROTEIN}$$

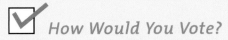

How Would You Vote?

Ricin is difficult to disperse through the air and is unlikely to be used in a large-scale terrorist attack. However, ricin powder did turn up in a Senate office building. Scientists are working to develop a vaccine against ricin. If mass immunizations were to be offered, would you sign up to be vaccinated? See BiologyNow for details, then vote online.

Key Concepts

INTRODUCTION

Life depends on enzymes and other proteins. All proteins consist of polypeptide chains. The chains are sequences of amino acids that correspond to genes—sequences of nucleotide bases in DNA. The path leading from genes to proteins has two steps: transcription and translation.

TRANSCRIPTION

During transcription, the two strands of the DNA double helix are unwound in a gene region. Exposed bases of one strand become the template for assembling a single strand of RNA. Only one type of RNA transcript encodes the message that gets translated into protein. It is called messenger RNA.   Section 14.1

CODE WORDS IN THE TRANSCRIPTS

The nucleotide sequence in DNA is read three bases at a time. Sixty-four base triplets correspond to specific amino acids and represent the genetic code.

The code words have been highly conserved through time. Only a few simple eukaryotes, prokaryotes, and prokaryote-derived organelles have slight variations on the genetic code.   Section 14.2

TRANSLATION

During translation, amino acids are bonded together into a polypeptide chain in a sequence specified by base triplets in messenger RNA. Transfer RNA delivers amino acids one at a time to ribosomes. An RNA component of ribosomes catalyzes the chain-building reaction.   Sections 14.3, 14.4

MUTATIONS IN THE CODE WORDS

Gene mutations introduce changes in protein structure, protein function, or both. The changes may lead to small or large variation in the shared traits that characterize individuals of a population.   Section 14.5

Links to Earlier Concepts

Once again you will meet up with the nucleic acids DNA and RNA (Section 3.7). Gene transcription has features in common with DNA replication, so you may wish to review Section 13.3 before you start. You will again consider how protein primary structure emerges (3.5), this time in the context of RNA interactions. The last section of this chapter will expand your knowledge of DNA repair mechanisms (13.3) and gene mutation (1.4, 3.6, 12.3, 12.8).

14.1 How Is RNA Transcribed From DNA?

LINKS TO
SECTIONS
3.7, 13.2, 13.3

*In **transcription**, the first step in protein synthesis, a sequence of nucleotide bases is exposed in an unwound region of a DNA strand. That sequence is the template upon which a single strand of RNA is assembled from adenine, cytosine, guanine, and uracil subunits.*

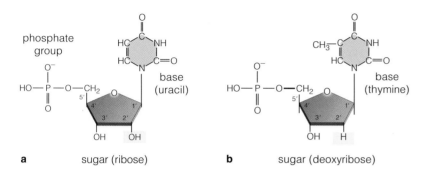

a   sugar (ribose)

b   sugar (deoxyribose)

c Example of base pairing between a DNA strand and a new RNA strand assembled on it during *transcription*:

DNA template

RNA transcript

d Example of base pairing between an old DNA strand and a new strand forming on it during *DNA replication*:

DNA template

new DNA strand

Figure 14.2 (**a**) Uracil, one of four ribonucleotides in RNA. The other three—adenine, guanine, and cytosine—differ only in their bases. Uracil compared with (**b**) thymine, a DNA nucleotide. (**c**) Base pairing of DNA with RNA during transcription, compared with (**d**) base pairing during DNA replication.

The chapter introduction may have left you with the impression that protein synthesis requires one class of RNA molecules. It actually requires three. When genes that specify proteins are transcribed, the outcome is **messenger RNA** (mRNA). *This is the only class of RNA that carries the protein-building codes.* **Ribosomal RNA** (rRNA) and **transfer RNA** (tRNA) are transcribed from different genes. The rRNA becomes a component of ribosomes, the structures in which polypeptide chains are assembled. The tRNA delivers amino acids one by one to ribosomes in the order specified by mRNA.

THE NATURE OF TRANSCRIPTION

An RNA molecule is almost but not quite like a single strand of DNA. It has four kinds of ribonucleotides, each with the five-carbon sugar ribose, one phosphate group, and one base. Three bases—adenine, cytosine, and guanine—are the same as those in DNA. In RNA, though, the fourth base is **uracil**, not thymine. Uracil, too, can pair with adenine, which means that a new RNA strand can base-pair with a DNA strand. Figure 14.2 is a simple way to think about this pairing.

Transcription *differs* from DNA replication in three respects. Only part of one DNA strand, not the whole molecule, is unwound and used as the template. The enzyme **RNA polymerase**, not DNA polymerase, adds ribonucleotides one at a time to the end of a growing strand of RNA. Also, transcription results in one free RNA strand, not a hydrogen-bonded double helix.

DNA contains many protein-coding regions. Each is transcribed separately, and each has its own START

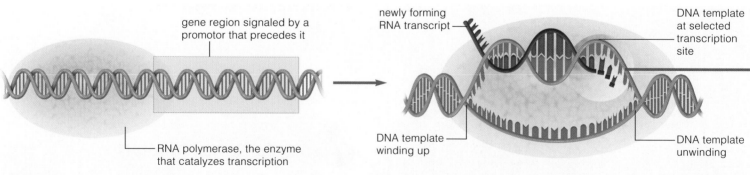

gene region signaled by a promotor that precedes it

newly forming RNA transcript

DNA template at selected transcription site

RNA polymerase, the enzyme that catalyzes transcription

DNA template winding up

DNA template unwinding

ⓐ RNA polymerase initiates transcription at a promoter in DNA. After binding to a promoter, RNA polymerases recognize a base sequence in DNA as a template for making a strand of RNA from free ribonucleotides, which have the bases adenine, cytosine, guanine, and uracil.

ⓑ All through transcription, the DNA double helix becomes unwound in front of the RNA polymerase. Short lengths of the newly forming RNA strand briefly wind up with its DNA template strand. New stretches of RNA unwind from the template (and the two DNA strands wind up again).

Figure 14.3 *Animated!* Gene transcription. By this process, an RNA molecule is assembled on a DNA template. (**a**) Gene region of DNA. The base sequence along one of DNA's two strands (not both) is used as the template. (**b**–**d**) Transcribing that region results in a molecule of RNA.

and STOP signal. A **promoter** is a START signal, a base sequence in DNA to which RNA polymerases bind and prepare for transcription. After binding, an RNA polymerase recognizes a gene region and moves along it. It uses the gene's base sequence as a template for covalently bonding free ribonucleotides together in a complementary sequence, as in Figure 14.3. When it reaches a sequence that signals "the end" of the gene region, the new RNA is released as a free transcript.

FINISHING TOUCHES ON THE mRNA TRANSCRIPTS

In eukaryotic cells, mRNA transcripts are modified before leaving the nucleus. Just as a dressmaker may snip off some threads or put bows on a dress before it leaves the shop, so do cells tailor their "pre-mRNA." For instance, some enzymes attach a modified guanine "cap" to the start of a pre-mRNA transcript. Others attach about 100 to 300 adenine ribonucleotides as a tail to the other end. Hence its name, poly-A tail.

Later, the pre-mRNA's cap will bind to a ribosome. Enzymes will nibble off the tail from the tip on back. Thus each tail's length dictates how long a particular protein-building message will last in the cytoplasm.

A transcript's message gets processed even before it leaves the nucleus. Eukaryotic genes contain **exons**: protein-coding base sequences that are interrupted by noncoding sequences, or **introns**. Both are transcribed, but all introns are snipped out before the transcript reaches the cytoplasm (Figure 14.4). Either all exons are retained in a mature mRNA transcript or some are

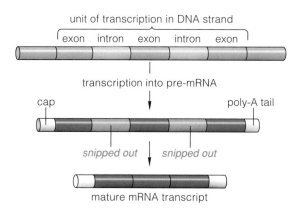

Figure 14.4
Animated! How pre-mRNA transcripts are processed into final form. Inside the nucleus, some or all introns are removed, and the transcript gets a cap and a tail.

removed and the rest are spliced together in various combinations. By this **alternative splicing**, one gene can specify two or more proteins that differ slightly in form and function! Cells use different combinations of exons at different times. Alternative splicing was once considered to be a rare event. However, it may occur in half (or all) genes of the human genome. It helps explain how human cells can make hundreds of thousands of proteins from only 21,500 or so genes.

In gene transcription, a sequence of exposed bases on one of the two strands of a DNA molecule serves as a template for synthesizing a complementary strand of RNA.

RNA polymerases assemble the RNA from four kinds of ribonucleotides that differ in their bases: A, U, C, and G.

Before leaving the nucleus, each new mRNA transcript, or pre-mRNA, undergoes modification into final form.

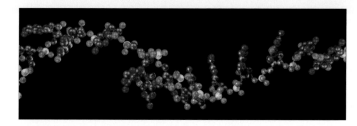

C What happened in the gene region? RNA polymerase catalyzed the covalent bonding of ribonucleotides to one another to form an RNA strand. The base sequence in the new strand is complementary to the exposed bases on the DNA as a template. Many other proteins assist in transcription; compare Section 13.3.

d At the end of the gene region, the last stretch of the new transcript is unwound and released from the DNA template. Shown below it is a model for a transcribed strand of RNA.

14.2 The Genetic Code

The correspondence between genes and proteins is encoded in protein-building "words" in mRNA transcripts. Three nucleotide bases make up each three-letter word.

Figure 14.5*a* shows a bit of mRNA transcribed from a DNA template. To translate it, you have to know how many letters (bases) make each word (amino acid). That is what Marshall Nirenberg, Philip Leder, Severo Ochoa, and Gobind Korana figured out. After mRNA has docked at a ribosome, its bases are "read" *three at a time*. The base triplets in mRNA are **codons**. Figure 14.5*b* shows how their sequence corresponds to the amino acid sequence in a growing polypeptide chain.

There are sixty-four different codons even though there are only twenty amino acids in proteins (Figure 14.6). Why so many? Think it through. If the codon were only one nucleotide, mRNA could specify only four kinds of amino acids. Codons of two nucleotides could code for sixteen kinds of amino acids—still not enough. Mixes of three nucleotides could code for sixty-four kinds—more than enough.

Certain codons actually do specify more than one kind of amino acid. For instance, both GAA and GAG specify glutamate. Also, in most species, the first AUG in the transcript is a START signal for translating "three-bases-at-a-time." It also means methionine is the first amino acid in all new polypeptide chains. UAA, UAG, and UGA do not specify any amino acid. They are STOP signals that block further additions of amino acids to a new chain.

The set of sixty-four different codons is the **genetic code**, and it has been highly conserved through time. Prokaryotes, a few organelles derived from them, and some protists of ancient lineages have a few slightly variant codons. For instance, a few unique codons give mitochondria their own "mitochondrial code." We can predict that they are outcomes of gene mutations that did not alter the mix of proteins in adverse ways. The near-universal use of the genetic code indicates that there is little tolerance for variation.

> *The genetic code is a set of sixty-four different codons, which are nucleotide bases in mRNA that are "read" in sets of three. Different codons (base triplets) specify different amino acids.*

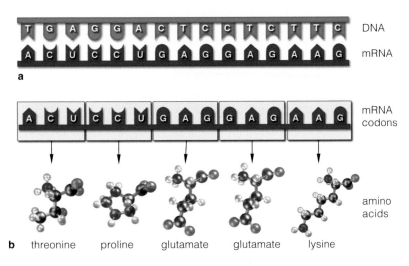

b threonine proline glutamate glutamate lysine

Figure 14.5 Example of the correspondence between genes and proteins. (**a**) An mRNA transcript of a gene region of DNA. Three nucleotide bases, equaling one codon, specify one amino acid. This series of codons (base triplets) specifies the sequence of amino acids shown in (**b**).

Figure 14.6 *Animated!* *Right*, the near-universal genetic code. Each codon in mRNA is a set of three ribonucleotide bases. Sixty-one of these base triplets encode specific amino acids. Three are signals that stop translation.

The *left* vertical column (*brown*) lists choices for the first base of a codon. The *top* horizontal row (*light tan*) lists the second choices. The *right* vertical column (*dark tan*) lists the third. To give three examples, reading left to right, the triplet U G G corresponds to tryptophan. Both U U U and U U C correspond to phenylalanine.

first base	second base				third base
	U	C	A	G	
U	phenylalanine	serine	tyrosine	cysteine	U
	phenylalanine	serine	tyrosine	cysteine	C
	leucine	serine	STOP	STOP	A
	leucine	serine	STOP	tryptophan	G
C	leucine	proline	histidine	arginine	U
	leucine	proline	histidine	arginine	C
	leucine	proline	glutamine	arginine	A
	leucine	proline	glutamine	arginine	G
A	isoleucine	threonine	asparagine	serine	U
	isoleucine	threonine	asparagine	serine	C
	isoleucine	threonine	lysine	arginine	A
	methionine (or START)	threonine	lysine	arginine	G
G	valine	alanine	aspartate	glycine	U
	valine	alanine	aspartate	glycine	C
	valine	alanine	glutamate	glycine	A
	valine	alanine	glutamate	glycine	G

14.3 The Other RNAs

Let's take stock. The codons in an mRNA transcript are the words in protein-building messages. Without translators, words that originated in DNA mean nothing; it takes the other two classes of RNA to synthesize proteins. Before getting into the mechanisms of translation, reflect on this overview of their structure and function.

Figure 14.7 shows the molecular structure for one of the tRNAs. All cells have pools of tRNAs and amino acids in their cytoplasm. Each tRNA has a molecular "hook," an attachment site for an amino acid. It has an **anticodon**, a ribonucleotide base triplet that can base-pair with a complementary codon in an mRNA transcript. When tRNAs bind to mRNA on a ribosome, the amino acid attached to each becomes positioned automatically in the order that the codons specify.

There are sixty-four codons but not as many kinds of tRNAs. How do tRNAs match up with more than one type of codon? According to base-pairing rules, adenine pairs with uracil, and cytosine with guanine. However, in codon–anticodon interactions, these rules

can loosen for the third base in a codon. This freedom in codon–anticodon pairing at a base is known as the "wobble effect." For example, AUU, AUC, and AUA specify isoleucine. All three codons can base-pair with one type of tRNA that hooks on to isoleucine.

Again, interactions between the tRNAs and mRNA take place at ribosomes. A ribosome has two subunits made of rRNA and structural proteins (Section 4.5 and Figure 14.8). In eukaryotic cells, they are built in the nucleus and moved to the cytoplasm. There, a large and small subunit converge as an intact, functional ribosome only when mRNA is to be translated.

LINK TO SECTION 4.5

Only mRNA carries DNA's protein-building instructions from the nucleus into the cytoplasm.

tRNAs deliver amino acids to ribosomes. Their anticodons base-pair with codons in the order specified by mRNA.

Polypeptide chains are built on ribosomes, each consisting of a large and small subunit made of rRNA and proteins.

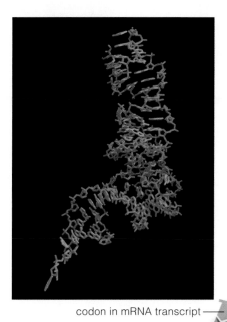

codon in mRNA transcript

anticodon in tRNA

amino acid

Figure 14.7 Model for a tRNA. The icon shown to the right is used in following illustrations. The "hook" at the lower end of this icon represents the binding site for a specific amino acid.

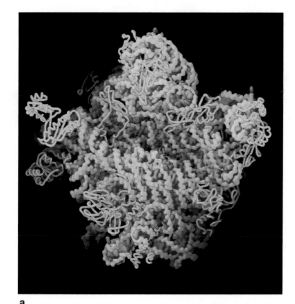

a

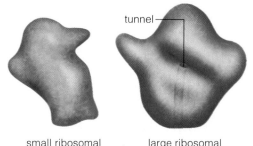

tunnel

small ribosomal subunit + large ribosomal subunit → intact ribosome

b

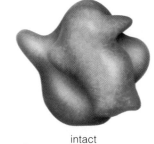

Figure 14.8 (**a**) Ribbon model for the large subunit of a bacterial ribosome. It has two rRNA molecules (*gray*) and thirty-one structural proteins (*gold*), which stabilize the structure. At one end of a tunnel through the large subunit, rRNA catalyzes polypeptide chain assembly. This is an ancient, highly conserved structure. Its role is so vital that the corresponding subunit of the eukaryotic ribosome, which is larger, may be similar in structure and function. (**b**) Model for the small and large subunits of a eukaryotic ribosome.

14.4 The Three Stages of Translation

An mRNA transcript that encodes DNA's information about a protein enters an intact ribosome. There, its codons are translated into a polypeptide chain—a protein's primary structure (Section 3.5). Translation of the protein-building message proceeds through three continuous stages called initiation, elongation, and termination.

Only one kind of tRNA can start the *initiation* stage of translation. It alone has the anticodon UAC—which is complementary to the START codon of every mRNA transcript. The anticodon and codon meet up when this initiator tRNA binds to a small ribosomal subunit. Next, a large ribosomal subunit joins with the small subunit. Together, the initiator tRNA, the ribosome, and the mRNA transcript form an initiation complex (Figure 14.9*a–c*). The next stage can begin.

During the *elongation* stage, a polypeptide chain is synthesized while the mRNA passes between the two ribosomal subunits, a bit like a thread being moved through the eye of a needle. Many tRNA molecules deliver amino acids to the ribosome, and each binds to the mRNA in the order specified by their codons. One region of an rRNA molecule located at the center of the large ribosomal subunit is highly acidic, and it functions as an enzyme. It catalyzes the formation of peptide bonds between amino acids (Figure 14.9*d–f*).

Figure 14.9*g* shows how one peptide bond forms between the most recently attached amino acid and the next one brought to the ribosome. Here, you might wish to look once more at Section 3.5, which includes a step-by-step description of peptide bond formation during protein synthesis.

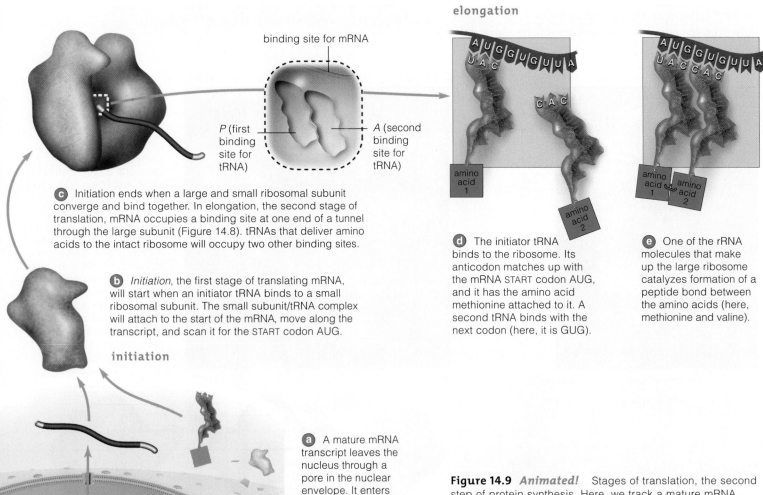

elongation

binding site for mRNA

P (first binding site for tRNA)

A (second binding site for tRNA)

c Initiation ends when a large and small ribosomal subunit converge and bind together. In elongation, the second stage of translation, mRNA occupies a binding site at one end of a tunnel through the large subunit (Figure 14.8). tRNAs that deliver amino acids to the intact ribosome will occupy two other binding sites.

b *Initiation*, the first stage of translating mRNA, will start when an initiator tRNA binds to a small ribosomal subunit. The small subunit/tRNA complex will attach to the start of the mRNA, move along the transcript, and scan it for the START codon AUG.

initiation

a A mature mRNA transcript leaves the nucleus through a pore in the nuclear envelope. It enters the cytoplasm, which has many free amino acids, tRNAs, and ribosomal subunits.

amino acid 1

amino acid 2

amino acid 1

amino acid 2

d The initiator tRNA binds to the ribosome. Its anticodon matches up with the mRNA START codon AUG, and it has the amino acid methionine attached to it. A second tRNA binds with the next codon (here, it is GUG).

e One of the rRNA molecules that make up the large ribosome catalyzes formation of a peptide bond between the amino acids (here, methionine and valine).

Figure 14.9 *Animated!* Stages of translation, the second step of protein synthesis. Here, we track a mature mRNA transcript that formed inside the nucleus of a eukaryotic cell. It passes through pores across the nuclear envelope and enters the cytoplasm, which contains pools of many free amino acids, tRNAs, and ribosomal subunits.

During *termination*, the last stage of translation, the mRNA's STOP codon enters the ribosome. No tRNA has a corresponding anticodon. Proteins called release factors bind to the ribosome. Binding triggers enzyme activity that detaches the mRNA *and* the polypeptide chain from the ribosome (Figure 14.9*i–k*).

In cells that are quickly using or secreting proteins, you often see many clusters of ribosomes (polysomes) on an mRNA transcript, all translating it at the same time. This is what happens in unfertilized eggs, which usually stockpile mRNA transcripts in the cytoplasm in preparation for the cell divisions that lie ahead.

Many newly formed polypeptide chains carry out their functions in the cytoplasm. Others have a special sequence of amino acids. The sequence is a shipping label that gets them into ribosome-studded, flattened sacs of rough ER (Section 4.6). In the organelles of the endomembrane system, the chains will take on final form before shipment to their ultimate destinations as structural or functional proteins.

Translation is initiated when a small ribosomal subunit and an initiator tRNA arrive at an mRNA transcript's START codon, and a large ribosomal subunit binds to them.

tRNAs deliver amino acids to a ribosome in the order dictated by the linear sequence of mRNA codons. A polypeptide chain lengthens as peptide bonds form between the amino acids.

Translation ends when a STOP codon triggers events that cause the polypeptide chain and the mRNA to detach from the ribosome.

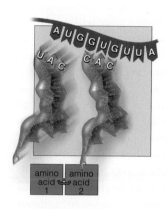

f The first tRNA is released, and the ribosome moves to the next codon position.

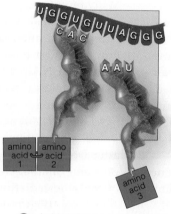

g A third tRNA binds with the next codon (here it is UUA). The ribosome catalyzes peptide bond formation between amino acids 2 and 3.

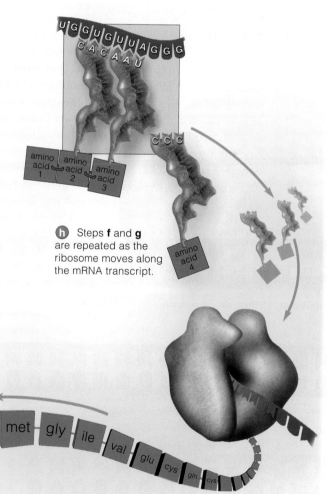

h Steps **f** and **g** are repeated as the ribosome moves along the mRNA transcript.

termination

i A STOP codon moves into the area where the chain is being built. It is the signal to release the mRNA transcript from the ribosome.

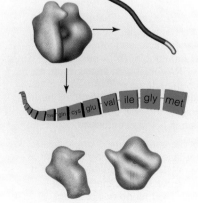

j The new polypeptide chain is released from the ribosome. It is free to join the pool of proteins in the cytoplasm or to enter rough ER of the endomembrane system.

k The two ribosomal subunits now separate, also.

14.5 Mutated Genes and Their Protein Products

LINKS TO
SECTIONS
2.3, 3.6, 7.1

When a cell taps its genetic code, it is making proteins with precise structural and functional roles that keep it alive. If a gene changes, the mRNA transcribed from it may change and specify an altered protein. If the protein has a crucial role, the outcome will be a dead or abnormal cell.

Gene sequences can change. Sometimes one base gets substituted for another in the nucleotide sequence. At other times, an extra base is inserted or one is lost. Such small-scale changes in the nucleotide sequence of a DNA molecule are **gene mutations**, and they can alter the message that becomes encoded in mRNA. Cells have some leeway, because more than one codon can specify the same amino acid. For example, if UCU replaced UCC in an mRNA transcript, this might not be bad, because both codons specify serine. However, as the next examples show, many mutations result in proteins that function in an altered way or not at all.

COMMON GENE MUTATIONS

During DNA replication, recall, the wrong nucleotide may become paired with an exposed base on the DNA template and slip by proofreading and repair enzymes (Section 13.3). This type of mutation is a **base-pair substitution**. When the altered message is translated, it may call for the wrong amino acid or a premature STOP codon. Figure 14.10*b* shows how adenine *replaced* one thymine in the gene for beta hemoglobin, which can give rise to sickle-cell anemia (Section 3.6).

Figure 14.10*c* depicts another gene mutation, one in which a single base—thymine—was *deleted*. Again,

DNA polymerases read base sequences in blocks of three. A deletion is one of the *frameshift* mutations; it shifts the "three-bases-at-a-time" reading frame. An altered mRNA is transcribed from the mutant gene, so an altered protein is the result.

Frameshift mutations fall in the broader categories of **insertions** and **deletions**. One or more base pairs become inserted into DNA or are deleted from it.

Other mutations arise from transposable elements, or **transposons**, that can jump around in the genome. Geneticist Barbara McClintock found that these DNA segments or copies of them move spontaneously to a new location in a chromosome or even to a different chromosome. When transposons land in a gene, they alter the timing or duration of its activity, or block it entirely. Their unpredictability can give rise to odd variations in traits. Figure 14.11 gives an example.

HOW DO MUTATIONS ARISE?

Many mutations happen spontaneously while DNA is being replicated. This is not surprising, given the swift pace of replication (about twenty bases per second in humans and a thousand bases per second in certain bacteria). DNA polymerases and DNA ligases can fix most mistakes (Section 13.3). But sometimes they go on assembling a new strand right over an error. The bypass can result in a mutated DNA molecule.

Not all mutations are spontaneous. A number arise after DNA is exposed to mutation-causing agents. To give an example, x-rays and other high-energy forms of **ionizing radiation** break chromosomes into pieces (Figure 14.12). Ionizing radiation damages DNA indirectly, also. When it penetrates living tissues, it leaves behind a long trail of destructive free radicals. Doctors and

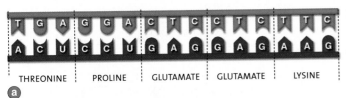

part of DNA template

mRNA transcribed from DNA

THREONINE   PROLINE   GLUTAMATE   GLUTAMATE   LYSINE   resulting amino acid sequence

a

base substitution in DNA

altered mRNA

THREONINE   PROLINE   VALINE   GLUTAMATE   LYSINE   altered amino acid sequence

b

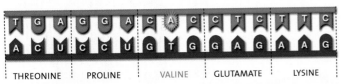

deletion in DNA

altered mRNA

THREONINE   PROLINE   GLYCINE   ARGININE   altered amino acid sequence

c

Figure 14.10 *Animated!* Example of gene mutation. (**a**) Part of a gene, the mRNA, and the specified amino acid sequence of the beta chain in hemoglobin. (**b**) A base-pair substitution in DNA replaces a thymine with an adenine. When the altered mRNA transcript is translated, valine replaces glutamate as the sixth amino acid of the new polypeptide chain. Sickle-cell anemia is the eventual outcome.

(**c**) Deletion of the same thymine would be a frameshift mutation. The reading frame for the rest of the mRNA shifts, a different protein product forms, and it causes thalassemia— a different type of red blood cell disorder.

Figure 14.11 Barbara McClintock, who won a Nobel Prize for her research. She proved that transposons slip into and out of different locations in DNA. The curiously nonuniform coloration of kernels in strains of Indian corn (*Zea mays*) sent her on the road to discovery.

Several genes govern pigment formation and deposition in corn kernels, which are a type of seed. Mutations in one or more of these genes produce yellow, white, red, orange, blue, and purple kernels. However, as McClintock realized, *unstable* mutations can cause streaks or spots in *individual* kernels.

All of a corn plant's cells have the same pigment-encoding genes. But a transposon invaded a pigment-encoding gene before the plant started growing from a fertilized egg. While a kernel's tissues were forming, its cells could not make pigment, but the same transposon jumped out of the pigment-encoding gene in some of its cells. Descendants of *those* cells could make pigment. The spots and streaks in individual kernels are visual markers for those cell lineages.

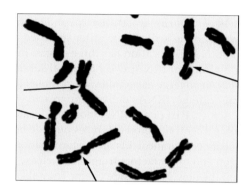

Figure 14.12 Chromosomes from a human cell after exposure to gamma rays, a form of ionizing radiation. We can expect such broken pieces (*arrows*) to be lost during interphase, when DNA is being replicated. The extent of the chromosome damage in an exposed cell typically depends on how much radiation it absorbed.

dentists both use the lowest possible doses of x-rays to minimize the damage to a patient's DNA.

Nonionizing radiation excites electrons to a higher energy level. DNA absorbs one form, ultraviolet (UV) light. Two nucleotide bases in DNA—cytosine and thymine—are most vulnerable to excitation that can change base-pairing properties. UV light can induce adjacent thymine bases in a DNA strand to pair *with each other*, as a bulky dimer (page 217). At least seven gene products interact as a DNA repair mechanism to remove the dimer, which wrinkles the DNA. If DNA polymerase encounters a thymine dimer, it will make replication errors. Exposing unprotected skin to the sun invites thymine dimer formation in skin cells.

When thymine dimers are not repaired, they cause DNA polymerases to make even more errors during the next replication cycle. They are the original source of mutations that lead to certain cancers.

Natural and synthetic chemicals accelerate rates of gene mutations. For instance, **alkylating agents** can transfer charged methyl or ethyl groups to reactive sites in DNA. At these sites, DNA is more vulnerable to mistakes in base pairing and to mutation. Cancer-causing agents in cigarette smoke and many other substances exert their effects by alkylating DNA.

THE PROOF IS IN THE PROTEIN

When a mutation arises in a somatic cell of a sexually reproducing individual, its good or bad effects will not endure; it is not passed on to offspring. If it arises in a germ cell or a gamete, however, it may enter the evolutionary arena. It also may do so when it is passed on to offspring by asexual reproduction. Either way, *the protein product of such heritable mutations will have harmful, neutral, or beneficial effects on the individual's capacity to function in the prevailing environment.* The effects of uncountable mutations in millions of species have had spectacular evolutionary consequences—and that is a topic of later chapters.

A gene mutation is a permanent change in one or more bases in the nucleotide sequence of DNA. The most common types are base-pair substitutions, insertions, and deletions.

Exposure to harmful radiation and to chemicals in the environment can cause mutations in DNA.

A protein specified by a mutated gene may have harmful, neutral, or beneficial effects on the individual's capacity to function in the environment.

Summary

Introduction   All enzymes and other proteins that are essential for life consist of polypeptide chains. Each chain, a linear sequence of amino acids, corresponds to nucleotide base sequences in DNA that form genes. The path from genes to proteins has two steps: transcription and translation (Figure 14.13).

Section 14.1   In eukaryotic cells, genes are transcribed in the nucleus and then translated cytoplasm. Both steps occur in the cytoplasm of prokaryotic cells, which have no nucleus. Enzymes unwind the two strands of a DNA double helix in a specific gene region. RNA polymerases covalently bond ribonucleotides one after another into a new RNA transcript, in an order complementary to the exposed bases on the DNA template. Adenine, guanine, cytosine, and uracil are the bases in ribonucleotides.

The mRNA transcript gets modified before it leaves the nucleus. Its 5′ end gets capped, and its 3′ end gets a poly-A tail, which paces how long the mRNA will stay intact in the cytoplasm. The introns between exons (the protein-coding portions of genes) are snipped out. The exons can be spliced together in different combinations.

Biology ⊘ Now
Learn how genes are transcribed and transcripts are processed with the animation on BiologyNow.

Sections 14.2, 14.3   Only messenger RNA (mRNA) carries the protein-building information in DNA to ribosomes for translation. Its genetic message is written in codons, or sets of three nucleotides along an mRNA strand that specify an amino acid. There are sixty-four codons, a few of which act as START or STOP signals for translation. That set constitutes a highly conserved genetic code. A few variations in code words evolved among prokaryotes and prokaryote-derived organelles (e.g., mitochondria) and in a few ancient lineages of single-celled eukaryotes.

Translation requires three classes of RNAs. Transfer RNA (tRNA) molecules have anticodons that can bind briefly to complementary codons in mRNA. They also have a binding site for a free amino acid, which they deliver to ribosomes during protein synthesis. Different tRNAs reversibly bind different amino acids. Ribosomal RNA (rRNA) and proteins that stabilize it make up the two subunits that form ribosomes.

Biology ⊘ Now
Explore the genetic code with the interaction on BiologyNow.

Section 14.4   During translation, peptide bonds form between amino acids in the order specified by codons in mRNA. Translation has three stages. In initiation, an initiator tRNA, two ribosomal subunits, and an mRNA converge as an initiation complex. In the elongation stage, tRNAs deliver amino acids to the intact ribosomes. Part of an rRNA molecule located in the ribosome's central region catalyzes peptide bond formation between amino acids. In the termination stage, a STOP codon and other factors trigger the release of mRNA and the new polypeptide chain. They also cause the ribosome's subunits to separate from each other.

Biology ⊘ Now
Observe the translation of an mRNA transcript with the animation on BiologyNow.

Section 14.5   Gene mutations are heritable, small-scale changes in the base sequence of DNA. Major types are base-pair substitutions, insertions, and deletions. Many arise spontaneously as DNA is being replicated. Some arise after transposons jump to new locations in chromosomes; others arise after DNA is exposed to ionizing radiation or to chemicals in the environment. Mutations may cause changes in protein structure, protein function, or both.

Biology ⊘ Now
Investigate the effects of mutation with the animation on BiologyNow.

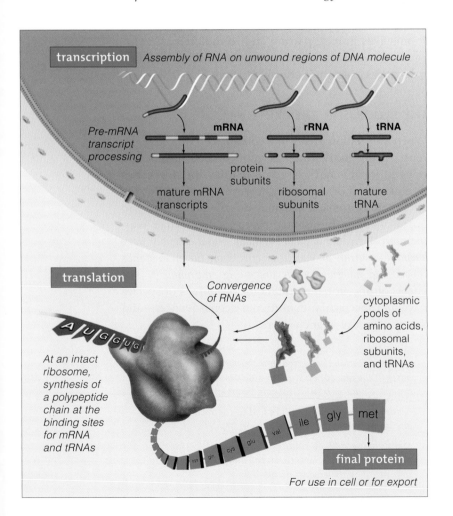

transcription | *Assembly of RNA on unwound regions of DNA molecule*

Pre-mRNA transcript processing
mRNA   **rRNA**   **tRNA**

protein subunits

mature mRNA transcripts   ribosomal subunits   mature tRNA

translation

Convergence of RNAs

cytoplasmic pools of amino acids, ribosomal subunits, and tRNAs

At an intact ribosome, synthesis of a polypeptide chain at the binding sites for mRNA and tRNAs

AUGGUG

cys gln glu val ile gly met

final protein

For use in cell or for export

Figure 14.13 *Animated!*   Summary of protein synthesis in eukaryotic cells. DNA is transcribed into RNA in the nucleus. RNA is translated in the cytoplasm. Prokaryotic cells do not have a nucleus; transcription and translation proceed in their cytoplasm.

1. DNA contains many different gene regions that are transcribed into different _____ .
   a. proteins
   c. mRNAs, tRNAs, rRNAs
   b. mRNAs only
   d. all of the above

2. An RNA molecule is typically _____ .
   a. a double helix
   c. double-stranded
   b. single-stranded
   d. triple-stranded

3. An mRNA molecule is synthesized by _____ .
   a. replication
   c. transcription
   b. duplication
   d. translation

4. Each codon specifies a(n) _____ .
   a. protein
   c. amino acid
   b. polypeptide
   d. mRNA

5. _____ different codons represent a near-universal genetic code.
   a. Twelve
   c. Thirty-four
   b. Twenty
   d. Sixty-four

6. Anticodons pair with _____ .
   a. mRNA codons
   c. RNA anticodons
   b. DNA codons
   d. amino acids

7. _____ can cause gene mutations.
   a. replication errors
   d. non-ionizing radiation
   b. transposons
   e. b and c are correct
   c. ionizing radiation
   f. all of the above

8. Match the terms with the most suitable description.
   ____ alkylating agent
   a. protein-coding parts of a mature mRNA transcript
   ____ chain elongation
   b. base triplet for amino acid
   c. second stage of translation
   ____ exons
   d. base triplet; pairs with codon
   ____ genetic code
   e. one environmental agent that induces mutation in DNA
   ____ anticodon
   ____ introns
   f. set of 64 codons for mRNA
   ____ codon
   g. noncoding part of pre-mRNA transcript, removed before translation

Additional questions are available on Biology⊛Now™

Critical Thinking

1. Using Figure 14.6, translate this nucleotide sequence in part of an mRNA transcript into an amino acid sequence:

   5′—GGTTTCTTCAAGAGA—3′

2. Briefly review Section 13.3. Now suppose that DNA polymerase made a wrong base pairing while a crucial gene region of DNA was being replicated. DNA repair mechanisms did not kick in to fix the mistake. Here is the part of the DNA strand that contains the error:

...AATTCCGACTCCTATGG
...TTAAGGTTGAGGATACC

After the DNA molecule is replicated, two daughter cells form. One daughter cell is carrying the mutation and the other cell is normal. Develop a hypothesis to explain this observation.

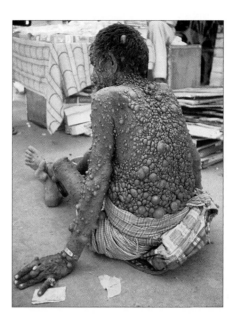

Figure 14.14 Soft skin tumors on an individual affected by the autosomal dominant disorder called neurofibromatosis.

3. *Neurofibromatosis* is a human autosomal dominant disorder caused by mutations in the *NF1* gene. It is characterized by the formation of soft, fibrous tumors in the peripheral nervous system and skin as well as abnormalities in muscles, bones, and internal organs (Figure 14.14).

Because the mutant allele is dominant, an affected child usually has an affected parent. Yet in 1991, scientists reported that a boy developed neurofibromatosis even though his parents did not. When they examined both copies of the boy's *NF1* gene, they found that the gene on the chromosome he inherited from his father contained a transposon. Neither father nor mother had a transposon in any of the copies of their *NF1* genes. Explain the cause of neurofibromatosis in the boy and how it arose.

4. Cigarette smoke is mostly carbon dioxide, nitrogen, and oxygen. The rest contains at least fifty-five different chemicals identified as carcinogenic, or cancer-causing, by the International Agency for Research on Cancer (IARC). When these carcinogens enter the bloodstream, enzymes convert them to a series of chemical intermediates that are easier to excrete. Some of the intermediates bind irreversibly to DNA. Propose one mechanism by which smoking cigarettes can cause cancer.

5. *Antisense drugs* may help us fight cancer and viral diseases, including SARS. These short mRNA strands are complementary to mRNAs that have been linked to these illnesses. Speculate on how these drugs work.

6. In some cases, the termination of transcription of prokaryotic DNA depends on the structure of the newly forming RNA transcript. The terminal end of an mRNA transcript often folds back tightly on itself and makes a hairpin-looped structure, like the one shown at right.

Why do you suppose that a "stem-loop" structure such as this stops transcription of prokaryotic DNA when the RNA polymerases reach it?

```
              C
          U—C
          G—C
          A—U
          C—G
          C—G
          G—C
          C—G
          C—G
...CCCACAG—CAUUUUU...
```

Between You and Eternity

You are in college, your whole life ahead of you. Your risk of developing cancer is as remote as old age, an abstract statistic that is easy to forget.

"There is a moment when everything changes—when the width of two fingers can suddenly be the total distance between you and eternity." Robin Shoulla wrote those words after being diagnosed with breast cancer. She was seventeen. At an age when most young women are thinking about school, parties, and potential careers, Robin was dealing with *radical mastectomy*—the removal of a breast, all lymph nodes under the arm, and skeletal muscles in the chest wall under the breast. She was pleading with her oncologist not to use her jugular vein for chemotherapy and wondering if she would survive through the next year (Figure 15.1).

Robin became an annual statistic—one of 10,000 or so females and, to a lesser extent, males who develop breast cancer before they are forty years old. About 180,000 new cases are diagnosed each year in the United States.

Cancers are as diverse as their underlying causes, but several gene mutations predispose individuals to developing certain kinds. Either the mutant genes are inherited or they mutate spontaneously in individuals after attacks by environmental agents, including some viruses, toxic chemicals, and ultraviolet radiation.

One gene on chromosome 17 encodes *ERBB2*, a type of membrane receptor. *ERBB2* is part of a control pathway that governs the cell cycle—that is, when and how often cells divide. It also is one of the proto-oncogenes. When such genes are mutated or overexpressed, they help trigger cancerous transformations. The cells of about 25 percent of breast cancer patients have too many of these receptors or extra copies of the gene itself. The cells do not stop dividing, and abnormal masses of cells are the outcome.

Two different genes, *BRCA1* and *BRCA2*, encode two of the proteins that can act as tumor suppressors. They help prevent the formation of benign or cancerous cell masses, as explained in Section 9.5. Such proteins usually function as part of DNA repair mechanisms. That is why mutation in *BRCA1* or *BRCA2* compromises the cell's capacity to fix damaged DNA. Other mutations are free to accumulate throughout the DNA and set the stage for cancer.

BRCA1 and *BRCA2* are called *breast cancer genes*, because their mutated forms often occur in cancerous breast cells. If a *BRCA* gene mutates in one of three especially dangerous ways, the individual has about an 80 percent chance of developing breast cancer before reaching age seventy.

Robin Shoulla survived. Although radical mastectomy is rarely performed today—a modified procedure is just as

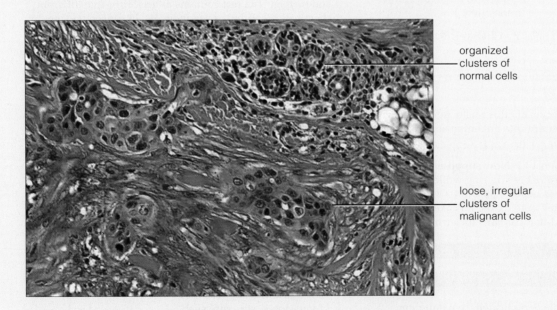

organized clusters of normal cells

loose, irregular clusters of malignant cells

Figure 15.1 Breast cancer. This light micrograph shows irregular clusters of carcinoma cells that infiltrated the ducts in breast tissue. On the facing page, Robin Shoulla. Diagnostic tests revealed cells like this in her body.

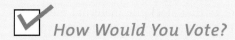
Watch the video online!

effective and less disfiguring—it is the only option when cancer cells infiltrate muscles under the breast. It was Robin's only option. She may never know which mutation caused her cancer. Thirteen years later, she has what she calls a normal life—a career, husband, children. Her goal is to grow very old with gray hair and spreading hips, smiling.

Robin's story lends immediacy to the world of **gene controls**. By these molecular mechanisms, all cells control when and how fast specific genes will be transcribed and translated, and whether gene products will be switched on or silenced. You will consider the impact of such controls in chapters throughout the book—and in many chapters of your life.

How Would You Vote?

Some females at high risk of developing breast cancer opt for prophylactic mastectomy, the surgical removal of one or more breasts even before cancer develops. Many of them would never have developed cancer. Should the surgery be restricted to cancer treatment? See BiologyNow for details, then vote online.

Key Concepts

OVERVIEW OF THE CONTROLS

Control mechanisms govern when, how, and to what extent an individual's genes are expressed. They respond to signaling molecules and to changing conditions.

Diverse control elements work before, during, and after gene transcription or translation. They interact with DNA, RNA, and protein products.

In multicelled species, long-term controls guide the stage-by-stage development of new individuals. Selective gene expression in embryos results in cell differentiation, whereby different cell lineages become specialized in composition, structure, and function.   Section 15.1

EXAMPLES FROM EUKARYOTES

Precise controls govern an embryo's development. As examples, the orderly, regional expression of certain genes causes animal organs and limbs to form where they are supposed to. In the embryos of mammals, controls also compensate for sex chromosome imbalances. A female inherits twice as many X chromosomes as a male, but controls shut down most of the genes on one of her two X chromosomes in each cell.   Section 15.2

CASE STUDY: FRUIT FLY DEVELOPMENT

Drosophila research revealed how a complex body plan emerges as different cells in a developing embryo activate or suppress shared genes in different ways.   Section 15.3

WHAT ABOUT THE PROKARYOTES?

Prokaryotic gene controls deal mainly with short-term changes in nutrient availability and other aspects of the environment. The main gene controls bring about fast adjustments in rates of transcription.   Section 15.4

Links to Earlier Concepts

You will be applying your knowledge of the organization of chromosomal DNA (Section 9.1) and of mRNA transcript processing (14.1). You may wish to review the mechanism that governs the activity of the key enzyme in tryptophan synthesis (6.4). Your understanding of the characteristics of autosomal recessive inheritance (12.3) and of the basis of sex determination in humans (12.1, 12.5) will come in handy.

15.1 When Controls Come Into Play

LINKS TO
SECTIONS
6.4, 9.1, 14.1

*Ultimately, **gene expression** refers to controls over the kinds and amounts of proteins that are in a cell in any specified interval. Tremendous coordination goes into synthesizing, stockpiling, using, exporting, and degrading thousands of types of proteins.*

SOME CONTROL MECHANISMS

Regulatory elements interact with DNA, RNAs, new polypeptide chains, and final proteins. Different kinds respond to shifts in concentrations of substances or to outside signals, such as hormones. Many responses exert *negative* control; they slow or stop some activity. Others exert *positive* control; they enhance it.

For instance, **promoters** are short stretches of base sequences in DNA where regulatory proteins gather and control transcription of specific genes, often in response to a hormonal signal. **Enhancers** are binding sites where such proteins increase transcription rates.

Chemical modification also can exert control. Many methyl groups (—CH₃) are "painted" on parts of newly replicated DNA to block access to genes. Acetyl groups (—CH₃CO⁻) are attached to DNA to make genes accessible. **Methylation** and **acetylation** are the names for the addition of such groups to DNA or any other molecule.

When, how, and to what extent any of these controls come into play depends on the type of cell, its functions, its chemical environment, and signals from the outside. Later chapters provide rich examples. For now, become familiar with the points at which control is exerted.

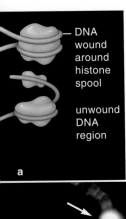

DNA wound around histone spool

unwound DNA region

a

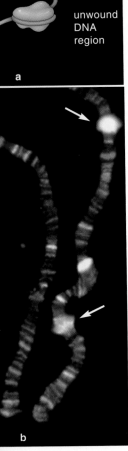

Figure 15.2 Examples of gene control mechanisms. (**a**) Loosening a chromosome's DNA–histone units may expose genes for transcription. Attaching an acetyl group to a histone makes it loosen its grip on the DNA wound around it. Transcription enzymes attach and detach these groups.

(**b**) *Drosophila* polytene chromosomes. To sustain a rapid growth rate, *Drosophila* larvae eat continuously and use a lot of saliva. Giant chromosomes in their salivary glands form by repeated DNA replications. Each has hundreds or thousands of the same DNA molecule, aligned side by side.

An insect hormone, ecdysone, serves as a regulatory protein; it promotes gene transcription. In response to the hormonal signal, these chromosomes loosen and puff out in regions where genes are being transcribed. Puffs are largest and most diffuse where transcription is most intense (*arrows*).

b

POINTS OF CONTROL

Controls Before Transcription The *access* to genes is under control. Remember how histones and other proteins help keep eukaryotic chromosomes organized (Section 9.1)? Where a DNA molecule is wound up tightly, polymerases cannot access genes. Acetylation can make histones loosen their grip (Figure 15.2a). As another example, a maternal *or* paternal allele at any locus in a diploid cell may become methylated, which can block the gene's influence on a trait.

Controls also affect *how* a gene will be transcribed. For instance, some gene sequences can be rearranged or multiplied. In immature amphibian eggs and gland cells of certain insect larvae, the chromosomal DNA is copied repeatedly in interphase. The copying results in *polytene* chromosomes, which contain hundreds or thousands of side-by-side copies of genes. The repeats allow these cells to churn out copious amounts of the gene products necessary for survival (Figure 15.2b).

Control of Transcript Processing Many controls influence mRNA transcript processing. Remember, the pre-mRNA transcripts are modified in ways that affect whether, when, and how they are translated (Section 14.1). Consider what happens in different muscle cells. Exons of the gene for troponin, a contractile protein, are put together in different combinations in different muscle cells. As a result, each type of muscle cell gets mRNA transcripts that are unique in a small region. The structure and functioning of the troponin product vary in subtle ways among them.

Also, the nuclear envelope helps control when the mRNA transcript reaches a ribosome. The transcript cannot pass through a nuclear pore complex unless proteins become attached to it. A base sequence in the untranslated end of mRNA is like a zip code. Controls "read" the code and attach proteins to it. The bound proteins help move the transcript to the region where it is supposed to be translated or stored. Destinations are vital. Which mRNAs—and, in time, gene products—end up in different regions of an immature egg's cytoplasm are "maternal messages" on how to start to construct the body plan of a new embryo.

Unfertilized eggs that stockpile maternal messages keep them silent with the help of controls called Y-box proteins. When phosphorylated, Y-box proteins bind and help stabilize mRNA. When many of the proteins bind to a transcript, they block its translation. In other words, phosphorylation of Y-box proteins is a control mechanism in mRNA inactivation. You will read about such controls in later sections of the book.

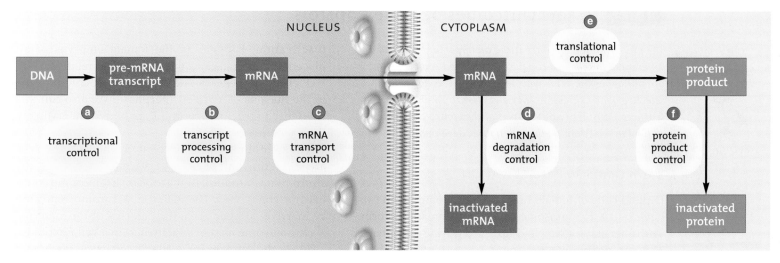

a DNA may be chemically modified, which can influence access to genes. In some species, genes become duplicated or rearranged.

b Pre-mRNA spliced in alternative ways can lead to different forms of a protein. Other modifications affect whether a transcript reaches the cytoplasm.

c Transport protein binding determines whether an mRNA will be delivered to a specific region of cytoplasm for local translation.

d How long an mRNA lasts depends on the proteins that are attached to it and the length of its poly-A tail.

e Translation can be blocked. mRNA cannot attach to a ribosome when proteins bind to it. Initiation factors can be inactivated.

f A new protein may be inactivated or activated. Control of enzymes and other proteins influences many cell activities.

Figure 15.3 *Animated!* Controls that influence whether, when, and how a gene in eukaryotic DNA will be expressed.

Control of Translation Many kinds of molecules function in coordinated ways during translation, and each is controlled independently. Some controls work on initiation factors and ribosome components. Others work through mRNA transcript stability. The longer a transcript lasts, the more times it can be translated. Enzymes start nibbling at the poly-A tail of a mature mRNA transcript within minutes of its appearance in the cytoplasm. How fast they digest it depends on the tail's length, its base sequences, and the proteins that have become attached to it (Section 14.1).

Controls After Translation Control is exerted over new enzymes and other proteins. For instance, Y-box proteins become activated only when enzymes attach a phosphate group to them. Other controls activate, inhibit, and stabilize diverse molecules that take part in protein synthesis. Allosteric control of tryptophan synthesis is a case in point (Section 6.4).

SAME GENES, DIFFERENT CELL LINEAGES

Later in the book, you will read about how complex organisms develop. For now, tentatively accept this premise: All cells of your body started out life with the same genes, because every one arose by mitotic cell divisions from the same fertilized egg. They all transcribe many of the same genes and are alike in most aspects of structure and housekeeping activities.

In other ways, however, *nearly all of your body cells became specialized in composition, structure, and function.*

This process—**cell differentiation**—is central to the development of all multicelled species. By selecting particular subsets of genes, specialized cells and their descendants give rise to different tissues and organs.

Here is an example: Cells generally transcribe the genes coding for enzymes of glycolysis all the time. But immature red blood cells alone transcribe genes for hemoglobin. Liver cells transcribe genes required to make enzymes that neutralize some toxins, but they are the only ones that do so. While your eyes formed, certain cells accessed genes necessary for synthesizing crystallin. No other cells in your body can activate the genes for this protein, which makes up the transparent fibers of the lens in each eye.

Figure 15.3 summarizes the main control points over gene expression in eukaryotic cells.

Gene expression is controlled by regulatory elements that interact with one another, with control elements built into the DNA, with RNA, and with newly synthesized proteins. Different forms of controls work before, during, and after transcription and translation.

Control also is exerted through chemical modifications that activate, inactivate, or restrict access to specific gene regions in DNA.

During development of all multicelled organisms, cells become different in composition, structure, and function as genes are activated and suppressed in selective ways.

15.2 A Few Outcomes of Gene Controls

LINKS TO
SECTIONS
12.1, 12.2. 12.5

The preceding section introduced an important idea. All differentiated cells in a complex, multicelled body use most of their genes the same way, but each type engages in selective gene expression that gives rise to its distinctive features. Consider two examples of the controls that guide the selections during embryonic development.

X CHROMOSOME INACTIVATION

Diploid cells of female humans and female calico cats have two X chromosomes. One is in threadlike form. The other stays scrunched up, even during interphase. This scrunching is a programmed shutdown of about 75 percent of the genes on *one* of two homologous X chromosomes. That shutdown, called **X chromosome inactivation**, happens in the female embryos of all placental mammals and their marsupial relatives.

Figure 15.4*a* shows one condensed X chromosome in the nucleus of a cell at interphase. We also call this condensed structural form a Barr body (after Murray Barr, who first identified it).

An X chromosome is inactivated when XX embryos are still a tiny ball of cells. In placental mammals, the shutdown is random, in that *either* chromosome could become condensed. The maternal X chromosome may be inactivated in one cell; the paternal or the maternal X chromosome may be inactivated in a cell next to it.

Once the random molecular selection is made in a cell, all of that cell's descendants make the exact same selection as they go on dividing to form tissues. What is the outcome? *A fully developed female has patches of tissue where genes of the maternal X chromosome are being expressed and patches of tissue where genes of the paternal X chromosome are being expressed.* She is a "mosaic" for the expression of X-linked genes.

When alleles on two homologous X chromosomes are not identical, patches of tissues through the body often show variation. Mosaic tissues can be observed in women who are heterozygous for a rare mutant allele that causes an absence of sweat glands. Sweat glands form in some patches of skin only. Where sweat glands are absent, the mutant allele is on the active X chromosome. The mosaic effect is especially apparent in females affected by *anhidrotic ectodermal dysplasia* (Figure 15.4*b*). Abnormalities in the skin and structures derived from it, including teeth, hair, nails, and sweat glands, are signs of this heritable disorder.

A different mosaic tissue effect shows up in female calico cats, of the sort shown in Figure 15.5. These cats are heterozygous for a certain coat color allele on their X chromosomes.

According to the theory of **dosage compensation**, the shutdown is not an accident of evolution; it is a gene control mechanism. In mammals, recall, males are XY, which means that females have twice as many X chromosome genes (Section 12.5). Inactivating one of their two X chromosomes balances gene expression

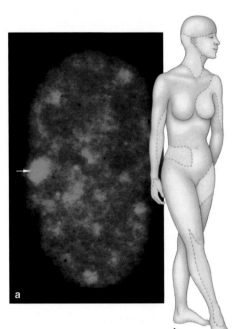

Figure 15.4 (**a**) In the somatic cell nucleus of a human female, a condensed X chromosome, also called a Barr body (*arrow*). The X chromosome in cells of human males is not condensed this way.

(**b**) A mosaic tissue effect that becomes apparent in women who are affected by anhidrotic ectodermal dysplasia. Some patches of skin have sweat glands, but other patches (color-coded *yellow*) have none.

Figure 15.5 *Animated!* Why is this female cat "calico"? In her body cells, one of her two X chromosomes has a dominant allele for the brownish-black pigment melanin. Expression of the allele on her other X chromosome codes for orange fur. When this cat was still an embryo, one X chromosome was inactivated at random in each cell that had formed by then. Patches of different colors reflect which allele was shut down in cells that formed a given tissue region. (White patches are an outcome of an interaction that involves a different gene, the product of which blocks melanin synthesis.)

petal
carpel
stamen
sepal

(a) Wild-type flower

(b) The abnormal flowers of four mutant plants

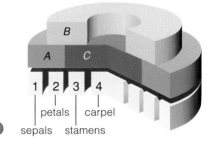

B
A   C

1 / 2 / 3 / 4
petals | carpel
(c) sepals   stamens

Figure 15.6 *Animated!* Controls over formation of flowers, based on mutations in *Arabidopsis thaliana*. This plant's flowers have four sepals, four petals, six stamens, and two fused carpels. The pattern of expression of floral identity genes affects whorls of cells that differentiate near the tip of newly forming flowers. Whorl 1 becomes sepals; whorl 2, petals; whorl 3, stamens; and whorl 4, the carpel. The above model represents the relative locations of tissues in which each set of floral identity genes—*A*, *B*, or *C*—is expressed. Gene products "tell" cells in each whorl what to do.

between the sexes. The normal development of female embryos depends on this type of control.

How, in a single nucleus, does only one of two X chromosomes get shut down? Methylation of histones and the action of *XIST*, an X-linked gene, do the trick. The *XIST* product, a large RNA molecule, sticks like masking paint to chromosomal DNA. Although we do not know why, the *XIST* gene on only one of the two chromosomes is active. That chromosome and its genes get painted with RNA. The other one remains paint-free; its genes remain available for transcription —sort of. Be sure to read *Critical Thinking* question 9 on page 241. It puts a twist on this generalized picture of X chromosome inactivation.

GENE CONTROL OF FLOWER FORMATION

Plants, too, offer fine examples of gene controls. For instance, when some plant shoots put on new growth, young plant cells right behind the tips differentiate in ways that produce flowers. Whorls of the new tissues become sepals, petals, stamens, and carpels (Figure 15.6). Studies of mutations in the common wall cress plant, *Arabidopsis thaliana*, support an **ABC model** for how all of the specialized parts of a flower develop in a predictable pattern. Three sets of master genes—*A*, *B*, and *C*—guide the process. As you will see, they are like the genes that control how body parts of animal embryos form in predictable patterns.

The cells dividing at the tip of a floral shoot form whorls of tissue, one over the other, like onion layers. What will the cells in each whorl become? It depends

on which genes of the *ABC* group are activated. In the outermost whorl, only *A* genes are switched on, and their products trigger events that cause sepals to form. Moving inward, cells in the next whorl express *A* and *B* genes; they give rise to petals. Farther in are cells that express *B* and *C* genes; they give rise to stamens (male floral structures). Cells of the innermost whorl express *C* genes only; they give rise to a fused carpel (a female floral structure).

Support for the model comes from mutations in genes of the *ABC* group (Figure 15.6*b*). Mutation in an *A* group gene alters the two outermost whorls. The flower that forms has stamens and carpels but no petals. Mutation in a *B* group gene affects the second and third whorls; sepals replace petals, and a carpel replaces stamens. Mutation in a *C* group gene alters the innermost whorls. The resulting flower is sterile (no stamens, no carpel) but has a profusion of petals.

The product of a different gene, *Leafy*, controls the activation of the sets of *ABC* genes. Certain mutations in *Leafy* keep flowers from forming on shoots where we normally expect to see them. And what switches on *Leafy*? Evidence points to a steroid hormone.

Dosage compensation in mammals is an example of gene control in eukaryotes. Most of the genes on one of the X chromosomes in females (XX) are inactivated so that early development proceeds the same as it does in males (XY).

As another example, selective expression of ABC genes controls how flowers develop.

15.3 There's a Fly in My Research

Patterns in the body plan emerge as an embryo develops, and they are both beautiful and fascinating. Researchers have correlated many of the patterns with expression of specific genes at particular times, in particular tissues. Tiny fruit flies yielded big clues to the connection.

For about a hundred years, *Drosophila melanogaster* has been the fly of choice for laboratory experiments. Why? It costs almost nothing to feed this fruit fly, which is only about 4.6 centimeters (1.8 inches) long and can live in bottles. Also, *D. melanogaster* reproduces fast and has a short life cycle, and disposing of dozens of spent bodies after an experiment is a snap. We now know how all of its 13,601 genes are distributed along the length of its four pairs of chromosomes.

Anatomical, cytological, biochemical, and genetic studies of *Drosophila* continue to reveal gene controls over development. In addition, they yield insights into evolutionary connections among groups of animals.

Discovery of Homeotic Genes   Like fruit flies, most eukaryotic species have **homeotic genes**, a class of master genes that contain information about mapping out the basic body plan. The genes code for regulatory proteins that include a "homeodomain," a sequence of about sixty amino acids. This sequence binds to control elements in promoters and enhancers.

Different homeotic genes are transcribed in specific parts of a developing embryo, so their products become concentrated in local tissue regions. Body parts form as the products interact with one another and with control elements to switch on other genes along the length of the body's main axis, according to an inherited plan.

Researchers discovered homeotic genes in mutant fruit flies that had body parts growing out of the wrong places. As an example, the *antennapedia* gene is supposed to be transcribed only in embryonic tissues that give rise to a thorax, complete with legs. This gene normally is not transcribed in cells of all other tissue regions. But Figure 15.7*a* shows what happened after a mutation altered

some control over transcription and the gene was wrongly transcribed in the tissue destined to become a head.

Plants, too, have master genes. You already read about how they control floral development. Similarly, a different master gene helps leaf veins in corn plants form straight, parallel lines. When the gene mutates, the veins twist.

More than 100 homeotic genes have been identified in diverse eukaryotes—and the same mechanisms control their transcription. Many of the genes are functionally interchangeable among species as evolutionarily distant as yeasts and humans, so we can expect that they evolved in the most ancient eukaryotic cells. Their protein products often differ only in modest substitutions. In other words, one amino acid has replaced another, but its chemical properties are still similar.

Knockout Experiments   *Drosophila* researchers made more discoveries about how embryos develop. For instance, with **knockout experiments**, a wild-type gene is mutated in a way that prevents its transcription or translation. If genetically engineered knockout individuals differ in form or behavior from wild-type individuals, this may be a clue to the function of the missing gene. Such experiments have yielded insights into the functions of many hundreds of genes in different organisms.

Researchers tend to name the genes based on what happens in their absence. For instance, *eyeless* is a control gene expressed in fruit fly embryos. In its absence, no eyes form. *Dunce* is a regulatory gene required for learning and memory. *Wingless*, *wrinkled*, and *minibrain* genes are self-explanatory. *Tinman* is necessary for heart development. Among other things, *groucho* prevents overproduction of whisker bristles. Figure 15.7 shows a few of the mutants.

In other experiments, researchers add special promoters to a gene so that they can control its expression with an external cue, such as temperature. They delete genes from one part of the *Drosophila* genome and put them back someplace else. This molecular sleight of hand revealed that expression of the *eyeless* gene can induce an eye to form not only on the fruit fly head, but also on the wings and legs (Figure 15.7*c*).

Figure 15.7   (**a**) Experimental evidence of controls over where body parts develop. In *Drosophila* larvae, activation of genes in one group of cells normally results in antennae on the head. A mutation that affects *antennapedia* gene transcription puts legs on the head. This is one of the genes controlled by regulatory proteins with homeodomains. (**b**) Model for a homeodomain binding to a transcriptional control site in DNA. (**c**) More *Drosophila* mutations.

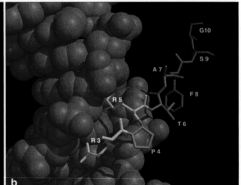

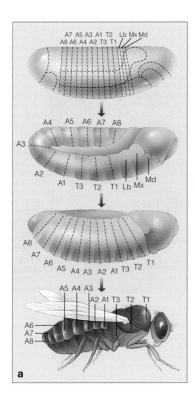

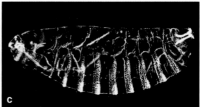

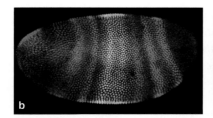

Figure 15.8 Genes and *Drosophila*'s segmented body plan. (**a**) Fate map for the surface of a *Drosophila* zygote. Such maps indicate where each differentiated cell type in the adult originated. The pattern starts with the polar distribution of maternal mRNA and proteins in the unfertilized egg. This polarity dictates the future body axis. A series of segments will develop along this axis. Genes specify whether legs, wings, eyes, or some other body parts will develop on a particular segment.

Briefly, here is how it happens: Maternal gene products prompt expression of gap genes. Different gap genes become activated in regions of the embryo with higher or lower concentrations of different maternal gene products. Gap gene products influence each other's expression as well. They form a primitive spatial map.

Depending on where they occur relative to concentrations of gap gene products, embryonic cells express different pair-rule genes. Products of pair-rule genes accumulate in seven transverse stripes that mark the onset of segmentation (**b**). They activate other genes, the products of which divide the body into units (**c**). These interactions influence the expression of homeotic genes, which collectively govern the structural and functional identity of each segment.

The *eyeless* gene even has counterparts in humans (the *PAX6* gene), mice (*Pax-6*), and squids (also *Pax-6*). Humans who have no functional *PAX6* genes have eyes with malformed irises. *PAX-6* inserted into any tissue of an eyeless mutant fly induces an eye to form wherever it is expressed. This kind of molecular evidence points to a shared ancestor among animals as evolutionarily distant as insects, cephalopods, and mammals.

Filling In Details of Body Plans Let's take stock. As an embryo develops, cells in different body regions become organized in different ways. Cells divide and differentiate. They migrate or stick to cells of the same type in tissues. They live or die after performing their function. These are genetically programmed events that fill in details of the body in orderly patterns, in keeping with the expression of master genes. Those genes are switched on in specific tissues, at specific stages of development. The products of those genes deal mainly with transcription. In effect, they form a three-dimensional map along the main body axis.

Depending on where undifferentiated cells are relative to the map, they are the start of specialized tissues and organs. That is how the sequential expression of master genes along the body axis gives rise to the body segments of fruit flies (Figure 15.8).

Pattern formation is the name for the emergence of embryonic tissues and organs in orderly patterns, at times and in places where we expect them to be. Section 43.5 offers a closer look at the controlled gene interactions that fill in details of the animal body plan.

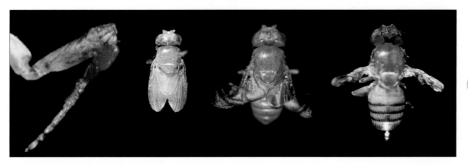

c A few more *Drosophila* mutations that yielded clues to gene function *Left to right,* an eye that formed on a leg, yellow miniature, curly wings, vestigial wings, and a double thorax.

15.4 Prokaryotic Gene Control

LINK TO
SECTION
8.2

In prokaryotic cells, gene controls deal mainly with quickly slowing down and starting up transcription in response to short-term shifts in environmental conditions. A diversity of long-term controls is not required; none of these species slowly develops into a complex, multicelled form (Table 15.1).

When nutrients are plentiful and when other external conditions also favor growth and reproduction, all prokaryotic cells rapidly transcribe genes that specify all of the enzymes required for nutrient absorption and other growth-related tasks. Genes that are tapped most often occur one after the other as a set of genetic information in the DNA. They all can be transcribed together, which yields a single RNA strand.

NEGATIVE CONTROL OF THE LACTOSE OPERON

With this bit of background, consider an example of how one kind of prokaryote responds to the presence or absence of lactose. *Escherichia coli* lives in the gut of mammals, where it dines on nutrients traveling past. Milk typically nourishes mammalian infants. It does not contain glucose, the sugar of choice for *E. coli*. It does contain lactose, a different sugar.

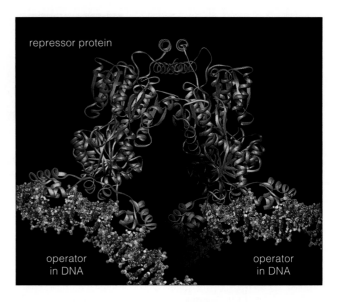

Figure 15.9 Model for the repressor protein of the lactose operon when it is bound to two operators in DNA.

Table 15.1	Prokaryotic Versus Eukaryotic Gene Control

Prokaryotic Gene Control

1. Control mechanisms adjust enzyme-mediated reactions in response to short-term changes in nutrient availability and other environmental conditions.

2. Operons control the expression of more than one gene at a time.

3. Transcriptional controls are *reversibly* inhibited when conditions do not favor growth and reproduction.

4. Translation starts immediately; prokaryotic RNA transcripts have no introns, no processing controls.

Eukaryotic Gene Control

1. Some control mechanisms adjust enzyme-mediated reactions in response to short-term changing conditions.

2. In multicelled species, other controls activate sets of genes at different times, in different tissues. They induce generally *irreversible* events that are part of a long-term program of growth and development.

3. Diverse controls operate during gene transcription and translation, and on the gene products. mRNA transcripts are processed in the nucleus; controls govern the timing and rate of their translation in the cytoplasm.

After being weaned, infants of most species drink little (if any) milk. Even so, *E. coli* cells can still use lactose if and when it shows up in the gut. They can activate a set of three genes for lactose-metabolizing enzymes. In *E. coli* DNA, a promoter precedes all three genes, and two operators flank it. Each **operator** is a binding site for a type of regulatory protein known as repressor, which stops transcription (Figure 15.9).

Any arrangement in which a promoter and a set of operators control access to more than one prokaryotic gene is called an **operon**.

In the absence of lactose, a repressor molecule binds to the set of operators. Binding causes the DNA region that contains the promoter to twist into a loop, as in Figure 15.10. RNA polymerase, the workhorse that transcribes genes, is not able to bind to a looped-up promoter. The result is that operon genes are not used when they are not required.

When lactose *is* in the gut, *E. coli* converts some of it to allolactose. This sugar binds to the repressor and changes its molecular shape. The altered repressor cannot bind to operators. The looped DNA unwinds and RNA polymerase transcribes the genes, so lactose-degrading enzymes are produced when required.

POSITIVE CONTROL OF THE LACTOSE OPERON

E. coli cells pay far more attention to glucose than to lactose. Even when lactose is in the gut, the lactose operon is not used much—unless there is no glucose.

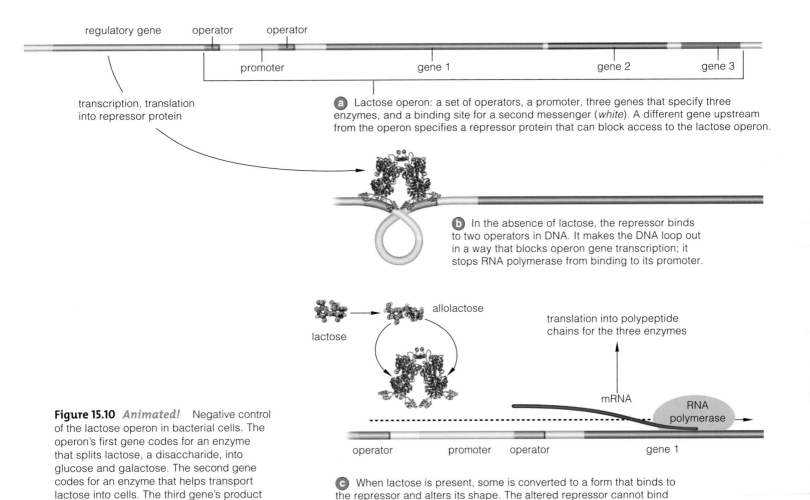

regulatory gene   operator   operator

promoter   gene 1   gene 2   gene 3

transcription, translation into repressor protein

a Lactose operon: a set of operators, a promoter, three genes that specify three enzymes, and a binding site for a second messenger (*white*). A different gene upstream from the operon specifies a repressor protein that can block access to the lactose operon.

b In the absence of lactose, the repressor binds to two operators in DNA. It makes the DNA loop out in a way that blocks operon gene transcription; it stops RNA polymerase from binding to its promoter.

allolactose

lactose

translation into polypeptide chains for the three enzymes

mRNA

RNA polymerase

operator   promoter   operator   gene 1

Figure 15.10 *Animated!* Negative control of the lactose operon in bacterial cells. The operon's first gene codes for an enzyme that splits lactose, a disaccharide, into glucose and galactose. The second gene codes for an enzyme that helps transport lactose into cells. The third gene's product helps metabolize certain sugars.

c When lactose is present, some is converted to a form that binds to the repressor and alters its shape. The altered repressor cannot bind to operators, so RNA polymerase is free to transcribe the operon genes.

When that happens, an activator called CAP (short for catabolite activator protein) exerts positive control over the lactose operon. It makes a promoter far more inviting to RNA polymerase. But CAP cannot issue the invitation until it has already become bound to a chemical messenger: cAMP (short for cyclic adenosine monophosphate). When the activator and cAMP join together as a complex to the promoter, they make it much easier for RNA polymerase to start transcribing genes. Such complexes are called transcription factors.

When glucose is plentiful, ATP forms by glycolysis (Section 8.2), but synthesis of an enzyme necessary to synthesize cAMP is blocked. The blocking ends when glucose is scarce and lactose becomes available. cAMP accumulates, the CAP–cAMP complexes form, and the lactose operon genes are transcribed fast. The gene products allow lactose to be converted to glucose, the preferred sugar of *E. coli*.

Unlike cells of *E. coli*, many of us develop *lactose intolerance*. Cells making up the lining of our small intestine make and then secrete lactase into the gut. As many people age, however, concentrations of this lactose-digesting enzyme decline. Lactose accumulates and is moved on to the large intestine, or colon. It promotes population explosions of resident bacteria. As the bacterial cells busily digest the lactose, a gaseous metabolic product accumulates, the gas distends the colon's wall and causes pain. The short fatty acid chains that form during bacterial metabolism cause diarrhea, which can be severe.

Prokaryotic cells do not require extensive controls over long-term development of complex bodies; they are small, fast reproducers. The main controls guide the transcription of enzyme-coding genes in response to short-term shifts in nutrient availability and other outside conditions.

Summary

Section 15.1 Gene expression within a cell changes in response to chemical conditions and signals from the outside. In complex multicelled species, it is subject to long-term controls over growth and development.

Control mechanisms govern whether, when, and how a gene is expressed. Hormones, activator proteins, and other regulatory elements interact with one another before, during, and after transcription and translation.

With negative control mechanisms, regulatory elements slow or stop a cell activity. With positive control mechanisms, they promote it.

Control is exerted before, during, and after gene transcription and translation. Transcription is a major control point for most eukaryotic genes because so many of the participating molecules can be controlled independently of the others.

Diverse controls guide embryonic development of multicelled eukaryotes. Each cell in an embryo inherits the same genes, but they start selectively activating and suppressing some in unique ways. The outcome of selective gene expression is called cell differentiation: cell lineages become unique in one or more aspects of composition, structure, and function. Those lineages are the start of specialized tissues and organs.

Biology⊗Now
Review the control points for gene expression with the animation on BiologyNow.

Section 15.2 Two examples are given of eukaryotic gene controls, one in mammals, the other in plants:

In the embryos of all female mammals, X chromosome inactivation is an outcome of the interactions between a product of the *XIST* gene and control elements in one of the two X chromosomes in cells. This control mechanism maintains a required balance of gene expression between the sexes while mammalian embryos are developing.

Studies of mutations in *Arabidopsis thaliana* support an ABC model for the formation of flowers. Three sets of master genes (*A, B,* and *C*) guide the differentiation of whorls of cells into sepals, petals, stamens, and carpels.

Biology⊗Now
Observe how eukaryotic gene controls influence development with the animation on BiologyNow.

Section 15.3 Many gene controls were identified through experiments with mutant forms of *Drosophila melanogaster*. Others were discovered by knockout experiments in which individual genes are deactivated before a new wild-type fly develops.

While embryos of this fruit fly and of most other eukaryotes develop, homeotic genes are activated in sequence. The protein products of this class of master genes become more or less concentrated along the body's main axis, which maps out the basic body plan. Cells differentiate according to their location along the map. Their descendants fill in the details of the body plan by forming specialized tissues and organs in patterns where we expect them to be.

Section 15.4 Prokaryotic cells do not have great structural complexity and do not undergo development. Most of the gene controls reversibly adjust transcription rates in response to environmental conditions, especially nutrient availability. Bacterial operons are examples of prokaryotic gene controls. The lactose operon controls three genes, the three products of which digest lactose. It has a promoter region in DNA (binding sites for RNA polymerase). Two operators flank it and are binding sites for a repressor protein that can block transcription.

Biology⊗Now
Explore the structure and function of the bacterial lactose operon with the animation on BiologyNow.

Self-Quiz
Answers in Appendix II

1. The expression of a given gene depends on the _____ .
   a. type and function of cell   c. environmental signals
   b. chemical conditions   d. all of the above

2. Control mechanisms adjust gene expression in response to changing _____ .
   a. nutrient availability   c. signals from other cells
   b. solute concentrations   d. all of the above

3. Regulatory elements interact with _____ .
   a. DNA   c. gene products
   b. RNA   d. all of the above

4. At _____ in DNA, regulatory proteins gather and control transcription of specific genes.
   a. promoters   c. operators
   b. enhancers   d. both a and b

5. Eukaryotic gene controls govern _____ .
   a. transcription   e. mRNA degradation
   b. RNA processing   f. gene products
   c. translation   g. a through e
   d. RNA transport   h. all of the above

6. Eukaryotic genes guide _____ .
   a. fast short-term activities   c. development
   b. overall growth   d. all of the above

7. Cell differentiation _____ .
   a. occurs in all complex multicelled organisms
   b. requires unique genes in different cells
   c. involves selective gene expression
   d. both a and c
   e. all of the above

8. During X chromosome inactivation _____ .
   a. many genes are shut down c. sweat glands form
   b. RNA paints chromosomes   d. both a and b

9. A cell with a Barr body is _____ .
   a. prokaryotic   c. from a female mammal
   b. from a male mammal   d. infected by Barr virus

10. Homeotic gene products _____ .
   a. are binding sites that flank a bacterial operon
   b. map out a developing embryo's body plan
   c. control X chromosome inactivation
   d. both a and c

11. Knockout experiments mutate _____ genes.
   a. bacterial   c. engineered
   b. wild-type   d. both a and c

12. A(n) _____ is a promoter and a set of operators that control access to two or more prokaryotic genes.
    a. lactose molecule   c. dosage compensator
    b. operon   d. both b and c

13. Match the terms with the most suitable description.
    ____ ABC model   a. a big RNA is its product
    ____ *XIST* gene   b. binding site for repressor
    ____ operator   c. cells become specialized in
    ____ Barr body   composition, function, etc.
    ____ process of cell   d. inactivated X chromosome
        differentiation   e. how flowers develop
    ____ methylation   f. —CH$_3$ additions to DNA

Additional questions are available on Biology 🅔 Now™

Critical Thinking

1. Do all transcriptional controls operate in prokaryotic as well as eukaryotic cells? Why or why not?

2. If all cells in your body start out life with the same inherited information on how to build proteins, then what caused the differences between a red blood cell and a white one? Between a white blood cell and a nerve cell?

3. Unlike most rodents, guinea pigs are well developed at the time of birth. Within a few days, they can eat grass, vegetables, and other plant material.

Suppose a breeder decides to separate baby guinea pigs from their mothers three weeks after they were born. He wants to raise the males and the females in different cages. However, he has trouble identifying the sex of young guinea pigs. Suggest how a quick look through a microscope can help him identify the females.

4. Calico cats are almost always female. A male calico cat is usually sterile. Find out why.

5. Reflect on the mutant *Arabidopsis thaliana* flowers in Figure 15.6. Small changes in the structure of control genes brought about those changes. Would you predict that such changes figured in the evolution of more than 295,000 kinds of plants, each with distinctive flowers?

Also reflect on the *Drosophila melanogaster* mutants shown in Figure 15.7. Would you predict that homeotic gene mutations figured in the evolution of the more than 1.5 million known species of animals?

6. *Duchenne muscular dystrophy*, a genetic disorder, affects boys almost exclusively. Muscles begin to atrophy (waste away) in affected children, who typically die in their teens or early twenties (Section 12.7).

Muscle biopsies of a few women who carry an allele that is associated with the disorder identified some body regions of atrophied muscle tissue. They also showed that muscles adjacent to a region of atrophy were normal or even larger and more chemically active, as if to compensate for the weakness of the adjoining region.

Form a hypothesis about the genetic basis of Duchenne muscular dystrophy that includes an explanation of why the symptoms might appear in some body regions but not others.

7. Figure 15.11 shows seven "spots" that emerge in the wings of a developing moth larva. The spots identify where seven distinct eyespots will appear on the wings

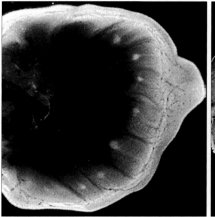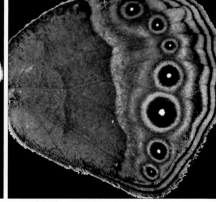

Figure 15.11 *Left*, seven spots in the embryonic wing of a moth larva identify the presence of a gene product that will induce the formation of seven "eyespots" in the wing of the adult (*right*).

of adult moths. What is the name of the class of genes responsible for mapping out such details of the body plan of developing embryos, including insect larvae?

8. Geraldo isolated an *E. coli* strain in which a mutation has hampered the capacity of CAP to bind to a region of the lactose operon, as it would do normally. How will this mutation affect transcription of the lactose operon when the *E. coli* cells are exposed to the following conditions? Briefly state your answers:
    a. Lactose and glucose are both available.
    b. Lactose is available but glucose is not.
    c. Both lactose and glucose are absent.

9. About 300 million years ago, before mammals began their great adaptive radiation, their X and Y chromosomes were about the same in size. When paired, the two sex chromosomes typically synapsed and exchanged alleles along their length. The X chromosome now carries 1,141 genes. Over time, however, the Y chromosome lost most of itself and now contains only 255 genes. Its big claim to fame is ownership of the *SRY* gene, the master of sex determination.

Think about the dosage compensation theory, sketched out in Section 15.2. According to this theory, *X chromosome inactivation* is nature's way of compensating for a double dose of X-linked genes in XX embryos, because there are not enough genes left on the puny Y chromosome to balance out their expression. And yet, about 15 percent of the genes on an inactivated X chromosome escape being painted to varying degrees—which means women make more copies of certain proteins than men do.

Besides this, another 10 percent of the X-linked genes might or might not get painted in individual embryos—which means women differ significantly from one another in which X-linked genes are active.

Now consider this: Human and chimpanzee genomes differ by 1.5 percent. Women differ from men by 1 percent! Go ahead and let your brain chew on that one.

You may wish to start with this recent article: L. Carel and H. Willard, "X-Inactivation Profile Reveals Extensive Variability in X-Linked Gene Expression in Females," *Nature* 2005; 434(7031):400–404.

Golden Rice, or Frankenfood?

Not too long ago, the World Health Organization made a conservative estimate that 124 million children around the world show vitamin A deficiencies. Their skin, eyes, and mucous membranes are dry and vulnerable to infection. They do not grow and develop as they should, and they show signs of mental impairment. Each year at least a million die of malnutrition, and about 350,000 end up permanently blind.

Ingo Potrykus and Peter Beyer wanted to help. As they knew, beta-carotene is a yellow pigment in all plant leaves, and it also is a precursor for vitamin A. These geneticists borrowed three genes from garden daffodils (*Narcissus pseudonarcissus*) and a bacterium, and transferred them to rice plants. The plants transcribed the genes and did something they could not do before. They made beta-carotene not only in their leaves but also in their *seeds*—the grains of Golden Rice (Figure 16.1).

Why rice? Rice is the main food for 3 billion people in impoverished countries. There, the poor cannot afford leafy vegetables and other sources of beta-carotene. Getting beta-carotene into rice grains would be the least costly way to deliver the vitamin to those who need it the most, but doing so was beyond the scope of conventional breeding practices. Research continues, and the amount of beta-carotene in SGR1, a more recent version of Golden Rice, is twenty-three times higher than the prototype.

No one wants children to suffer or die. However, many people oppose the idea of genetically modified (GM) foods, including golden rice. Possibly they are unaware of the history of agrarian societies, because it is not as if our ancestors were twiddling their green thumbs. For thousands of years, their artificial selection practices coaxed new plants and new breeds of cattle, cats, dogs, and birds from wild ancestral stocks. Meatier turkeys, huge watermelons, big juicy corn kernels from puny hard ones—the list goes on (Figure 16.1).

And we are newcomers at this! During the 3.8 billion years before we even made our entrance, nature busily conducted uncountable numbers of genetic experiments by way of mutation, crossing over, and gene transfers between species. These processes introduced changes in the molecular messages of inheritance, and today we see their outcomes in the sweep of life's diversity.

Perhaps the unsettling thing about the more recent human-directed changes is that the pace has picked up, hugely. We are getting much better at tinkering with the genetics of many organisms. We do this for pure research and for useful, practical applications.

a   b

Figure 16.1 Where one genetic engineering success story started: (**a**) Researchers transferred genetic information from ordinary daffodils into rice plants, which then used it to stockpile beta-carotene in their seeds—rice grains. (**b**) Two successive generations of Golden Rice compared with grains from a regular rice plant at lower left. *Facing page*, an artificial selection success story—a big kernel from a modern strain of corn next to tiny kernels of an ancestral corn species discovered in a prehistoric cave in Mexico.

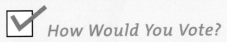

For instance, many crop plants, including corn, beets, and potatoes, have been modified. They are now widely planted. They are less temperamental about their living conditions than rice plants are, and they have not run rampant through ecosystems. After a decade-long study in the United Kingdom, researchers concluded that the new crop plants being monitored were doing no harm. Throughout Arizona, farmers grow cotton plants that are genetically engineered for pest resistance. The plantings have not put the environment at risk and might even be less disruptive compared to current agricultural practices. University of Arizona entomologist Bruce Tabashnik, who is monitoring cotton fields, notes that farmers have cut applications of chemical pesticides by 75 percent.

Take stock of how far you have come in this unit. You started with cell division mechanisms that allow parents to pass on DNA to new generations. You moved to the chromosomal and molecular basis of inheritance, then on to gene controls that guide life's continuity. The sequence parallels the history of genetics. And now, you have arrived at the point in time where geneticists hold molecular keys to the kingdom of inheritance. What they are unlocking is already having impact on life in the biosphere.

How Would You Vote?

Nutritional labeling is required on all packaged food in the United States, but genetically modified food products may be sold without labeling. Should food distributors be required to label all products made from genetically modified plants or livestock? See BiologyNow for details, then vote online.

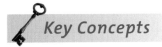

Key Concepts

MAKING RECOMBINANT DNA

Researchers routinely make recombinant DNA molecules. They use restriction enzymes to isolate, cut, and join gene regions from DNA of different species. They use plasmids and other vectors to insert the recombinant molecule into target cells.   Section 16.1

ISOLATING AND AMPLIFYING DNA FRAGMENTS

Researchers isolate and make many copies of genes that interest them. PCR is now the gene amplification method of choice. The genes are copied in amounts large enough for research and practical applications.   Section 16.2

DECIPHERING DNA FRAGMENTS

Sequencing methods reveal the linear order of bases in a sample of DNA. Automated methods complete the task with impressive speed.   Sections 16.3, 16.4

MAPPING AND ANALYZING WHOLE GENOMES

Genomics is concerned with mapping and sequencing of the genomes of humans and other species. Comparative genomics yields evidence of evolutionary relationships among groups of organisms.   Section 16.5

USING THE NEW TECHNOLOGIES

Genetic engineering results in transgenic organisms, which incorporate genes from another species. With gene therapy, a mutated or altered gene is isolated, modified, and copied. Copies are inserted back into the individual to cover the gene's function. The new technologies raise social, legal, ecological, and ethical questions.   Sections 16.6–16.10

Links to Earlier Concepts

This chapter builds on earlier explanations of the molecular structure of DNA (Sections 3.7, 13.2), and DNA replication and DNA repair (13.4). You may wish to review quickly the nature of mRNA transcript processing (14.1) and controls over gene transcription (14.1). You will come across more uses for radioisotopes (2.2) and fluorescent light (6.6). You will be reminded of why it is useful to know about membrane proteins (5.2). You will see why the lactose operon is not necessarily of obscure interest (15.4).

16.1 A Molecular Toolkit

Analysis of genes starts with manipulation of DNA. With molecular tools, researchers can cut DNA from different sources, then splice the fragments together.

LINKS TO
SECTIONS
13.1, 13.3, 14.1

THE SCISSORS: RESTRICTION ENZYMES

In 1970, Hamilton Smith was studying viral infection of *Haemophilus influenzae*. This bacterium protects itself from infection by cutting up viral DNA before it can get inserted into the bacterial chromosome. Smith and his colleagues isolated one of the bacterial enzymes that cuts viral DNA. It was the first known **restriction enzyme**. In time, several hundred strains of bacteria and a few eukaryotic cells yielded thousands more.

A restriction enzyme cuts double-stranded DNA at a specific base sequence between four and eight base pairs in length. Most of these recognition sites contain the same nucleotide sequence, in the 5'→3' direction, on both strands of the DNA. For instance, the enzyme

*Eco*RI recognizes and cuts GAATTC (Figure 16.2). It makes staggered cuts that produce a "sticky end," or single-stranded "tail,"on the DNA fragments. The tail can base-pair with a tail of another fragment cut by the same enzyme, because the sticky ends of both will match up as base pairs. Tiny nicks remain when DNA fragments base-pair. Remember **DNA ligases** (Section 13.3)? They seal the nicks, which yields a recombinant molecule (Figure 16.2). We define **recombinant DNA** as any molecule consisting of base sequences from two or more organisms of the same or different species.

CLONING VECTORS

Bacterial cells, recall, have only one chromosome—a circular DNA molecule. But many also have plasmids. A **plasmid** is a small circle of extra DNA with just a few genes (Figure 16.3*a*). It gets replicated along with the bacterial chromosome. Bacteria normally can live without plasmids. Even so, certain plasmid genes are useful, as when they confer resistance to antibiotics.

Under favorable conditions, bacteria divide often, so huge populations of genetically identical cells form swiftly. Before each division, replication enzymes copy chromosomal DNA *and* plasmid DNA, in some cases repeatedly. This gave researchers the idea of inserting DNA fragments into a plasmid to see if a bacterial cell would replicate them right along with the plasmid.

A plasmid that has accepted foreign DNA and can slip into a host bacterium, yeast, or some other cell is a **cloning vector**. Most vectors have been engineered to incorporate multiple cloning sites, which are unique restriction enzyme sequences in one part of the vector (Figure 16.3). As you will see, cloning vectors contain genes that help researchers identify which cells take them up. Viruses also are used as cloning vectors.

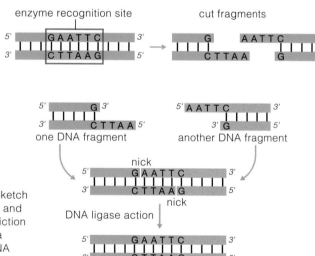

Figure 16.2 Sketch of the formation and splicing of restriction fragments into a recombinant DNA molecule.

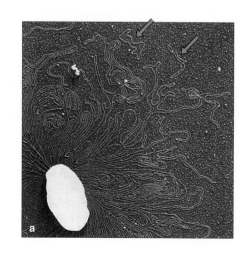

Figure 16.3 (**a**) Plasmids (*arrows*) from a ruptured *Escherichia coli* cell. (**b**) A commercially available cloning vector. Its useful restriction enzyme sites are listed at right. This vector includes antibiotic resistance genes (*blue*) and the bacterial *lacZ* gene (*red*). Researchers can check for the expression of these genes as a way to identify the bacterial cells that take up recombinant molecules.

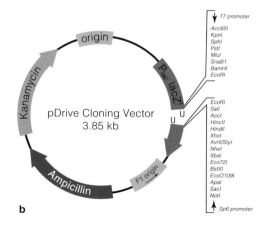

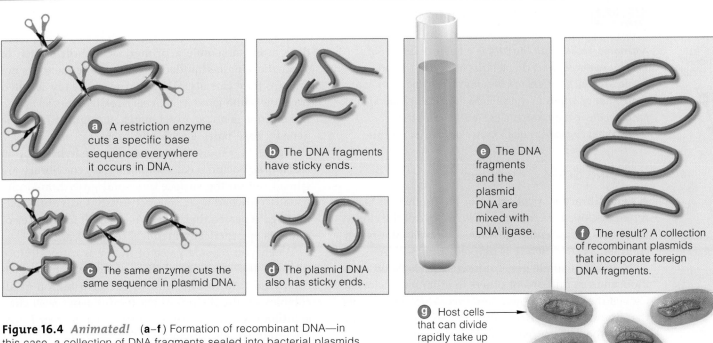

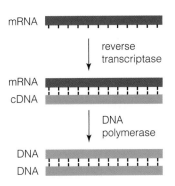

Figure 16.4 *Animated!* (**a–f**) Formation of recombinant DNA—in this case, a collection of DNA fragments sealed into bacterial plasmids. (**g**) Recombinant plasmids are inserted into host cells that can rapidly make multiple copies of the foreign DNA of interest.

a A restriction enzyme cuts a specific base sequence everywhere it occurs in DNA.

b The DNA fragments have sticky ends.

c The same enzyme cuts the same sequence in plasmid DNA.

d The plasmid DNA also has sticky ends.

e The DNA fragments and the plasmid DNA are mixed with DNA ligase.

f The result? A collection of recombinant plasmids that incorporate foreign DNA fragments.

g Host cells that can divide rapidly take up the recombinant plasmids.

A cell that takes up a cloning vector may give rise to a huge population of descendant cells, each with an identical copy of the vector and the foreign DNA inserted into it. Collectively, the identical cells hold many "cloned" copies of the foreign DNA.

Such DNA cloning is a tool that helps researchers amplify and harvest unlimited amounts of particular DNA fragments for their studies (Figure 16.4).

cDNA CLONING

Remember those introns in eukaryotic DNA (Section 14.1)? Bacterial cells cannot remove introns from RNA, as eukaryotic cells do. That is why researchers often use mature mRNA transcripts. The introns already have been removed, and protein-coding sequences and a few sequences that are identifiable signals are left. Researchers also may use mRNA to study gene expression, because the cells that are actively using a gene obviously contain mRNA transcribed from it.

Restriction enzymes will not cut single-stranded molecules, so they will not cleave mRNA (which is single-stranded). However, mRNA can be cloned if it is first transcribed—in reverse. Replication enzymes isolated from viruses or bacterial cells can transcribe mRNA inside a test tube. **Reverse transcriptase** is one viral enzyme that can catalyze the bonding of free nucleotides into one strand of *complementary* DNA, or **cDNA** on an mRNA template (Figure 16.5). Base pairs

Figure 16.5 How to make cDNA. Reverse transcriptase catalyzes the assembly of a single DNA strand on an mRNA template, forming an mRNA–cDNA hybrid molecule. Next, DNA polymerase replaces the mRNA with another DNA strand. The result is double-stranded DNA.

of cDNA get hydrogen-bonded to those of mRNA, forming a hybrid molecule. Next, DNA polymerase is added to the mix. It strips RNA bases from the hybrid molecule while it copies the first strand of cDNA into a second strand. The result, a double-stranded DNA copy of the original mRNA, may be used for cloning.

Molecular biologists manipulate DNA and RNA. Restriction enzymes cut DNA from individuals of different species or the same species. DNA ligases glue the fragments into plasmids.

A recombinant plasmid is a cloning vector. It can slip into bacteria, yeast, or other cells that divide rapidly. The host cells make multiple, identical copies of the foreign DNA.

Reverse transcriptase, a viral enzyme, uses a single strand of mRNA as the template to make cDNA for cloning.

16.2 From Haystacks to Needles

LINKS TO
SECTIONS
2.1, 10.1, 13.2

*A **genome**, recall, is all the DNA in a haploid number of the chromosomes that characterize a species. To study or modify any gene, researchers must first find it among thousands of others in the genome, and it's like searching for a needle in a haystack. Once found, it must be copied many times to make enough material for experiments.*

ISOLATING GENES

A **gene library** is a collection of host cells that house different cloned fragments of DNA. We call the cloned fragments of an entire genome a *genomic* library. By contrast, a *cDNA library* is derived from mRNA.

How can a single gene of interest be isolated from thousands or millions of others in a library of clones? Clones that have the gene are mixed up with others that do not. Researchers might decide to use a **probe** to find the gene. Probes are short stretches of DNA that are complementary to a gene of interest and that are tagged with a label, such as a radioisotope, that devices can detect (Section 2.1). Probes base-pair with DNA in a gene region, then researchers pinpoint the gene by detecting the label on the probe. Any base pairing between DNA (or RNA) from more than one source is known as **nucleic acid hybridization**.

How do researchers make a probe? If they already know the gene sequence of interest, they can use it to design and assemble a **primer**, or a short stretch of synthetic, single-stranded DNA. If the sequence is not known, they can use DNA that was already isolated from the same gene in a closely related species. Even if the probe is not an exact match, it might still tag the gene by base-pairing with part of it.

Figure 16.6 shows steps of one probe hybridization technique. Bacterial cells containing a gene library are spread out on the surface of a solid growth medium, usually enriched agar, in a petri dish. Individual cells undergo repeated divisions, which result in colonies of millions of genetically identical bacterial cells.

When you press a piece of nylon or nitrocellulose filter on top of the petri dish, some cells from each colony stick to it. They mirror the distribution of all colonies on the dish. Soaking the filter in an alkaline solution ruptures the cells, which releases their DNA. The solution also denatures DNA—which separates into single strands that stick to the filter in the spots where the colonies were. When the probe is washed over the filter, it hybridizes with (sticks to) only the DNA with the targeted sequence.

The hybridized probe can be detected with x-ray film or computerized imaging devices. Its position on the film pinpoints the position of the original colony on the petri dish. Cells from that colony alone can be cultured to isolate the cloned gene of interest.

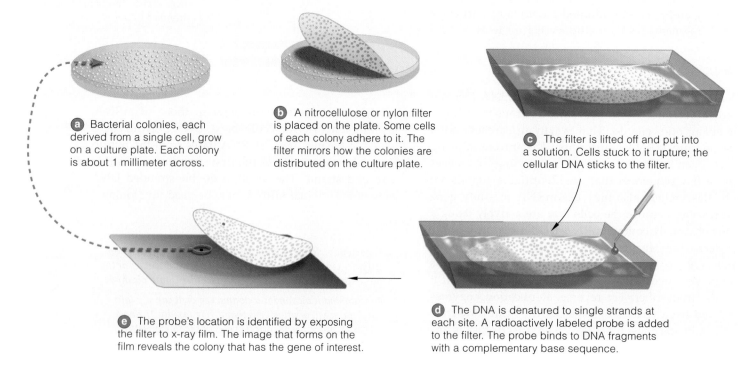

a Bacterial colonies, each derived from a single cell, grow on a culture plate. Each colony is about 1 millimeter across.

b A nitrocellulose or nylon filter is placed on the plate. Some cells of each colony adhere to it. The filter mirrors how the colonies are distributed on the culture plate.

c The filter is lifted off and put into a solution. Cells stuck to it rupture; the cellular DNA sticks to the filter.

d The DNA is denatured to single strands at each site. A radioactively labeled probe is added to the filter. The probe binds to DNA fragments with a complementary base sequence.

e The probe's location is identified by exposing the filter to x-ray film. The image that forms on the film reveals the colony that has the gene of interest.

Figure 16.6 *Animated!* How a radioactive probe helps identify a bacterial colony that contains a targeted gene.

Figure 16.7 *Animated!* Two rounds of the polymerase chain reaction, or PCR. A bacterium, *Thermus aquaticus*, is the source for the *Taq* polymerase. Thirty or more cycles of PCR may yield a billionfold increase in the number of starting DNA molecules that serve as templates.

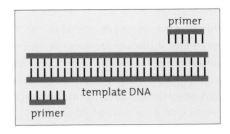

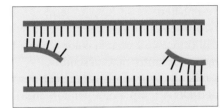

a Primers, free nucleotides, and DNA templates are mixed with heat-tolerant DNA polymerase.

BIG-TIME AMPLIFICATION—PCR

Researchers may replicate a gene, or part of it, with **PCR** (*Polymerase Chain Reaction*). PCR uses primers and a heat-tolerant polymerase for a hot–cold cycled reaction that replicates targeted DNA fragments. The technique can replicate the fragments by a billionfold. It can transform one needle in a haystack, that one-in-a-million DNA fragment, into a huge stack of needles with a little hay in it.

Figure 16.7 shows the reaction steps. The primers are designed to base-pair with particular nucleotide sequences on either end of the fragment of interest. Usually they are between ten and thirty bases long.

In a PCR reaction, researchers mix primers, DNA polymerase, nucleotides, and the DNA that will serve as a template for replication. Then they expose the mixture to cycles of high and low temperatures that are repeated again and again. At high temperature, the two strands of a DNA double helix separate. When the mixture is cooled, some of the primers hybridize with the DNA template.

The elevated temperatures required to separate the DNA strands destroy typical DNA polymerases. But the heat-tolerant DNA polymerase employed for PCR reactions is from *Thermus aquaticus*, a bacterium that lives in hot springs (Chapter 21). Like all other DNA polymerases, it recognizes primers bound to DNA as places to start synthesis. The temperature is raised to the optimum for this enzyme (72°C). Then synthesis occurs along the DNA template until the temperature cycles up and the DNA strands are separated again.

When the temperature cycles down, the primers rehybridize, and the reactions run once more. With each round of temperature cycling, the number of copies of targeted DNA can double. PCR quickly and exponentially amplifies even a tiny bit of DNA.

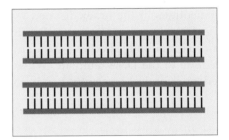

b When the mixture is heated, the DNA denatures. When it is cooled, some primers hydrogen-bond to the DNA templates.

c *Taq* polymerase uses the primers to initiate synthesis. The DNA templates are copied. The first round of PCR is completed.

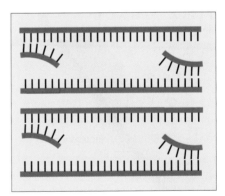

d The mixture is heated again. This denatures all the DNA into single strands. When the mixture is cooled, some of the primers hydrogen-bond to the DNA.

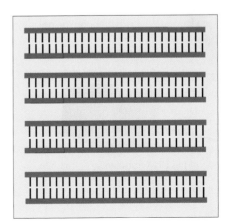

e *Taq* polymerase uses the primers to initiate synthesis, copying the DNA. The second round of PCR is complete. Each successive round of synthesis can double the number of DNA molecules.

Probes may be used to help identify one particular gene among many in gene libraries.

The polymerase chain reaction (PCR) is a method of rapidly and exponentially amplifying the number of particular DNA fragments.

16.3 Automated DNA Sequencing

LINK TO
SECTION
6.6

Sequencing reveals the order of nucleotides in DNA. This technique uses DNA polymerase to partially replicate a DNA template. Automated techniques have largely replaced manual methods.

Automated DNA sequencing can reveal the sequence of a stretch of cloned or PCR-amplified DNA in just a few hours. Researchers use four standard nucleotides (T, C, A, and G). They also use four modified versions, which we represent here as T*, C*, A*, and G*. Each form of modified nucleotide has been labeled with a pigment that will fluoresce a certain color when a laser beam hits it. Each will halt strand assembly.

Researchers mix all eight kinds of nucleotides with a single-stranded DNA template, a primer, and DNA polymerase. The polymerase uses the primer to copy the template DNA into new strands of DNA. One by one, it adds nucleotides in the order dictated by the sequence of the DNA template (Figure 16.8a). Every time, the polymerase randomly attaches a standard *or* a modified nucleotide to the DNA template. When one of the modified nucleotides covalently bonds to the forming DNA strand, no more can be added. After enough time passes, there will be some new strands that stop at each base in the DNA template sequence.

Eventually the mixture holds millions of copies of DNA fragments, all fluorescent-tagged on one end. These fragments are separated by **gel electrophoresis**, a technique that sorts fragments as they move through a semisolid slab (of polyacrylamide) in response to an electric field.

Depending on their lengths, the fragments migrate at different rates through the gel. The gel hinders the migration of longer ones more than shorter ones. By analogy, elephants running through the forest in India cannot move between the trees as fast as tigers can.

The shortest fragments migrate fastest and are first to arrive at the end of the gel. The longest fragment is last. Fragments of the same length move through the gel at the same speed, and they gather into bands.

A laser beam shines on each band when it passes through the end of the gel. The modified nucleotides attached to the fragments fluoresce in response to the light, and the sequencer detects and records the color of each band. Because each color designates one of the four particular nucleotides, the order of colored bands reveals the DNA sequence. The machine itself rapidly assembles the sequence data.

Figure 16.8b shows the partial results from one run through an automated DNA sequencer. Each peak in the tracing represents the detection of one fluorescent color as the fragments reached the end of the gel. The sequence is shown beneath the graph line.

> *DNA sequencing rapidly reveals the order of nucleotides in a cloned or amplified DNA fragment.*

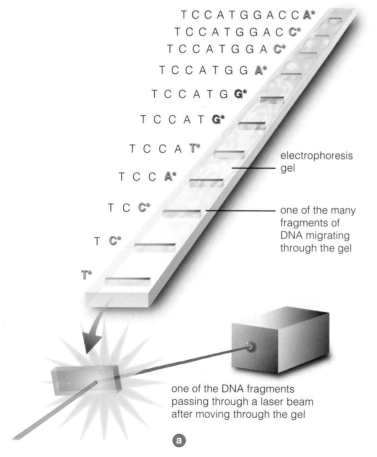

T C C A T G G A C C A*
T C C A T G G A C C*
T C C A T G G A C*
T C C A T G G A*
T C C A T G G*
T C C A T G*
T C C A T*
T C C A*
T C C*
T C*
T*

electrophoresis gel

one of the many fragments of DNA migrating through the gel

one of the DNA fragments passing through a laser beam after moving through the gel

a

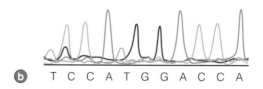

b T C C A T G G A C C A

Figure 16.8 *Animated!* Automated DNA sequencing. (**a**) Researchers synthesize DNA fragments by using a template and fluorescent nucleotides. Gel electrophoresis sorts out the fragments by length. (**b**) The order of the fluorescent bands that appear in the gel is detected by the sequencer. That order indicates the template DNA sequence. Today, researchers throughout the world use sequence databases that can be accessed via the Internet.

16.4 Analyzing DNA Fingerprints

Except for identical twins, no two people have exactly the same sequence of bases in their DNA. One individual can be distinguished from all others on the basis of this molecular fingerprint.

Each human has a unique set of fingerprints. In addition, like other sexually reproducing species, each also has a **DNA fingerprint**—a unique array of DNA sequences that are inherited from parents in a Mendelian pattern. More than 99 percent of the DNA is the same in all humans, but the other fraction of 1 percent is unique to each individual. Some of these unique stretches of DNA are sprinkled through the human genome as **tandem repeats**—many copies of the same short base sequences, positioned one after the other along a DNA molecule.

For example, one person's DNA might contain four repeats of the bases TTTTC in a certain location. Another person's DNA might have them repeated fifteen times in the same location. One person might have ten repeats of CGG, and another might have fifteen. Such repetitive sequences slip spontaneously into DNA during replication, and their numbers grow or shrink over time. The mutation rate is relatively high in these regions.

DNA *fingerprinting* reveals differences in the tandem repeats among individuals. A restriction enzyme cuts their DNA into an assortment of fragments. The sizes of those fragments are unique to the individual. They reveal genetic differences between individuals, and they can be detected as RFLPs (*Restriction Fragment Length Polymorphisms*).

The fragments can be subjected to gel electrophoresis to form distinct bands according to their length. The banding pattern of genomic DNA fragments is the DNA fingerprint unique to the individual. For all practical purposes, it is identical only between identical twins. The odds of two unrelated people sharing an identical DNA fingerprint are 1 in 3,000,000,000,000.

PCR can be used to amplify tandem-repeat regions. Again, differences in the size of DNA fragments amplified by this technique can be detected by gel electrophoresis. A few drops of blood, semen, or cells from a hair follicle at a crime scene or on a suspect's clothing yield enough DNA to amplify with PCR, and then generate a fingerprint.

DNA fingerprints help forensic scientists identify criminals, victims, and innocent suspects. Figure 16.9 shows some tandem repeat RFLPs that were separated by gel electrophoresis. Those samples of DNA had been taken from seven people and from a bloodstain left at a crime scene. One of the DNA fingerprints matched.

Defense attorneys initially challenged the use of DNA fingerprinting as evidence in court. Today, however, the procedure has been firmly established as accurate and unambiguous. DNA fingerprinting is routinely submitted as evidence in disputes over paternity, and it is being widely used to convict the guilty and to exonerate the innocent. At this writing, DNA evidence has helped release well over 100 innocent people from prison.

DNA fingerprint analysis has even wider application. For instance, it confirmed that human bones exhumed from a shallow pit in Siberia belonged to five individuals of the Russian imperial family, all shot to death in secrecy in 1918. More recently, it was used to identify the remains of those who died in the World Trade Center on September 11, 2001.

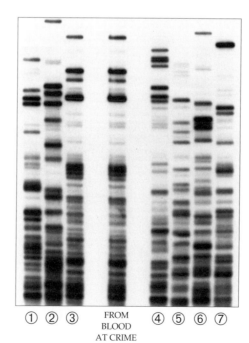

① ② ③   FROM BLOOD AT CRIME SCENE   ④ ⑤ ⑥ ⑦

Figure 16.9 One case of a damning comparison of the DNA fingerprints from a bloodstain left behind at a crime scene and from blood samples of seven suspects (the series of circled numbers).

Can you point out which of the seven DNA fingerprints is an exact match?

16.5 The Rise of Genomics

LINKS TO
SECTIONS
3.7, 13.2, 14.5

The potential benefits of sequencing and analyzing the thousands of genes in the genome of selected organisms—say, the human genome—soon became apparent. Automated gene sequencing techniques were developed in response.

THE HUMAN GENOME PROJECT

By 1986, scientists were arguing about sequencing the 3 billion bases of the human genome. Many insisted that benefits for medicine and pure research would be incalculable. Others insisted that the mapping would divert funds from other work that was more urgent and had a better chance of success.

Automated sequencing had just been invented, as had PCR, the polymerase chain reaction. At the time, both techniques were cumbersome, expensive, and far from standardized, but many sensed their potential. Waiting for faster methods seemed the most efficient approach to sequencing the human genome—but who would decide when the technology was fast enough?

Several independent organizations launched their own versions of the Human Genome Project. Walter Gilbert started one company and declared he would sequence and patent the human genome. In 1988, the National Institutes of Health (NIH) annexed the entire Human Genome Project by hiring James Watson as its head and providing 200 million dollars per year to researchers. A public consortium formed between the NIH and institutions working on different versions of

the project. Watson set aside 3 percent of the funding for studies into ethical and social issues arising from the research. He then resigned in 1992 because of a disagreement with the NIH about patenting partial gene sequences. Francis Collins replaced him in 1993.

Amid ongoing squabbles over patent issues, Craig Venter started Celera Genomics (Figure 16.10). Venter cheekily declared that his new company would be the first to finish and patent the genome sequence. This prompted the public consortium to move its gene sequencing efforts into high gear.

Sequencing of the human genome was officially completed in 2003—fifty years after the discovery of the structure of DNA. About 99 percent of the coding regions in human DNA have been deciphered with a high degree of accuracy. A number of other genomes also have been fully sequenced.

What do we do with this vast amount of data? The next step is to investigate questions about precisely what each sequence means—what the genes do, what the control mechanisms are, and how they operate.

At this writing, 19,438 are confirmed as genes, and another 2,188 are probably genes. This does not mean that geneticists have learned what the genes encode.

Among the bizarre discoveries: Protein-encoding genes make up less than 2 percent of our genome. Millions of transposable elements repeated over and over make up more than half of it. There are almost as many *pseudogenes*—inactivated, nonfunctional copies of genes—as there are genes!

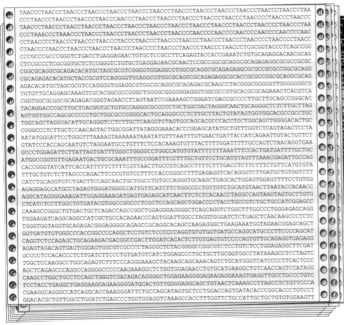

Figure 16.10 Some of the bases of the human genome—and a few of the supercomputers used to sequence it—at Celera Genomics in Maryland.

Figure 16.11 Complete yeast genome array on a DNA chip about 19 millimeters (3/4 inch) across. *Green* spots pinpoint genes that are active during fermentation. *Red* pinpoints the genes used in aerobic respiration, and *yellow*, the ones that are active in both pathways.

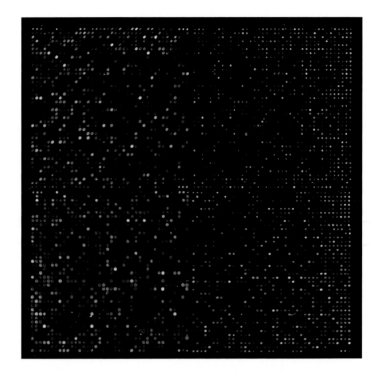

GENOMICS

Research into genomes of humans and other species has converged into a new research field—**genomics**. *Structural* genomics focuses on actual mapping and sequencing of the genomes of individuals. *Comparative* genomics sifts through the maps for similarities and differences that point to evolutionary connections.

Comparative genomics has practical applications as well as potential for research. The basic premise is that the genomes of all existing organisms are derived from common ancestors. For instance, pathogens share some conserved genes with human hosts even though they are only remotely related. Shared gene sequences, how they are organized, and where they differ might hold essential clues to where our immune defenses against pathogens are strongest or the most vulnerable.

Genomics has potential for **human gene therapy**—the transfer of one or more normal or modified genes into a person's body cells to correct a genetic defect or boost resistance to disease. However, even though the human genome is fully sequenced, it still is not easy to manipulate within the context of a living individual.

Today, experimenters use stripped-down viruses as vectors that inject genes into human cells. Some gene therapies deliver modified cells into a patient's tissue. In many cases, therapies make a patient's symptoms subside even when the modified cells are producing just a small amount of a required protein.

A caveat: No one can yet predict whether a virus-injected gene will be delivered to the right tissues and whether cellular mechanisms will maintain it.

DNA CHIPS

Analysis of genomes is now advancing at a stunning pace. Researchers pinpoint which genes are silent and which are being expressed with the use of **DNA chips**. These are microarrays of thousands of gene sequences representing a large subset of an entire genome—all stamped onto a glass plate that is about the size of a small business card.

A cDNA probe is built by using mRNA from, say, cells of a cancer patient. The free nucleotides used to synthesize the complementary strand of DNA have been labeled with a fluorescent pigment. Only genes that are expressed at the time the cells are harvested are making mRNA, so those genes alone make up the resulting probe population. The labeled probe is then incubated along with a chip made from genomic DNA. Wherever the probe binds with complementary base sequences on the chip, there will be a spot that glows under fluorescent light. Analysis of which spots on the chip are glowing reveals which of the thousands of genes inside the cells are active and which are not.

DNA chips are being used to compare different gene expression patterns between cells. Examples are yeasts grown in the presence and absence of oxygen, and different types of cells from the same multicelled individual. RNA from one set of cells is transformed into green fluorescent cDNA, and RNA from the other set into red fluorescent cDNA. The cDNAs are mixed and incubated with a genomic DNA chip. Green or red fluorescence indicates expression of genes in the different cell types. Yellow is a mixture of both red and green, and it indicates that both genes were being expressed at the same time in a cell (Figure 16.11).

In genomics, automated gene sequencing, the use of DNA chips, and other techniques let researchers rapidly evaluate and compare genome-spanning expression patterns.

16.6 Genetic Engineering

Genetic engineering is the deliberate modification of an individual's genome. Genes from another species may be transferred to an individual. Conversely, the individual may have its own genes isolated, modified and copied, and then receive copies of the modified genes.

Genetic engineering started with bacterial species, so consider them first. The kinds that take up plasmids are now widely used in basic research, agriculture, medicine, and industry. Plasmids, again, function as vectors for transferring fragments of foreign or modified DNA into an organism.

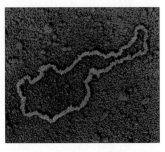

For instance, like you, bacterial cells have the metabolic machinery to make complex organic compounds. Genetically engineered types can be employed to transcribe genes that have been transferred to plasmids and synthesize desired proteins. Immense populations do this; they make useful amounts of medically valued proteins in huge stainless steel vats. *E. coli* cells were the first to transcribe and translate synthetic genes for human insulin. Their descendants were the first large-scale, cost-effective bacterial factory for proteins. In addition to insulin, vats of microbes churn out human somatotropin (growth hormone), hemoglobin, blood-clotting factors, interferon, and a variety of drugs and vaccines that we have come to depend upon.

Certain bacteria also hold potential for industry and for cleaning up environmental messes—that is, for *environmental remediation*. In nature, they break down organic wastes as part of their metabolic activities and help cycle nutrients through ecosystems. Modified types digest crude oil into less harmful compounds. When sprayed on oil spills, as from a shipwrecked supertanker, they can help mop up oil. Other species sponge up excess phosphates, heavy metals, and other pollutants, even radioactive wastes.

> *Genetic engineering refers to the directed alteration of an individual's genome. Microbes were the first targets.*
>
> *In some cases, DNA is transferred between individuals of different species, the outcome being a transgenic organism.*
>
> *In other cases, genes or gene regions from an individual are isolated, modified, then copied and inserted into the same individual.*

16.7 Designer Plants

Think back on those Golden Rice plants described in the chapter introduction. They are a prime example of genetic engineering that can produce valuable transgenic plants. There is some urgency surrounding much of this work, as you will now read.

As crop production expands to keep pace with human population growth, it puts unavoidable pressure on ecosystems everywhere. Irrigation leaves mineral and salt residues in soils. Tilled soil erodes, taking topsoil with it. Runoff clogs rivers, and fertilizer in it causes algae to grow so much that fish suffocate. Pesticides harm humans, other animals, and beneficial insects.

Pressured to produce more food at lower cost and with less damage to the environment, some farmers are turning to genetically engineered crop plants.

Cotton plants with a built-in insecticide gene kill only the insects that eat it, so farmers that grow them are not required to use as many pesticides. Certain transgenic tomato plants can grow, develop, and bear fruit in salty soils that would wither other plants. They also absorb and store excess salt in their leaves, thus purifying saline soil for future crops.

The cotton plants in Figure 16.12*a* were genetically engineered for resistance to a relatively short-lived herbicide. Spraying fields with this herbicide will kill all weeds—but not the engineered cotton plants. As you read in the chapter's introduction, the practice means that farmers can use reduced amounts of less toxic chemicals. They do not have to till the soil as much to control weeds, so river-clogging runoff can be reduced. As another example, Figure 16.12*b* shows transgenic aspen seedlings that grow well and do not make as much lignin. Lignin-deficient trees are better for making paper and other forest products.

Engineering plant cells starts with vectors that can carry genes into plant cells. *Agrobacterium tumefaciens* is a bacterial species that infects eudicots, including beans, peas, potatoes, and other major crops plants. Genes in its plasmids cause tumors to form on these plants; hence the name Ti plasmid (*Tumor-i*nducing). Researchers use the Ti plasmid to transfer foreign or modified genes into plants.

Researchers excise the tumor-inducing genes, then insert a desired gene into the plasmid (Figure 16.13). Some plant cells cultured with the modified plasmid may take it up. Whole plants may be regenerated.

Modified *A. tumefaciens* bacteria deliver genes into monocots that also are food sources, including wheat, corn, and rice. Researchers can even transfer genes into plants by way of electric shocks, chemicals, and blasts of microscopic particles coated with DNA.

Figure 16.12 (**a**) *Left,* control cotton plant. *Right,* cotton plant genetically engineered for herbicide resistance. Both plants were sprayed with a weed killer that is widely applied in cotton fields.

(**b**) Control plant (*left*) and four genetically engineered aspen seedlings. Vincent Chiang and coworkers suppressed a control gene involved in a lignin biosynthetic pathway. The modified plants synthesized normal lignin, but not as much. Lignin synthesis dropped by as much as 45 percent—yet cellulose production increased 15 percent. Root, stem, and leaf growth were greatly enhanced. Plant structure did not suffer. Wood harvested from such trees might make it easier to manufacture paper and some clean-burning fuels, such as ethanol. Lignin, a tough polymer, strengthens secondary cell walls of plants. Before paper can be made from wood, the lignin must be chemically extracted.

a   b

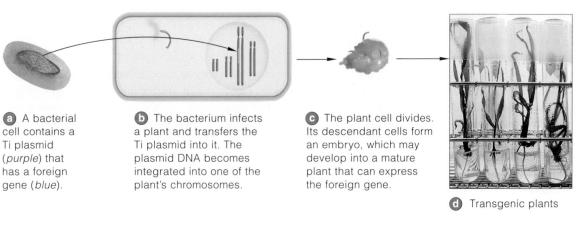

a A bacterial cell contains a Ti plasmid (*purple*) that has a foreign gene (*blue*).

b The bacterium infects a plant and transfers the Ti plasmid into it. The plasmid DNA becomes integrated into one of the plant's chromosomes.

c The plant cell divides. Its descendant cells form an embryo, which may develop into a mature plant that can express the foreign gene.

d Transgenic plants

e Example of a young plant with a fluorescent gene product.

Figure 16.13 *Animated!* (**a–d**) Ti plasmid transfer of an *Agrobacterium tumefaciens* gene to a plant cell. (**e**) A transgenic plant expressing a firefly gene for the enzyme luciferase.

Consider another compelling reason for modifying plant species: The food supply for most of the human population is extremely vulnerable. Farmers usually want to plant crops that give them the highest yields. Over time, genetically similar varieties have replaced the more diverse, older varieties. However, genetic uniformity makes food crops far more vulnerable to many pathogenic fungi, viruses, and bacteria.

That is why botanists comb the world for seeds of the older, diverse varieties of plants and of the wild ancestors of potatoes, corn, and other crop plants. They send their prizes—seeds with genes of a plant's lineage—to **seed banks**. These safe storage facilities are designed to preserve genetic diversity. They are now being tapped by genetic engineers as well as by traditional plant breeders.

Crop vulnerability is a huge problem. At one time, *Southern corn leaf blight* destroyed much of the United States corn crop. All of the plants carried the gene that conferred susceptibility to the fungal pathogen. Ever since that devastating epidemic, seed companies have been much more attentive to offering genetically diverse corn seeds. They tap seed banks, the treasure houses of plant genes.

Transgenic plants help farmers grow crops more efficiently and with less impact on the environment.

Genetic engineers as well as traditional plant breeders are tapping seed banks, which are safe storage facilities designed to preserve genetic diversity of plants.

16.8 Biotech Barnyards

LINKS TO
SECTIONS
5.2, 15.3

Laboratory mice were the first mammals to be genetically engineered. Today, featherless chickens, drug-producing goats, and transgenic pigs are part of the biotech barnyard.

TRANSGENIC ANIMALS

Traditional cross-breeding practices have produced unusual animals, including the featherless chicken in Figure 16.14. Now transgenic types are on the scene. The first ones arrived in 1982. Researchers isolated a gene for human somatotropin (growth hormone) and inserted it into a plasmid. They injected copies of the recombinant plasmids into fertilized mouse eggs that were later implanted into female mice. A third of the offspring of the surrogate mothers grew much larger than their littermates (Figure 16.15). The rat gene had become integrated into the host DNA and was being expressed in the transgenic mice.

Transgenic animals are used routinely for medical research. The functions of many gene products and how they can be controlled have been discovered by inactivating genes in "knockout mice" and analyzing the effect on phenotype (Section 15.3). Strains of mice, genetically modified mice to be susceptible to human diseases, help researchers study both the diseases and potential cures without experimenting on humans.

Genetically engineered animals also are sources of medically valued proteins. As a few examples, goats synthesize quantities of CFTR protein to treat cystic fibrosis and TPA protein to counter the bad effects of heart attacks. Rabbits make human interleukin-2, a protein that triggers divisions of immune cells called T lymphocytes. Cattle, too, may soon produce human

Figure 16.15 Evidence of a successful gene transfer. Two ten-week-old mouse littermates. *Left*, This one weighed 29 grams. *Right*, This one weighed 44 grams. It grew from a fertilized egg into which a gene for human somatotropin had been inserted.

collagen, which can be used to repair cartilage, bone, and skin. Goats make spider silk protein that might be used to make bullet-proof vests, medical supplies, and equipment for use in space. Different goats make human antithrombin, which is used to treat people with blood-clotting disorders (Figure 16.14*b*).

Genetic engineers have developed pigs that make environmentally friendlier manure. They have made freeze-resistant salmon, low-fat pigs, heftier sheep, and cows that are resistant to mad cow disease. Within a few years, they may give us allergen-free cats.

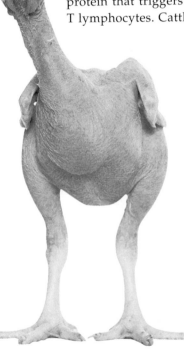

Figure 16.14 Genetically modified animals. (**a**) Featherless chicken developed by traditional cross-breeding methods in Israel. Such chickens survive in hot deserts where cooling systems are not an option. Chicken farmers in the United States have lost millions of feathered chickens in extremely hot weather. (**b**) Mira, a goat transgenic for human antithrombin III, an anticlotting factor. (**c**) Inquisitive transgenic pig at the Virginia Tech Swine Research facility.

Tinkering with the genetics of animals for the sake of human convenience does raise ethical questions. However, transgenic animal research may be viewed as an extension of thousands of years of acceptable barnyard breeding practices. Techniques have changed, but not the intent. Humans continue to have a vested interest in improving livestock.

KNOCKOUT CELLS AND ORGAN FACTORIES

Each year, about 75,000 people are on waiting lists for an organ transplant, but human donors are in short supply. There is talk of harvesting organs from pigs (Figure 16.14c), because pig organs function a lot like ours do. Transferring an organ from one species into another is called **xenotransplantation**.

The human immune system battles anything that it recognizes as "nonself." It rejects a pig organ at once, owing to a glycoprotein on the plasma membrane of cells that make up the blood vessels in pig organs. Antibodies circulating in human blood swiftly latch on to the sugar component and call for a response. In less than a few hours, blood inside the vessels coagulates massively and dooms the transplant. Drugs suppress this immune response, but a side effect is serious: the drugs make organ recipients vulnerable to infections.

Pig DNA contains two copies of *Ggta1*, the gene for an enzyme that catalyzes a key step in biosynthesis of alpha-1,3-galactose. This is the pig sugar that human antibodies recognize. Researchers have knocked out both copies of the *Ggta1* gene in transgenic piglets. Without the gene product, and the sugar, a pig tissue or organ may be less prone to rejection by the human immune system. Tissues and organs from such animals could help millions of people, including the ones with organs that have been severely damaged as a result of diabetes and Parkinson's disease.

Critics of xenotransplantation are concerned that, among other things, pig–human transplants would invite pig viruses to cross a species barrier and infect humans, perhaps catastrophically. Their concerns are not unfounded. In 1918, an influenza pandemic killed twenty million people worldwide. It originated with a swine flu virus—in pigs.

Many years have passed since the first transfer of foreign DNA into a plasmid. That transfer ignited an ongoing debate about potential dangers of transgenic organisms entering the environment before rigorous testing.

In 1972, Paul Berg and his associates were the first to make recombinant DNA. Researchers knew that DNA was not toxic, but they could not predict what would happen every time they fused genetic material from different organisms into the recombinant molecules. Would they accidentally make superpathogens? Could they create a new form of life by the fusion of DNA from two normally harmless organisms? What if their creation escaped from the laboratory and transformed other organisms in the natural environment?

In a remarkably quick and responsible display of self-regulation, scientists reached a consensus on the safety guidelines for DNA research. Adopted at once by the NIH, their guidelines listed precautions for laboratory procedures. They covered the design and use of host organisms that could survive only under the narrow range of conditions inside the laboratory. Researchers stopped using DNA from pathogenic or toxic organisms for recombination experiments until proper containment facilities were developed.

As added precautions, "fail-safe" genes are now built into genetically engineered bacteria. They remain silent unless the bacteria escape and are exposed to environmental conditions—whereupon the genes get activated, with lethal results for the cell. Suppose that the package has a *hok* gene next to a promoter of the lactose operon (Section 15.4). Sugars are plentiful in the environment. If they were to activate the *hok* gene in a bacterial cell that escaped, the gene's product would destroy membrane function and the escapee.

Even so, does Murphy's law also apply to genetic engineering? As with any human endeavor, things can go wrong. After rabbits started taking over much of Australia, researchers were tinkering with a rabbit-killing virus in a containment laboratory on an island. Maybe the virus escaped in flying insects. However it happened, the virus is out and about, and killing lots of rabbits (Section 46.10). It is an example of why researchers are expected to expect the unexpected.

Genetic engineering started more than two decades ago. The kinds of animals being sought are beyond the scope of traditional breeding practices. Pigs engineered as donors for human organs are among the more startling cases.

Rigorous safety guidelines for DNA research have been in place for decades in the United States. They have been adopted by the NIH, and researchers are expected to comply with their stringent standards.

16.10 Modified Humans?

We as a society continue to work our way through the ethical implications of applying the new DNA technologies. Even as we are weighing the risks and benefits, however, the manipulation of individual genomes has begun.

WHO GETS WELL?

Human gene therapy is often cited as one of the most compelling reasons for embracing the new research. We already have identified more than 15,500 genetic disorders. Many are rare in the population at large. Collectively, however, they show up in 3 to 5 percent of all newborns, and they cause 20 to 30 percent of all infant deaths every year. They account for about half of mentally impaired patients and nearly a fourth of all hospital admissions. They contribute to many age-related disorders that await all of us.

Rhys Evans, shown below, was born with a severe immune deficiency known as SCID-X1, which stems from mutations in gene *IL2RG*. Children affected by this disorder can live only in germ-free isolation tents, a "bubble," because they cannot fight infections.

In 1998, doctors withdrew stem cells from the bone marrow of eleven SCID-X1 boys. Stem cells, recall, are forerunners of other cell types, including white blood cells of the immune system. The doctors used a virus to insert nonmutated copies of *IL2RG* into each boy's stem cells, which they then infused back into his bone marrow. Months later, ten of the boys left isolation tents for good; gene therapy had successfully repaired their immune system. Since then, other gene therapy trials have freed many other SCID-X1 patients from life in a bubble. Rhys Evans is one of them.

In 2002, to the shock of researchers, two boys from the 1998 trial developed leukemia and one died. The researchers had anticipated that any cancer related to the therapy would be extremely rare. The very gene targeted to do the repair work—*IL2RG*—may be a problem, especially when combined with the viral vector that delivered the gene into stem cells. One other child who took part in a gene therapy experiment for SCID-X1 has developed leukemia. That it developed at all is evidence that our understanding of the human genome lags behind our ability to modify it.

WHO GETS ENHANCED?

When all is said and done, the idea of using human gene therapy to cure genetic disorders seems like a socially acceptable goal to most of us. Now see if your comfort level can move one step further. Would it also be acceptable to modify genes of some individual who falls within the normal range if he or she simply would like to minimize or enhance a particular trait?

We have already crossed the threshold of a brave new world. Researchers who are adept at transferring genes have already engineered strains of mice with enhanced memory and improved learning abilities. Perhaps their work is a beacon to those whose very lives have been turned upside down by Alzheimer's disease. Perhaps it draws others who are enchanted with the idea of simply getting more brain power.

The idea of selecting the most desired human traits is referred to as *eugenic engineering*. Yet who decides which forms of traits are most desirable? Realistically, cures for many severe but rare genetic disorders will not happen, because the payback for research is not financially attractive. Eugenics, however, might turn a profit. Just how much would potential parents pay to engineer tall or blue-eyed or fair-skinned children? Would it be okay to engineer "superhumans" with breathtaking strength or intelligence? How about an injection that would help you lose that extra weight and keep it off permanently? Where exactly is the line between interesting and abhorrent?

In a survey conducted in the United States, more than 40 percent of those interviewed said it would be fine to use gene therapy to make smarter and cuter babies. In one poll of British parents, 18 percent would be willing to use genetic enhancement to keep their child from being aggressive, and 10 percent would use it to keep a child from growing up to be homosexual.

Some argue that we must never alter the DNA of anything. The concern is that we just do not have the wisdom to bring about any genetic changes without causing irreparable damage to ourselves and nature.

One is reminded of our peculiar human tendency to leap before we look. And yet, something about the human experience gave us the capacity to imagine wings of our own making, a capacity that carried us to the frontiers of space. It gave one individual the dream of enhancing the rice plant genome to keep millions of children from going blind.

In this brave new world, two questions are before you: Should we be more cautious, because the risk takers may go too far? And what do we stand to lose if risks are not taken?

Be engaged; our understanding of the meaning of the human genome is changing even as you read this.

Summary

Section 16.1   Recombinant DNA technology uses restriction enzymes that can cut DNA into fragments. DNA ligases can splice the fragments into plasmids or some other cloning vector. Recombinant plasmids may be taken up by rapidly dividing cells, such as bacteria. When the host cells replicate, they make multiple, identical copies of the foreign DNA as well.

Bacteria cannot correctly express eukaryotic genes, which contain introns. Reverse transcriptase, a viral enzyme, can make a complementary DNA strand on mRNA. The hybrid molecule can then be converted to cDNA for cloning.

Biology⊜Now
Explore the tools used to make recombinant DNA with the animation on BiologyNow.

Section 16.2   A gene library is a mixed collection of cells that have taken up cloned DNA. Researchers can isolate a gene of interest from a library by using a probe, a short stretch of DNA that can base-pair with the gene and that is traceable (it is labeled with a detectable tag, such as a radioisotope). Probes can help researchers locate and base-pair with one clone among millions. Such base pairing between nucleotide sequences from different sources is known as nucleic acid hybridization.

The polymerase chain reaction (PCR) is a technique for rapidly copying DNA fragments. A sample of a DNA template is mixed with nucleotides, primers, and a heat-resistant DNA polymerase. Each round of PCR proceeds through a series of temperature changes that amplifies the number of DNA molecules exponentially.

Biology⊜Now
Learn how researchers isolate and copy genes with the interaction on BiologyNow.

Section 16.3   Automated DNA sequencing rapidly reveals the order of nucleotides in DNA fragments. As DNA polymerase copies a template DNA, progressively longer fragments stop growing as soon as one of four different fluorescent nucleotides becomes attached to them. Electrophoresis separates the labeled fragments into bands according to length. The order of the colored bands as they migrate through the gel reflects which fluorescent base was added to the end of each fragment, and so indicates the template DNA base sequence.

Biology⊜Now
Investigate DNA sequencing with the animation on BiologyNow.

Section 16.4   Tandem repeats are multiple copies of a short DNA sequence that follow one another along a chromosome. The number and distribution of tandem repeats, unique in each person, can be revealed by gel electrophoresis; they form a DNA fingerprint.

Biology⊜Now
Observe the process of DNA fingerprinting with the animation on BiologyNow.

Section 16.5   The entire human genome has been sequenced and is now being analyzed. Genomes of other organisms also have been fully sequenced.

The new field of genomics is concerned with the mapping and analysis of genomes. One branch, called comparative genomics, uses similarities and differences between DNA sequences of major groups of organisms to identify their evolutionary relationships.

DNA chips are microarrays used to compare patterns of gene expression within a genome.

Sections 16.6–16.8   Recombinant DNA technology and the mapping and analysis of genomes is the basis for genetic engineering. Genetic engineering is the directed modification of the genetic makeup of an organism, often to modify its phenotype. Researchers insert normal or modified genes from one organism into another of the same or different species. Gene therapies insert copies of modified genes into individuals to cover the functions of a mutant or altered gene.

Genetically engineered bacteria that contain plasmid vectors have diverse uses in basic research, medicine, agriculture, industry, and ecology. Transgenic crop plants help farmers use less toxic pesticides and produce food more efficiently. Genetic engineering of animals allows commercial production of human proteins, as well as research into genetic disorders.

Biology⊜Now
See how the Ti plasmid is used to genetically engineer plants with the animation on BiologyNow.

Section 16.9   There is always a risk that genetically modified experimental organisms can escape from the laboratory. Typically, potentially dangerous types have fail-safe genes built into their genome that will destroy them when exposed to conditions that exist anywhere except in the laboratory. Rigorous tests for safety must precede the release of any modified organism into the environment.

Section 16.10   The goal of human gene therapy is to transfer normal or modified genes into body cells to correct genetic defects. As with any new technology, the benefits must be weighed against potential risks.

Self-Quiz   *Answers in Appendix II*

1. Researchers can cut DNA molecules at specific sites by using _____ .
    a. DNA polymerase   c. restriction enzymes
    b. DNA probes   d. reverse transcriptase

2. Fill in the blank: A _____ is a small circle of bacterial DNA that contains only a few genes and is separate from the bacterial chromosome.

3. By reverse transcription, _____ is assembled on a(n) _____ template.
    a. mRNA; DNA   c. DNA; ribosome
    b. cDNA; mRNA   d. protein; mRNA

4. PCR stands for _____ .
   a. polymerase chain reaction
   b. polyploid chromosome restrictions
   c. polygraphed criminal rating
   d. politically correct research

5. Automated DNA sequencing relies on _____ .
   a. supplies of standard and labeled nucleotides
   b. primers and DNA polymerases
   c. gel electrophoresis and a laser beam
   d. all of the above

6. By gel electrophoresis, fragments of DNA can be separated according to _____ .
   a. sequence   b. length   c. species

7. _____ can be used to insert genes into human cells.
   a. PCR   c. Xenotransplantation
   b. Modified viruses   d. DNA microarrays

8. For each species, all _____ in a haploid number of chromosomes is the _____ .
   a. genomes; phenotype   c. mRNA; start of cDNA
   b. DNA; genome   d. cDNA; start of mRNA

9. Match the terms with the most suitable description.
   _____ DNA fingerprint   a. selecting "desirable" traits
   _____ Ti plasmid   b. mutations, crossovers
   _____ nature's genetic   c. used in some gene transfers
          experiments   d. a person's unique collection
   _____ nucleic acid   of tandem repeats
          hybridization   e. base pairing of nucleotide
   _____ eugenic   sequences from different
          engineering   DNA or RNA source

Additional questions are available on Biology ⬙ Now™

Critical Thinking

1. Lunardi's Market put out a bin of tomatoes having vine-ripened redness, flavor, and texture. A sign identified them as genetically engineered produce. Most shoppers selected unmodified tomatoes in the adjacent bin even though those tomatoes were pale pink, mealy textured, and tasteless. Which tomatoes would you pick? Why?

2. Biotechnologists envision a new Green Revolution. As they see it, designer plants hold down food production costs, reduce dependence on pesticides and herbicides, enhance crop yields, offer improved flavor and nutritional value, and often produce plants with salt tolerance and drought tolerance. Fruits and vegetables can be designed for flavor, nutritional value, and extended shelf life.

Genetically engineered food crops are widespread in the United States. At least 45 percent of cotton crops, 38 percent of soybean crops, and 25 percent of corn crops have been modified to withstand weedkillers or make their own pesticides. For years, modified corn and soybeans have been used in tofu, cereals, soy sauce, vegetable oils, beer, and soft drinks. They are fed to farm animals.

In Europe especially, public resistance to modified food runs high. Besides arguing that modified foods might be toxic and have lower nutritional value, many people worry that designer plants might cross-pollinate wild plants and produce "superweeds." The chorus of critics in Europe has forced American farmers to keep genetically engineered crops separated from traditional crops. Traditional crops only are exported to Europe. Such separation is both costly and difficult.

Read up on scientific research related to this issue and form your own opinions. The alternatives are to be swayed either by media hype (the term Frankenfood, for instance) or by sometimes biased reports from groups (such as chemical manufacturers), which have their own agendas.

3. The sequencing of the human genome is completed, and knowledge about many genes is being used to detect genetic disorders. Many insurance companies will pay for their female subscribers to take advantage of genetic testing for breast cancer and are willing to allow them to keep the results confidential.

Explain how a health insurance company might benefit financially if it were to encourage its subscribers to take confidential tests for breast cancer susceptibility.

4. Scientists at Oregon Health Sciences University produced Tetra, the first primate clone. They also made the first transgenic primate by inserting a jellyfish gene into a fertilized egg of a rhesus monkey. (The gene codes for a bioluminescent protein that fluoresces green; refer to Section 6.6). The egg was implanted in a surrogate monkey's uterus, where it developed into a male.

The long-term goal of this gene transfer project is not to make glowing-green monkeys. It is the transfer of human genes into primates whose genomes are most like ours. Transgenic primates could yield insight into genetic disorders. That insight might lead to the development of cures for those who are affected and of vaccines for those who are at risk.

Something more controversial is at stake. Will the time come when foreign genes can be inserted into human embryos? Would it be ethical to transfer a chimpanzee or monkey gene into a human embryo to cure a genetic defect? To bestow immunity against a potentially fatal disease such as AIDS? Think about it.

Appendix I. Classification System

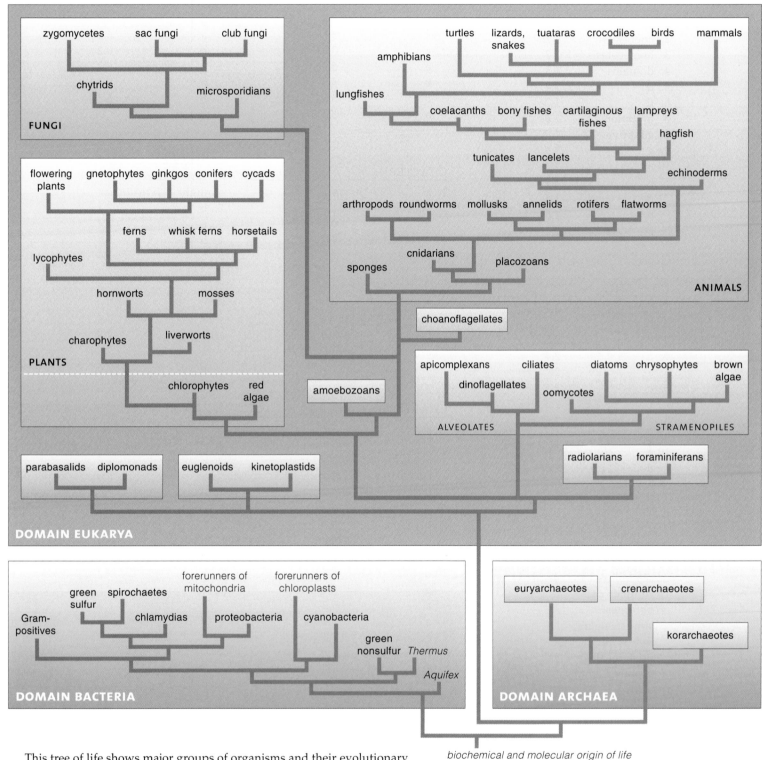

This tree of life shows major groups of organisms and their evolutionary connections. Each set of organisms (taxon) has living representatives. Each branch point represents the last common ancestor of the set above it. Small boxes within domains Archaea and Eukarya highlight taxa that are currently being recognized as the equivalent of kingdoms in earlier classification systems. This tree is based on morphological, genetic, and molecular comparisons of the most important groups. It is a work in progress and subject to refinements as more information comes in.

Appendix II. Answers to Self-Quizzes

CHAPTER 1

1. cell	1.1
2. energy	1.2
3. homeostasis	1.2
4. domains	1.3
5. d	1.2, 1.4
6. d	1.2
7. mutation	1.4
8. adaptive	1.4
9. a	1.6
10. c	1.1
e	1.4
d	1.5
b	1.5
a	1.5

CHAPTER 2

1. False	2.1
2. b	2.1
3. d	2.2
4. c	2.4
5. a	2.4
6. e	2.5
7. f	2.6
8. acid, base	2.6
9. c	2.6
10. e	*Introduction*
d	2.6
b	2.4
c	2.5
a	2.1

CHAPTER 3

1. complex carbo- hydrates:	
simple sugars	3.3
lipids; fatty acids or sterol rings	3.4
proteins; amino acids	3.5
nucleic acids; nucleotides	3.7
2. d	3.1
3. c	3.2
4. f	3.3
5. b	3.4
6. b	3.4
7. e	3.4
8. d	3.5, 3.7

9. d	3.6
10. d	3.7
11. b	3.7
12. c	3.5
e	3.7
b	3.4
d	3.7
a	3.3

CHAPTER 4

1. c	4.1
2. See Figure 4.15	
3. d	4.4
4. d	4.9
5. False	4.9
6. c	4.3
7. e	4.7
d	4.8
a	4.1, 4.4
b	4.6
c	4.6

CHAPTER 5

1. c	5.1
2. c	5.1
3. a	5.1
4. d	5.2
5. d	5.3
6. d	5.3
7. b	5.4
8. a	5.5
9. c	5.6
10. d	5.6
g	5.4
a	5.2
e	5.4
c	5.1
b	5.3
f	5.2

CHAPTER 6

1. c	6.1
2. d	6.1
3. b	6.1
4. d	6.3
5. d	6.4
6. b	6.5

7. c	6.2
g	6.2
a	6.2
d	6.2
e	6.2
b	6.3
f	6.2

CHAPTER 7

1. carbon dioxide, sunlight	*Introduction*
2. b	7.1
3. a	7.4
4. b	7.4
5. c	7.4
6. d	7.4
7. c	7.6
8. b	7.6
9. d	7.6
10. c	7.5
a	7.6
b	7.6

CHAPTER 8

1. d	8.1
2. c	8.2
3. b	8.4
4. c	8.4
5. See Figure 8.3	
6. c	8.5
7. b	8.5
8. d	8.6
9. b	8.2
c	8.5
a	8.3
d	8.4

CHAPTER 9

1. d	9.1
2. b	9.1
3. c	9.1
4. d	9.2
5. a	9.2
6. c	9.2
7. a	9.3
8. b	9.3
9. d	9.3

b	*9.3*	13.c	*12.9*	10.b	*15.3*		
c	*9.3*	e	*12.8*	11.b	*15.3*		
a	*9.3*	d	*12.9*	12.b	*15.4*		

CHAPTER 10

1. c	*10.1*	b	*12.8*	13.e	*15.2*
2. b	*10.2*	a	*12.2*	a	*15.2*
3. a	*10.2*	f	*12.9*	b	*15.4*
4. d	*10.1*			d	*15.2*
5. d	*10.2*	**CHAPTER 13**		c	*15.1*
6. b	*10.2*			f	*15.1*
7. d	*10.3*	1. c	*13.2*		

8. Sister chromatids
remain attached. *10.3*

2. d	*13.2*	**CHAPTER 16**	
9. d	*10.3*	3. c	*13.2*	
10.e	*10.4*	4. a	*13.3*	1. c	*16.1*
11.d	*10.2*	5. d	*13.3*	2. plasmid	*16.1*
a	*10.1*	6. b	*13.2*	3. b	*16.1*
c	*10.3*	7. b	*13.4*	4. a	*16.2*
b	*10.3*	8. c	*13.1*	5. d	*16.3*

e	*13.4*	6. b	*16.3*
a	*13.2*	7. b	*16.5*
b	*13.3*	8. b	*16.5*
d	*13.3*	9. d	*16.4*
f	*13.3*	c	*16.7*

CHAPTER 11

1. a	*11.1*			b	*Introduction*
2. b	*11.1*			e	*16.2*
3. a	*11.1*	**CHAPTER 14**		a	*16.10*
4. b	*11.1*				
5. c	*11.2*	1. c	*14.1*		
6. a	*11.2*	2. b	*14.1*		
7. d	*11.3*	3. c	*14.1*		
8. c	*11.5*	4. c	*14.1*		
9. a	*11.5*	5. d	*14.2*		
10.b	*11.3*	6. a	*14.3*		
d	*11.2*	7. f	*14.5*		
a	*11.1*	8. e	*14.5*		
c	*11.1*	c	*14.4*		
		a	*14.1*		
		f	*14.2*		
		d	*14.3*		
		g	*14.1*		
		b	*14.2*		

CHAPTER 12

1. d	*12.2*	**CHAPTER 15**	
2. c	*12.5*		
3. b	*12.3*		
4. b	*12.3*	1. d	*15.1*
5. b	*12.3*	2. d	*15.1*
6. False	*12.7*	3. d	*15.1, 15.4*
7. d	*12.7*	4. d	*15.1*
8. e	*12.8*	5. h	*15.1*
9. d	*12.9*	6. d	*15.1*
10.False	*12.9*	7. d	*15.1*
11.c	*12.9*	8. d	*15.2*
12.a	*12.10*	9. c	*15.2*

Appendix III. Answers to Genetics Problems

CHAPTER 11

1. a. Both parents are heterozygotes (*Aa*). Their children may be albino (*aa*) or unaffected (*AA* or *Aa*).

b. All are homozygous recessive (*aa*).

c. Homozygous recessive (*aa*) father, and heterozygous (*Aa*) mother. The albino child is *aa*, the unaffected children *Aa*.

2. Possible outcomes of an experimental cross between F_1 rose plants heterozygous for height (*Aa*):

	Ⓐ	ⓐ
Ⓐ	*AA* climber	*Aa* climber
ⓐ	*Aa* climber	*aa* shrubby

3:1 possible ratio of genotypes and phenotypes in F_2 generation

Possible outcomes of a testcross between an F_1 rose plant heterozygous for height and a shrubby rose plant:

Gametes F_1 hybrid:

	Ⓐ	ⓐ
ⓐ	*Aa* climber	*aa* shrubby
ⓐ	*Aa* climber	*aa* shrubby

Gametes shrubby plant:

1:1 possible ratio of genotypes and phenotypes in F_2 generation

3. a. *AB*

b. *AB, aB*

c. *Ab, ab*

d. *AB, Ab, aB, ab*

4. a. All offspring will be *AaBB*.

b. 1/4 *AABB* (25% each genotype)
1/4 *AABb*
1/4 *AaBB*
1/4 *AaBb*

c. 1/4 *AaBb* (25% each genotype)
1/4 *Aabb*
1/4 *aaBb*
1/4 *aabb*

d. 1/16 *AABB* (6.25% of genotype)
1/8 *AaBB* (12.5%)
1/16 *aaBB* (6.25%)
1/8 *AABb* (12.5%)
1/4 *AaBb* (25%)
1/8 *aaBb* (12.5%)
1/16 *AAbb* (6.25%)
1/8 *Aabb* (12.5%)
1/16 *aabb* (6.25%)

5. a. *ABC*

b. *ABC, aBC*

c. *ABC, aBC, ABc, aBc*

d. *ABC*
aBC
AbC
abC
ABc
aBc
Abc
abc

6. A mating of two M^L cats yields 1/4 *MM*, 1/2 M^LM, and 1/4 M^LM^L. Because M^LM^L is lethal, the probability that any one kitten among the survivors will be heterozygous is 2/3.

7. Yellow is recessive. Because F_1 plants have a green phenotype and must be heterozygous, green must be dominant over the recessive yellow.

8. a. *RR* and *rr*

b. all *Rr*

9. Because all F_1 plants of this dihybrid cross had to be heterozygous for both genes, then 1/4 (25%) of the F_2 plants will be heterozygous for both genes.

10. A mating between a mouse from a true-breeding, white-furred strain and a mouse from a true-breeding, brown-furred strain would provide you with the most direct evidence. Because true-breeding strains of organisms typically are homozygous for a trait being studied, all F_1 offspring from this mating should be heterozygous. Record the phenotype of each F_1 mouse, then let them mate with one another. Assuming only one gene locus is involved, these are possible outcomes for the F_1 offspring:

a. All F_1 mice are brown, and their F_2 offspring segregate: 3 brown : 1 white. *Conclusion*: Brown is dominant to white.

b. All F_1 mice are white, and their F_2 offspring segregate: 3 white : 1 brown. *Conclusion*: White is dominant to brown.

c. All F_1 mice are tan, and the F_2 offspring segregate: 1 brown : 2 tan : 1 white. *Conclusion*: The alleles at this locus show incomplete dominance.

11. The data reveal that these genes do not assort independently because the observed ratio is very far from the 9:3:3:1 ratio expected with independent assortment. Instead, the results can be explained if the genes are located close to each other on the same chromosome, which is called linkage.

12. Fred could use a testcross to find out if his pet's genotype is *WW* or *Ww*. He can let his black guinea pig mate with a white guinea pig having the genotype *ww*.

If any F_1 offspring are white, then the genotype of his pet is *Ww*. If the two guinea pig parents are allowed to mate repeatedly and all the offspring of the matings are black, then there is a high probability that his pet guinea pig is *WW*.

(For instance, if ten offspring are all black, then the probability that the male is *WW* is about 99.9 percent. The greater the number of offspring, the more confident Fred can be of his conclusion.)

13. **a.** <u>1/2</u> red   <u>1/2</u> pink   _____ white
    b. _____ red   <u>All</u> pink   _____ white
    c. <u>1/4</u> red   <u>1/2</u> pink   <u>1/4</u> white
    d. _____ red   <u>1/2</u> pink   <u>1/2</u> white

14. 9/16 walnut
   3/16 rose
   3/16 pea
   1/16 single

15. Because both parents are heterozygotes (Hb^AHb^S), the following are the probabilities for each child:

   a. 1/4 Hb^SHb^S

   b. 1/4 Hb^AHb^A

   c. 1/2 Hb^AHb^S

16. 2/3

17. The smooth rind/furrowed rind ratio is 230:230, which is exactly 1:1. The nonexplosive rind/explosive rind ratio is 227:233, which is close to a 1:1 ratio.

The overall ratio is close to a 1:1 ratio, which indicates that the genes are assorting independently.

18. See the percentages in the graph below. The varied colors in wheat kernels are due to the combined effects of incomplete dominance of alleles of two genes that influence the same phenotype.

CHAPTER 12

1. **a.** Human males (XY) inherit their X chromosome from their mother.

b. A male can produce two kinds of gametes. Half carry an X chromosome and half carry a Y chromosome. All the gametes that carry the X chromosome carry the same X-linked allele.

c. A female homozygous for an X-linked allele produces only one kind of gamete.

d. Fifty percent of the gametes of a female who is heterozygous for an X-linked allele carry one of the two alleles at that locus; the other fifty percent carry its partner allele for that locus.

2. Because Marfan syndrome is a case of autosomal dominant inheritance and because one parent bears the allele, the probability that any child of theirs will inherit the mutant allele is 50 percent.

3. **a.** Nondisjunction might occur during anaphase I or anaphase II of meiosis.

b. As a result of translocation, chromosome 21 may get attached to the end of chromosome 14. The new individual's chromosome number would still be 46, but its somatic cells would have the translocated chromosome 21 in addition to two normal chromosomes 21.

4. A daughter could develop this muscular dystrophy only if she inherited two X-linked recessive alleles— one from each parent. Males who carry the allele are unlikely to father children because they develop the disorder and die early in life.

5. In the mother, a crossover between the two genes at meiosis generates an X chromosome that carries neither mutant allele.

6. The phenotype appeared in every generation shown in the diagram, so this must be a pattern of autosomal dominant inheritance.

7. There is no scientific answer to this question, which simply invites you to reflect on the difference between a scientific and a subjective interpretation of this individual's condition.

Genotype	Phenotype	Number Displaying the Trait	Percent of Population
$A^1A^1B^1B^1$	Dark red	181	9.05
$A^1A^1B^1B^2$ or $A^1A^2B^1B^1$	Red	360	18.00
$A^1A^2B^1B^2$ or $A^1A^1B^2B^2$ or $A^2A^2B^1B^1$	Salmon	922	46.10
$A^1A^2B^2B^2$ or $A^2A^2B^1B^2$	Pink	358	17.90
$A^2A^2B^2B^2$	White	179	8.95
	Totals	2,000	100

Appendix IV. Periodic Table of the Elements

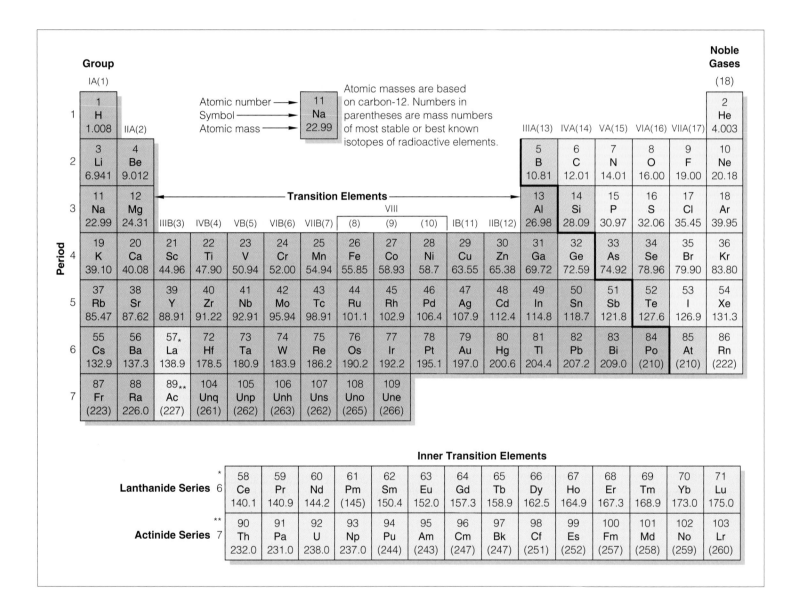

Appendix V. The Amino Acids

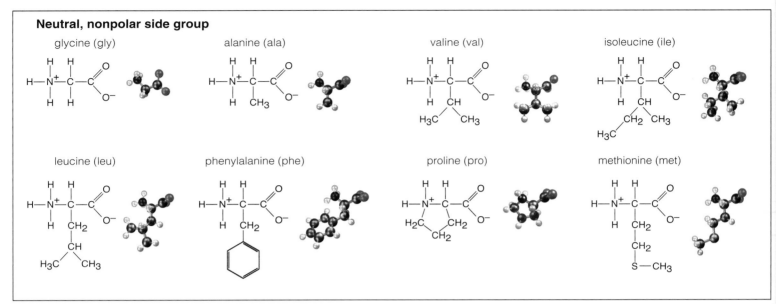

Neutral, nonpolar side group

glycine (gly)

alanine (ala)

valine (val)

isoleucine (ile)

leucine (leu)

phenylalanine (phe)

proline (pro)

methionine (met)

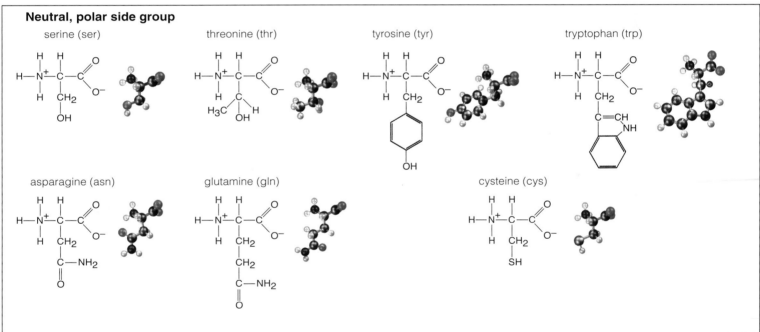

Neutral, polar side group

serine (ser)

threonine (thr)

tyrosine (tyr)

tryptophan (trp)

asparagine (asn)

glutamine (gln)

cysteine (cys)

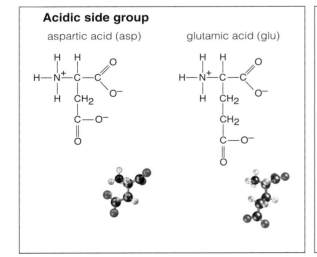

Acidic side group

aspartic acid (asp)

glutamic acid (glu)

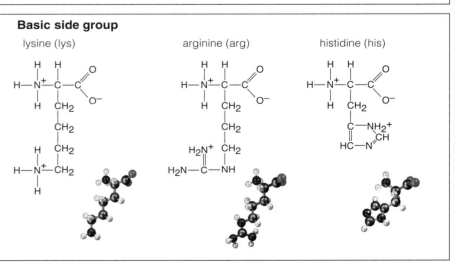

Basic side group

lysine (lys)

arginine (arg)

histidine (his)

Metric-English Conversions

Length

English		Metric
inch	=	2.54 centimeters
foot	=	0.30 meter
yard	=	0.91 meter
mile (5,280 feet)	=	1.61 kilometer

To convert	multiply by	to obtain
inches	2.54	centimeters
feet	30.00	centimeters
centimeters	0.39	inches
millimeters	0.039	inches

Weight

English		Metric
grain	=	64.80 milligrams
ounce	=	28.35 grams
pound	=	453.60 grams
ton (short) (2,000 pounds)	=	0.91 metric ton

To convert	multiply by	to obtain
ounces	28.3	grams
pounds	453.6	grams
pounds	0.45	kilograms
grams	0.035	ounces
kilograms	2.2	pounds

Volume

English		Metric
cubic inch	=	16.39 cubic centimeters
cubic foot	=	0.03 cubic meter
cubic yard	=	0.765 cubic meters
ounce	=	0.03 liter
pint	=	0.47 liter
quart	=	0.95 liter
gallon	=	3.79 liters

To convert	multiply by	to obtain
fluid ounces	30.00	milliliters
quart	0.95	liters
milliliters	0.03	fluid ounces
liters	1.06	quarts

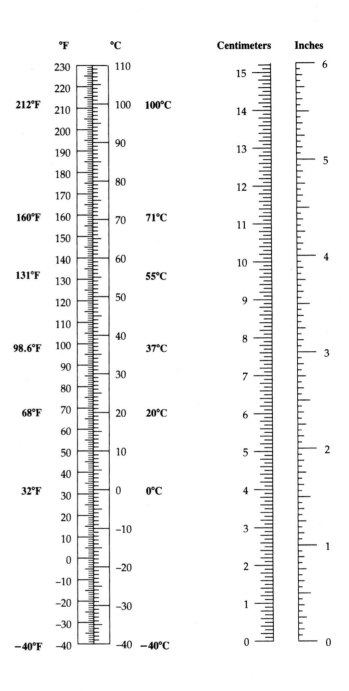

Appendix VII. Closer Look at Some Major Metabolic Pathways

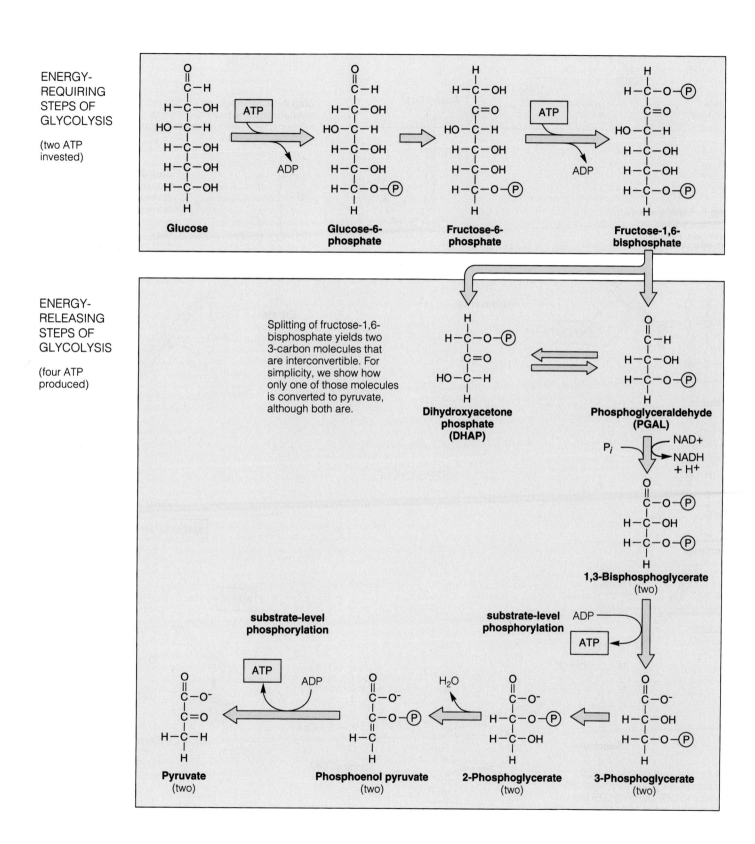

Figure A Glycolysis, ending with two 3-carbon pyruvate molecules for each 6-carbon glucose molecule entering the reactions. The *net* energy yield is two ATP molecules (two invested, four produced).

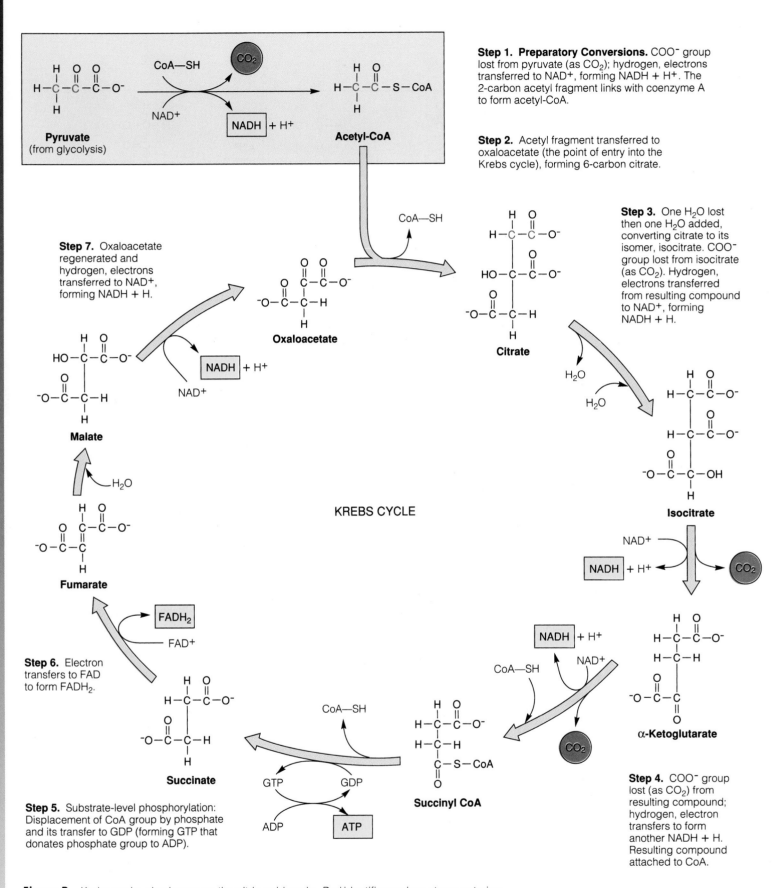

Step 1. Preparatory Conversions. COO^- group lost from pyruvate (as CO_2); hydrogen, electrons transferred to NAD^+, forming $NADH + H^+$. The 2-carbon acetyl fragment links with coenzyme A to form acetyl-CoA.

Step 2. Acetyl fragment transferred to oxaloacetate (the point of entry into the Krebs cycle), forming 6-carbon citrate.

Step 3. One H_2O lost then one H_2O added, converting citrate to its isomer, isocitrate. COO^- group lost from isocitrate (as CO_2). Hydrogen, electrons transferred from resulting compound to NAD^+, forming $NADH + H$.

Step 7. Oxaloacetate regenerated and hydrogen, electrons transferred to NAD^+, forming $NADH + H$.

KREBS CYCLE

Step 6. Electron transfers to FAD to form $FADH_2$.

Step 5. Substrate-level phosphorylation: Displacement of CoA group by phosphate and its transfer to GDP (forming GTP that donates phosphate group to ADP).

Step 4. COO^- group lost (as CO_2) from resulting compound; hydrogen, electron transfers to form another $NADH + H$. Resulting compound attached to CoA.

Pyruvate (from glycolysis), Acetyl-CoA, Citrate, Isocitrate, Oxaloacetate, Malate, Fumarate, Succinate, Succinyl CoA, α-Ketoglutarate

Figure B Krebs cycle, also known as the citric acid cycle. *Red* identifies carbon atoms entering the cyclic pathway (by way of acetyl-CoA) and leaving (by way of carbon dioxide). These cyclic reactions run twice for each glucose molecule that has been degraded to two pyruvate molecules.

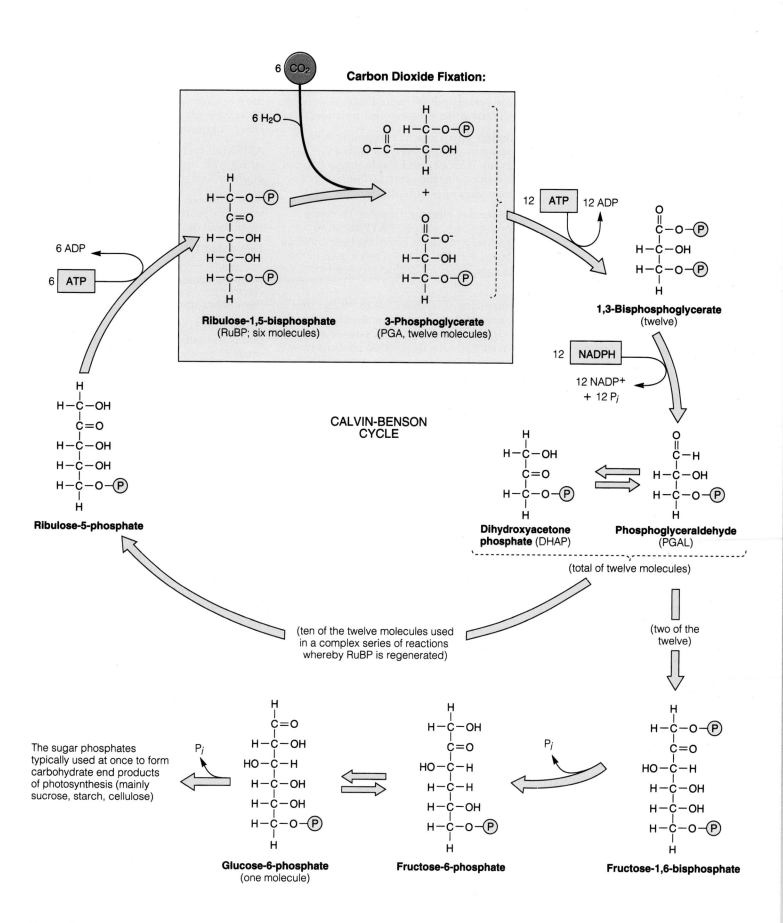

Figure C Calvin–Benson cycle of the light-independent reactions of photosynthesis.

Glossary

ABC model Proposed model for the genetic control of how flowers form; three groups of master genes, *A*, *B*, and *C*, guide the emergence of sepals, petals, and reproductive parts.

absorption spectrum Of chlorophylls and other photosynthetic pigments, the range of wavelengths that each type of pigment can absorb.

acetylation Attachment of an acetyl group to an organic compound.

acid Any substance that donates hydrogen ions (H^+) to other solutes or to water molecules.

activation energy Minimum amount of energy required to start a reaction; enzyme action lowers this energy barrier.

active site Crevice in an enzyme molecule where a specific reaction is catalyzed.

active transport Pumping of a specific solute across a cell membrane against its concentration gradient, through the interior of a transport protein that spans the lipid bilayer. Requires energy input.

adhesion protein Plasma membrane protein that helps cells locate cells of the same type and stick together.

aerobic respiration Oxygen-dependent pathway of ATP formation in which glucose is fully broken down to carbon dioxide and water in several stages: including glycolysis, the Krebs cycle and a few steps preceding it, and electron transfer phosphorylation. Occurs only in mitochondria. Typical net yield: 36 ATP.

alcohol Organic compound that has one or more hydroxyl groups (—OH) and dissolves easily in water (e.g, sugars).

alcoholic fermentation Anaerobic ATP-forming pathway that occurs only in the cytoplasm. NADH transfers electrons to acetaldehyde, forming ethanol. Reactions start with pyruvate from glycolysis and regenerate NAD^+. Net yield: 2 ATP.

alkylating agent Any substance that replaces a hydrogen atom with an alkyl (saturated organic) group in a biological molecule.

allele One of two or more molecular forms of a gene that arise by mutation and that specify slightly different versions of the same trait.

alternative splicing During mRNA transcript processing, the splicing of a gene's exons (protein-coding sequences) in various combinations; the same gene can specify two or more proteins of slightly different form and function.

amino acid Organic compound with an amino group (NH_2), a carboxylic acid group (COOH), and a side group bonded covalently to the same carbon atom. The monomer of polypeptide chains.

anaerobic electron transfer Of some prokaryotes, a type of ATP-forming pathway in which electrons flow through transfer chains in the plasma membrane. A compound other than oxygen is the final electron acceptor.

anaphase Of mitosis, stage when sister chromatids of each chromosome move to opposite spindle poles. Of anaphase I (meiosis), each duplicated chromosome and its homologue move to opposite poles of a microtubular spindle. Of anaphase II, sister chromatids of each chromosome move to opposite poles.

aneuploidy A condition in which body cells have more or fewer chromosomes relative to the parental chromosomal number.

animal Multicelled, motile heterotroph that ingests other organisms or their tissues. Nearly all animals have tissues and organ systems that arise from two or three primary tissue layers during early stages of embryonic development.

anthocyanin One of a class of accessory pigments in photosynthesis that reflect red to blue light.

anticodon Part of a tRNA molecule; a linear set of three nucleotide bases that can base-pair with an mRNA codon.

antioxidant Enzyme or cofactor that can neutralize free radicals.

Archaea Domain of prokaryotic species with molecular and biochemical traits that are distinct from bacteria; in some traits they are closer to eukaryotic cells. One of the first two lineages that evolved after the origin of life. All archaeans.

asexual reproduction Any reproductive mode by which genetically identical offspring arise from a single parent; e.g., prokaryotic fission, transverse fission, budding, and vegetative propagation.

atom Fundamental form of matter that has mass and takes up space and cannot be broken down into something else by everyday means.

ATP Adenosine triphosphate. A type of nucleotide; consists of the base adenine, the five-carbon sugar ribose, and three phosphate groups; the main carrier of energy between reaction sites in cells.

ATP/ADP cycle Alternating reactions in which ATP and ADP form as an outcome of phosphate group transfers.

ATP synthase Membrane-bound active transport protein that also functions as an enzyme of ATP formation.

automated DNA sequencing Robotic method of determining the nucleotide sequence of a region of DNA. Uses gel electrophoresis and laser detection of fluorescent tracers.

autosome Any chromosome of a type that is the same in males and females of sexually reproducing species.

autotroph A "self-feeding" organism; uses an environmental energy source and carbon atoms from CO_2 to make its own food. Photoautotrophs get energy from the sun's rays; chemoautotrophs use simple inorganic or organic compounds in the environment (e.g., methane).

Bacteria Domain of prokaryotic species; the first lineages after the origin of life. Most are chemoheterotrophs, including many parasites and pathogens.

bacteriophage One of a class of viruses that infect bacteria.

basal body An organelle that is located at the base of a flagellum or cilium, below the 9+2 array of microtubules; starts out as a barrel-shaped centriole.

base Any substance that accepts hydrogen ions (H^+) when dissolved in water, so forming hydroxyl ions (OH^-). Also, the nitrogen-containing component of a nucleic acid.

base-pair substitution Mutation in which one nucleotide is wrongly substituted for another during DNA replication.

bell curve Statistical distribution of the continuous variation for a specified trait in a population.

biofilm Large microbial populations anchored to lung epithelia and other surfaces by their own sticky, stiff polysaccharide secretions.

bioluminescence The production of fluorescent light by a living organism.

bipolar mitotic spindle Of eukaryotic cells, a dynamic array of microtubules that moves chromosomes with respect to its two poles during mitosis or meiosis.

buffer system A weak acid and the base that forms when it dissolves in water. The two work as a pair to counter slight shifts in pH.

bulk flow Mass movement of one or more substances in the same direction in response to pressure, gravity, or another external force.

C3 plant Type of plant in which three-carbon PGA is the first stable intermediate to form after carbon dioxide fixation in the second stage of photosynthesis.

C4 plant Type of plant in which makes four-carbon oxaloacetate is the first stable intermediate to form after carbon dioxide fixation in the second stage of photosynthesis.

calcium pump Active transport protein that pumps calcium ions across a cell membrane.

Calvin–Benson cycle Cyclic reactions that form sugar and regenerate RuBP in the second stage of photosynthesis. The reactions require carbon (from carbon dioxide). They use energy from ATP and hydrogens and electrons from NADPH, both of which form in the first stage.

CAM plant Type of plant that conserves water by opening stomata only at night, when it fixes carbon by repeated turns of the C4 pathway; stands for crassulacean acid metabolism.

cancer Malignant neoplasm; mass of abnormally dividing cells that can leave their home tissue and invade and form new masses in other parts of the body.

carbon fixation Process by which any autotrophic cell incorporates carbon atoms into a stable organic compound. Different cells get carbon dioxide from the air or dissolved in water.

carotenoid One of a class of accessory pigments in photosynthesis that reflect red, orange, and yellow light and color many fruits. One, beta-carotene, is a precursor of vitamin A.

cDNA DNA synthesized in laboratories from an mRNA transcript, through the use of the enzyme reverse transcriptase.

cell Smallest unit that still displays the properties of life; it has the capacity to survive and reproduce on its own.

cell cortex Three-dimensional mesh of actin filaments and other cytoskeletal proteins under the plasma membrane.

cell cycle Of eukaryotic cells, a series of events from the time a cell forms until that cell reproduces. Interphase, mitosis, and cytoplasmic division constitute one turn of the cycle.

cell differentiation In the developing embryos of multicelled organisms, the process by which different cell lineages selectively express the same genes and become specialized in their composition, structure, and function.

cell junction In multicelled species, any molecular structure that connects adjoining cells physically, chemically, or both at their plasma membranes.

cell plate formation Mechanism of cytoplasmic division in plant cells that involves formation of a cross-wall after nuclear division.

cell theory All organisms consist of one or more cells, the cell is the smallest unit of organization that still displays the properties of life, and the continuity of life arises directly from the growth and division of single cells.

cell wall Of many cells (not animal cells), a semirigid but permeable structure that surrounds the plasma membrane; helps a cell retain its shape and resist rupturing.

central vacuole In many mature, living plant cells, an organelle that stores amino acids, sugars, and toxic wastes; it gets larger during growth and forces the cell to enlarge and increase its surface area.

centriole A barrel-shaped structure that arises from a centrosome and organizes growing microtubules into a 9+2 array inside a cilium or flagellum.

centromere Constricted region of a eukaryotic chromosome that contains binding sites (kinetochores) to which spindle microtubules reversibly attach.

centrosome A small, dense patch of material in eukaryotic cells from which microtubules start growing; also known as a microtubule organizing center.

chemical bond A union between the electron structures of two or more atoms.

chemical equilibrium State at which the concentrations of reactants and products in a reversible chemical reaction show no net change.

chlorophyll *a* The main photosynthetic pigment in plants and algae; absorbs mainly red and violet light and reflects or transmits green light.

chlorophyll *b* An accessory pigment in photosynthesis; absorbs mainly blue and orange light.

chloroplast Organelle of photosynthesis in plants and algae. Two outer membranes enclose a semifluid interior (stroma). A third membrane forms a compartment inside that functions in ATP and NADPH formation; sugars form in the stroma.

chromatin All of the DNA molecules and associated proteins inside a nucleus.

chromosome In prokaryotic cells, a circular DNA double helix with a few proteins attached. In eukaryotic cells, a linear DNA double helix with many histones and other proteins attached.

chromosome number Sum total of all chromosomes in a given type of cell; e.g., diploid ($2n$) cells have two of each type, haploid cells (n) have one of each type.

cilium, plural cilia In some eukaryotic cells, a thin projection from the plasma membrane that is a motile structure or that has been modified into a sensory structure; cilia usually are shorter than flagella.

clone A genetically identical copy of DNA, a cell, or a multicelled organism.

cloning vector DNA molecule that can accept foreign DNA and be replicated inside a host cell.

codon A base triplet; a linear sequence of three nucleotides in mRNA that codes for an amino acid or a termination signal during translation.

coenzyme Small molecule that takes part in an enzymatic reaction and is reversibly modified during the reaction (e.g., a vitamin).

cohesion A capacity to resist rupturing when placed under tension (stretched).

communication protein Membrane protein that helps form an open channel between the cytoplasm of adjoining cells.

compound Molecule of two or more elements in proportions that do not vary, as they can do in mixtures.

concentration gradient Difference in the number of molecules or ions of any substance between two adjoining regions.

condensation reaction Type of chemical reaction in which two molecules become covalently bonded as a larger molecule; water often forms as a by-product.

consumer Type of heterotroph that feeds on the tissues of other organisms as its sources of carbon and energy.

continuous variation Of individuals of a population, a range of small differences in the phenotypic expression of a trait.

contractile ring mechanism The basis of cytoplasmic division of animal cells; a ring of contractile filaments beneath the plasma membrane shrinks in diameter and pinches the cytoplasm in two.

control group In experimental tests, a group used as a standard for comparison against one or more experimental groups.

covalent bond A sharing of one or more electrons between two atoms. In polar covalent bonds, atoms share the electrons unequally; in nonpolar covalent bonds, each gets an equal share of the electrons.

crossing over At prophase I of meiosis, reciprocal exchange of segments between two nonsister chromatids of a pair of homologous chromosomes; results in novel combinations of alleles.

cytoplasm All cell parts, particles, and semifluid substances between the plasma membrane and the nucleus (or, in the case of prokaryotic cells, the nucleoid).

cytoskeleton In eukaryotic cells, a dynamic system of protein filaments that structurally support, organize, and move the cell and its internal structures. Some prokaryotic cells contain a few similar protein filaments but these do not form a cytoskeleton.

deletion Loss of a chromosome segment through unequal crossovers, inversions, or chemical attack. Also, a mutation involving the loss of one or more bases of a DNA molecule.

denaturation Disruption of hydrogen bonds and other interactions holding a molecule in its three-dimensional shape, which changes. Heat, shifts in pH, and detergents can cause denaturation.

deoxyribonucleic acid See DNA.

diffusion Net movement of like ions or molecules from a region where they are most concentrated to an adjoining region where they are less concentrated (e.g., down a concentration gradient).

dihybrid experiment Type of experiment that starts with a cross between two true-breeding, homozygous parents that differ in two traits governed by alleles of two genes. The actual experiment is a cross between two of their F_1 offspring that are identically heterozygous for the two genes (e.g., *AaBb* x *AaBb*).

diploid chromosome number Of many sexually reproducing species, having two chromosomes of each type in body cells; the two pair as homologues at meiosis.

DNA Deoxyribonucleic acid. Double-stranded nucleic acid twisted into a helical shape; its base sequence encodes the primary hereditary information for all living organisms and many viruses.

DNA chip Microarray of thousands of gene sequences that represents a large subset of a genome; stamped onto a glass plate about the size of a business card and used to study gene expression.

DNA fingerprinting Method that can distinguish one individual from all others based on unique differences in parts of their DNA; fragments cut from the DNA (also called RFLPs) reveal genetic differences.

DNA ligase Type of enzyme that catalyzes the sealing of short stretches of DNA into a continuous strand during replication; also seals strand breaks.

DNA polymerase Type of enzyme that catalyzes the addition of free nucleotides to new DNA strands during replication; also proofreads and corrects mismatches.

DNA proofreading mechanism One of various enzyme-mediated processes that fixes most DNA replication errors and strand breaks.

DNA replication Process by which a cell duplicates its DNA molecules before it divides into daughter cells.

dosage compensation In mammals, a gene control mechanism that inactivates most of the genes on one of the two X chromosomes in females so that early development proceeds the same as it does in males (XY).

duplication Base sequence in DNA that has been repeated two or more times; can result from unequal crossovers.

egg Mature female gamete, or ovum.

electric gradient A difference in electric charge between adjoining regions.

electromagnetic spectrum All of the wavelengths of radiant energy less than 10^{-5} nanometers long to radio waves more than 10 kilometers long.

electron Negatively charged unit of matter; electrons occupy orbitals around the atomic nucleus.

electron transfer chain Organized array of enzymes and other molecules in a cell membrane that accept and then give up electrons in sequence; operation of chain permits the release and capture of energy in small, useful increments.

electron transfer phosphorylation Final stage of aerobic respiration; electron flow through electron transfer chains in inner mitochondrial membrane sets up H^+ concentration and electric gradient that drives ATP formation. Oxygen accepts electrons at the end of the chain and so keeps the chain clear for operation.

element A fundamental substance that is composed only one kind of atom.

emergent property With respect to life's levels of organization, a new property that emerges through interactions of entities at lower levels, none of which displays the property; e.g., living cells that emerge from "lifeless" molecules.

endocytosis Mechanism by which a patch of plasma membrane balloons inward and forms a vesicle, which sinks into the cytoplasm.

endomembrane system Endoplasmic reticulum, Golgi bodies, and transport vesicles concerned with modification of many new proteins, lipid assembly, and their transport within the cytoplasm or to the plasma membrane for export.

endoplasmic reticulum ER. Organelle that starts at the nuclear envelope and extends through the cytoplasm. Rough ER, with an abundance of ribosomes on its cytoplasmic side, functions in the modification of many new polypeptide chains. Assembly of membrane lipids, fatty acid breakdown, and inactivation of some toxins occurs in smooth ER.

energy A capacity to do work.

enhancer A small sequence in DNA that binds transcription-regulating molecules; enhances transcription rates.

entropy Measure of how much and how far a concentrated form of energy has been dispersed after an energy change.

enzyme Typically, a type of protein that catalyzes (speeds) a chemical reaction. Some RNAs also show catalytic activity.

epistasis Interaction among products of two or more gene pairs that influence the same trait.

ER See endoplasmic reticulum.

eukaryotic cell Type of cell that starts life with a nucleus and other membrane-bound organelles.

evaporation Process of conversion of a liquid to a gas; requires energy input.

evolution Heritable change in a line of descent. Outcome of microevolutionary events: gene mutation, natural selection, genetic drift, and gene flow.

exocytosis Movement of a vesicle from the cytoplasm to the plasma membrane, where it releases its contents outside the cell surface; the vesicle's membrane fuses with the plasma membrane and becomes part it.

exon One of the protein-coding base sequences in eukaryotic DNA; some or all of a gene's exons may be spliced together in various combinations for a mature mRNA transcript.

experiment, scientific A type of test that simplifies observation in nature or in the laboratory by manipulating and controlling the conditions under which observations are made.

experimental group A group upon which an experiment is performed, and compared with a control group.

FAD Flavin adenine dinucleotide. A nucleotide coenzyme; transfers electrons and unbound protons (H^+) from one reaction site to another.

fat Type of lipid with one, two, or three fatty acid tails attached to a glycerol head.

fatty acid Organic compound having a carboxyl group attached to a backbone with many as thirty-six carbon atoms; saturated types have single bonds only; unsaturated types have one or more double covalent bonds.

feedback inhibition Of cells, an activity causes a change in cellular conditions, and the change itself triggers a decrease or the shutting down of the activity.

fertilization Fusion of a sperm nucleus and an egg nucleus, the result being a single-celled zygote.

first law of thermodynamics A law of nature; states that the total amount of energy in the universe remains constant; energy can be converted from one form to another, but it cannot be created or destroyed.

flagellum, plural **flagella** A whip-like motile structure of many cells.

fluid mosaic model Explanation of cell membrane structure; the membrane has a mixed composition (mosaic) of lipids and proteins, the interactions and motions of which impart fluidity to it.

functional group An atom or a group of atoms with characteristic properties that is covalently bonded to the carbon backbone of an organic compound.

fungus, plural **fungi** Type of eukaryotic heterotroph that obtains nutrients by extracellular digestion and absorption; fungi are notable for being prolific spore producers and major decomposers.

Gel electrophoresis Procedure for separating DNA molecules by length or protein molecules by size and charge; the molecules separate from each other as they migrate through a gel matrix in response to an electric current.

gene Unit of information for a heritable trait in DNA, transmitted from parents to offspring.

gene control A molecular mechanism that governs if, when, or how a specific gene is transcribed or translated.

gene expression Conversion of heritable information in a gene into a product (e.g., DNA to mRNA to a protein).

gene library Mixed collection of host cells that contain cloned DNA fragments representing all or most of a genome.

gene mutation Small-scale change in the nucleotide sequence of a gene that can result in an altered protein product.

genetic code Correspondence between triplets of nucleotides in DNA (then in mRNA) and specific sequences of amino acids in a polypeptide chain; the near-universal language of protein synthesis in cells. A few variant code words occur in some species and in mitochondria.

genetic engineering Manipulation of an organism's DNA, usually to alter at least one aspect of phenotype.

genome All DNA in a haploid number of chromosomes for a species.

genomics The study of genes and gene function in humans and other organisms.

genus A group of related species; also the first name of the species epithet.

glycolysis Partial breakdown of glucose to two pyruvate molecules. First stage of the main energy-releasing pathways, including aerobic respiration, lactate fermentation, and alcoholic fermentation.

Golgi body Organelle of endomembrane system; its enzymes modify many new polypeptide chains, assemble lipids, and package both inside vesicles for secretion or for use inside cell.

growth factor In multicelled species, protein that stimulates increases in size (e.g., by promoting mitosis).

haploid chromosome number Total chromosome number in cells with one of each type of chromosome characteristic of the species.

helicase Type of enzyme that catalyzes the breaking of hydrogen bonds in DNA during replication, which allows the two strands of a double helix to unwind from each other.

heme Iron-containing functional group that reversibly binds oxygen.

heterotroph Organism unable to make its own organic compounds; feeds on autotrophs, other heterotrophs, or organic wastes.

heterozygous condition Having two different alleles at a particular locus on homologous chromosomes (e.g., *Aa*).

homeostasis State in which of physical and chemical aspects of the internal environment are being maintained in a range that is tolerable for cell activities.

homeotic gene One of a class of master genes; helps determine identity of body parts during embryonic development.

homologous chromosome One of a pair of chromosomes in body cells of diploid organisms; except for nonidentical sex chromosomes, they have the same size, shape, and gene sequence.

homozygous dominant Of diploid organisms, having a pair of dominant alleles at a gene locus on homologous chromosomes (e.g., *AA*).

homozygous recessive Of diploid organisms, having a pair of recessive alleles at a gene locus on homologous chromosomes (e.g., *aa*).

human gene therapy The transfer of one or more normal or modified genes into a person to correct a genetic defect, or to boost resistance to a disease.

hybrid Individual having a nonidentical pair of alleles for a trait being studied.

hydrocarbon An organic compound that has only hydrogen atoms bonded to its carbon backbone.

hydrogen bond A weak attraction that has formed between a covalently bonded hydrogen atom and an electronegative atom in a different molecule or in a different part of the same molecule.

hydrolysis A cleavage reaction in which an enzyme splits a molecule, and the components of water (—OH and —H) are attached to the fragments.

hydrophilic substance Any type of polar molecule that easily dissolves in water (e.g., glucose).

hydrophobic substance Any type of nonpolar molecule that resists dissolving in water (e.g., oil).

hydrostatic pressure Pressure exerted by a volume of fluid against a wall, membrane, or some other structure that contains it; also called turgor pressure.

hypertonic solution Of two fluids, the one with the higher solute concentration.

hypothesis, scientific An explanation of a phenomenon, one that has the potential to be proven false by experimental tests.

hypotonic solution Of two fluids, the one with the lower solute concentration.

independent assortment An outcome of random alignments at metaphase I of meiosis; each homologous chromosome and its partner are assorted into different gametes independently of other pairs. Crossing over can affect the outcome.

induced-fit model Explanation of how some enzymes speed a reaction; they change shape for a better fit with a bound substrate; the resulting tension destabilizes substrate bonds so that they break faster.

insertion Mutation involving insertion of one to a few extra bases into a DNA strand.

intermediate filament One of three major types of cytoskeletal elements; mechanically strengthens some cells.

interphase In a eukaryotic cell cycle, the interval between mitotic divisions; the cell grows in mass, roughly doubles the number of cytoplasmic components, and replicates its DNA.

introns One of the noncoding sequences in eukaryotic genes; removed from a pre-mRNA transcript before translation.

inversion Chromosomal change in which part of the sequence of DNA gets oriented in the reverse direction, with no molecular loss.

ion Atom that has a positive or negative electric charge as a result of an imbalance between its number of protons and electrons.

ionic bond Interaction between ions held together by the attraction of their opposite charges.

ionizing radiation Form of radiation that has enough energy to eject electrons from atoms.

isotonic solution Any fluid having the same solute concentration as another fluid to which it is being compared.

isotope One of two or more atoms of the same element (same number of protons) that differ in the number of neutrons.

karyotype Preparation of an individual's metaphase chromosomes arranged by length, centromere location, and shape.

kilocalorie A thousand calories of heat energy, the amount needed to raise the temperature of 1 kilogram of water by 1°C. Standard unit of measure for the energy content of foods.

kinase One of a class of enzymes that catalyze a phosphate-group transfer (e.g., a protein kinase).

knockout experiment An experiment in which researchers mutate a wild-type gene so that it cannot be transcribed or translated; the organism's phenotype may yield clues to the gene's function.

Krebs cycle The second stage of aerobic respiration in which pyruvate is broken down to carbon dioxide and water. Two ATP form. Occurs only in mitochondria.

lactate fermentation Anaerobic pathway of ATP formation. The pyruvate from glycolysis is converted to three-carbon lactate, and NAD^+ is regenerated. The net energy yield is 2 ATP.

light-dependent reactions First stage of photosynthesis. Pigments trap sunlight energy, which gets converted to chemical energy of ATP. In a noncyclic pathway, the reduced coenzyme NADPH also forms.

light-independent reactions Second stage of photosynthesis in which sugars are formed by using ATP and NADPH from the preceding stage. Involves carbon fixation and cyclic reactions in which sugars form and an organic compound—the entry point for the cyclic reactions—is regenerated.

linkage group All genes on the same chromosome.

lipid One of the nonpolar hydrocarbons; (e.g., a fat, oil, wax, sterol, phospholipid, or glycolipid).

lipid bilayer Structural basis of all cell membranes; primarily phospholipids arranged tail-to-tail in two layers, with the hydrophilic heads of one layer dissolved in cytoplasmic fluid and the other dissolved in fluid outside.

lysosome Specialized vesicle that functions in intracellular digestion.

meiosis Of eukaryotic cells, a nuclear division process that halves the parental chromosome number, to a haploid (n) number. Prerequisite to the formation gametes and sexual spores.

messenger RNA mRNA. Single-stranded ribonucleotide transcribed from DNA; the only RNA that carries protein-building information to ribosomes.

metabolic pathway Any sequence of enzyme-mediated reactions by which cells assemble and build or break down organic compounds.

metabolism All the controlled, enzyme-mediated chemical reactions by which cells acquire and use energy as they synthesize, store, degrade, and eliminate substances.

metaphase Of meiosis I, stage when all pairs of homologous chromosomes have become positioned at the equator of the bipolar microtubular spindle. Of mitosis or meiosis II, stage when all duplicated chromosomes are positioned at the spindle equator.

methylation Attachment of a methyl group to an organic compound; also a common gene control mechanism.

microfilament Thinnest of the cytoskeletal elements; consists of actin subunits. Functions in contraction, movements, and structural support.

microtubule Thickest of the cytoskeletal elements; consists of tubulin subunits. Contributes to cell shape, growth, and motion.

mimicry Looking like something else and confusing predators (or prey); confers a selective advantage upon the mimic, the model, or both.

mitochondrion, plural **mitochondria** Organelle that specializes in ATP formation; site of aerobic respiration's second and third stages.

mitosis In eukaryotic cells, a nuclear division mechanism that maintains the parental chromosome number for forthcoming daughter cells. Basis of growth, tissue repair, and often asexual reproduction.

mixture Two or more types of molecules intermingled in proportions that can and usually do vary.

model Theoretical explanation of something that has not been directly observed.

molecule Two or more atoms of the same or different elements joined by chemical bonds.

monohybrid experiment Type of experiment that starts with a cross between two true-breeding, homozygous parents that differ in one traits governed by alleles of one gene. The experiment is a cross between two of their F_1 offspring that are identically heterozygous for the two genes (e.g., *Aa* x *Aa*).

monomer Small molecule that is a repeating subunit in polymers, such as the sugar monomers of starch.

multiple allele system Three or more slightly different molecular forms of a gene that occur among the individuals of a population.

mutation Heritable change in DNA.

NAD+ Abbreviation for nicotinamide adenine dinucleotide. A nucleotide coenzyme; abbreviated NADH when reduced (carrying electrons and H+).

natural selection Microevolutionary process; the outcome of differences in survival and reproduction among individuals of a population that differ in the details of their heritable traits.

neoplasm Mass of cells (tumor) that lost control over growth and division.

neutron Type of subatomic particle in the nucleus of all atom except hydrogen; has mass but no charge.

nondisjunction The failure of sister chromatids or homologous chromosomes to separate from each other in meiosis or mitosis. Daughter cells end up with too many or too few chromosomes.

nonionizing radiation Form of radiation that carries enough energy to boost electrons to higher energy levels but not enough to eject them from an atom.

nuclear envelope A double membrane that is the outer boundary of the nucleus of eukaryotic cells.

nucleic acid Single-stranded or double-stranded molecule of nucleotides joined at phosphate groups (e.g., DNA, RNA).

nucleic acid hybridization Any base-pairing between DNA or RNA strands from different sources.

nucleoid Of prokaryotic cells, the region in which DNA is physically organized; not separated from the cytoplasm by a membrane boundary.

nucleolus In a nondividing cell, a mass of material in the nucleus from which the RNA subunits and proteins of ribosomes are assembled before shipment to the cytoplasm.

nucleosome Small stretch of eukaryotic DNA wound twice around a spool of proteins called histones.

nucleotide Small organic compound having a five-carbon sugar, a nitrogen-containing base, and a phosphate group.

nucleus Large organelle with an outer envelope of two pore-ridden lipid bilayers that separates eukaryotic DNA from the cytoplasm. Pores across the nuclear envelope selectively control the passage of substances across it.

operator Part of an operon; a binding site for a regulatory protein.

operon Promoter–operator sequence that controls transcription of more than one bacterial gene.

organelle One of the membrane-bound compartments that carry out specialized metabolic functions in eukaryotic cells (e.g., nucleus, mitochondria).

organic compound Any carbon-based molecule that also incorporates atoms of hydrogen and, often, oxygen, nitrogen, and other elements.

osmosis Diffusion of water across a selectively permeable membrane from a region where the water concentration is higher to a region where it is lower.

osmotic pressure Of a fluid, a measure of the tendency of water to follow its water concentration gradient and move into that fluid.

oxidation–reduction reaction Transfer of electrons between reactant molecules.

passive transport Diffusion of a solute across a cell membrane, through the interior of a transport protein.

pattern formation During development, the sculpting of embryonic cells into specialized tissues and organs at expected times, in expected regions.

PCR Polymerase chain reaction. A method that rapidly and enormously amplifies DNA fragments.

pedigree Chart of connections among individuals related by descent.

periodic table of the elements Tabular arrangement of elements based on their predictable chemical properties.

peroxisome Enzyme-filled vesicle that breaks down amino acids and fatty acids to hydrogen peroxide, which is then converted to harmless products.

pH scale Measure of the H+ concentration (or acidity) of blood, water, and other solutions. pH 7 is neutrality on the scale.

phagocytosis "Cell eating," a common endocytic pathway by which various cells engulf food bits, microbes, cellular debris.

phospholipids Type of lipid that has a phosphate group in its hydrophilic head. The main constituent of cell membranes.

phosphorylation Enzyme-mediated transfer of a phosphate group between molecules.

photon Unit of electromagnetic energy; has wave-like and particle-like properties.

photosynthesis Process by which organisms capture sunlight energy and use it first in the formation of ATP and NADPH, then in the formation of sugars from carbon dioxide and water. ATP donates energy that drives the sugar-building reactions, and NADPH donates electrons and hydrogen building blocks.

photosystem In photosynthetic cells, a cluster of membrane-bound, light-trapping pigments and other molecules.

phycobilin One of a class of accessory pigments in photosynthesis that reflects red to blue light.

pigment Any light-absorbing molecule.

plant A multicelled photoautotroph, most with well-developed roots and shoots (e.g., stems, leaves); plants are the primary producers for nearly all ecosystems on land.

plasma membrane Outermost cell membrane; structural and functional boundary between the cytoplasm and fluid surrounding the cell.

plasmid A small, circular molecule of bacterial DNA that carries a few genes and is replicated independently of the chromosome.

pleiotropy A case of alleles at a single gene locus having positive or negative impact on two or more traits.

polygenic inheritance Inheritance of multiple genes that affect the same trait.

polymer Large molecule of three to millions of monomers of the same or different kinds.

polypeptide chain Three or more amino acids linked by peptide bonds.

polyploidy Having three or more of each type of chromosome characteristic of a species.

prediction Statement, based on a hypothesis, about what you expect to observe in nature; the "if-then process."

pressure gradient Difference in pressure being exerted in two adjoining regions.

primary wall Of young plant cells, a thin wall that is pliable enough to permit mitotic division and shape changes; has cellulose, polysaccharides, glycoproteins.

primer Short nucleotide sequence that researchers design as a site of initiation for DNA synthesis on DNA or RNA.

probability The chance that each outcome of an event will occur is proportional to the number of ways in which the outcome can be reached.

probe Short nucleotide sequence, labeled with a tracer, designed to hybridize with part of a specific gene or mRNA.

producer Autotroph (self-feeder); nourishes itself using sources of energy and carbon from the environment. Photoautotrophs and chemoautotrophs are examples.

prokaryotic cell Archaean or bacterium; single-celled organism, most often walled; none has a nucleus or the other organelles seen in eukaryotic cells.

promoter Short stretch of DNA to which RNA polymerase binds and prepare for transcription.

prophase Of mitosis, the stage when all duplicated chromosomes in a cell start to condense, a microtubular spindle forms, and the nuclear envelope starts to break up. Duplicated pairs of centrioles move to opposite spindle poles. In prophase I of meiosis, crossing over also occurs.

"protist" Diverse single-celled, colonial, and multicelled species that structurally are the simplest eukaryotes; traditionally grouped in "kingdom Protista." Now being reclassified into major groupings that reflect evolutionary relationships.

proton Positively charged subatomic particle in the nucleus of all atoms.

pseudopod Flexible, transient lobe of membrane-enclosed cytoplasm that functions in motility and phagocytosis; amoebas, amoeboid cells, and many white blood cells form them.

Punnett-square method A simple way to predict the probable outcomes of a genetic cross by constructing and filling in a diagram of all possible combinations of genotypes, phenotypes, or both.

pyruvate Three-carbon compound that forms as the end product of glycolysis.

radioactive decay Natural, inevitable process by which an atom emits energy as subatomic particles and x-rays as its unstable nucleus spontaneously breaks apart; transforms one element into another in a predictable time span.

radiosotope Isotope with an unstable nucleus; too many or too few neutrons.

receptor, molecular A protein or some other molecule with a binding site for a specific signaling molecule. Receptor binding typically triggers alteration in some cell activity.

receptor, sensory Sensory cell or a specialized ending of one that detects a stimulus (a specific form of energy).

recognition protein One of a class of glycoproteins or glycolipids that project above the plasma membrane and that identify a cell *nonself* (foreign) or *self* (belonging to one's own body tissue).

recombinant DNA A DNA molecule that contains genetic material from more than one organism of the same species or different species.

repair enzyme Type of enzyme that helps repair some small-scale changes in a DNA strand; e.g., one specialized set of repair enzymes excises a mismatched base and replaces it with a suitable base.

reproduction Any process by which a parental cell or organism produces offspring. Among eukaryotes, asexual modes and sexual modes. Prokaryotes use prokaryotic fission only. Viruses cannot reproduce alone; host organisms are required for their replication cycle.

restriction enzyme A protein that can recognize and cut specific base sequences in double-stranded DNA. Bacteria use diverse kinds to cut up viral DNA before it can get inserted into the bacterial chromosome; also used in recombinant DNA technology.

reverse transcriptase Type of enzyme that assembles a single strand of DNA from free nucleotides on an RNA template; RNA viruses make them.

ribosomal RNA rRNA. Type of RNA that is a functional as well as structural component of ribosomes.

ribosome In all cells, the structure upon which polypeptide chains are built. An intact ribosome consists of two subunits of rRNA and proteins.

RNA Ribonucleic acid. Any of a class of single-stranded nucleic acids involved in transcription and translation; some RNAs show enzyme activity.

RNA polymerase Enzyme that catalyzes the addition of nucleotides to a growing strand of RNA (transcription).

rubisco RuBP carboxylase. Enzyme that fixes carbon in the C3 pathway of photosynthesis; catalyzes attachment of a carbon atom from carbon dioxide to RuBP, the first stable intermediate of the Calvin–Benson cycle.

salt Compound that releases ions (other than H⁺ and OH⁻) in solution.

scientific theory An explanation of the cause of a range of related phenomena; has not been disproved after rigorous testing by many diverse experiments and remains open to revision.

second law of thermodynamics A law of nature stating that the spontaneous direction of energy flow is from more concentrated forms to less concentrated forms; with each conversion, some energy is randomly dispersed in a form not as useful for doing work (e.g., heat).

secondary wall A rigid but permeable wall that forms inside the primary wall of many plant cells after the first growing season.

seed bank A safe storage facility where genes of diverse plant lineages are being preserved for botanists, plant breeders, and geneticists.

segregation, theory of Mendelian theory that applies to sexually reproducing organisms, which inherit pairs of genes on pairs of homologous chromosomes; the two genes of each pair are separated from each other at meiosis, and they later end up in separate gametes.

selective permeability Built-in capacity of a cell membrane to prevent or allow specific substances from crossing it at certain times, in certain amounts.

sex chromosomes Chromosomes that, in certain combinations, dictate the gender of a new individual.

sexual reproduction Production of genetically variable offspring by meiosis, gamete formation, and fertilization.

sister chromatids The two members of a duplicated chromosome, each the same DNA molecule and associated proteins; the two stay attached at the centromere until separated from each other during mitosis or meiosis.

sodium–potassium pump Cotransporter that, when energized, moves sodium and potassium ions across a cell membrane; as it actively transports sodium out of the cell, potassium passively diffuses in.

solute Any substance that is dissolved in a solution.

sperm Mature male gamete.

spore A haploid reproductive cell or body, often with a wall or coat, that is not a gamete and that does not take part in fertilization; transitional stage that gets the next generation through harsh conditions, aids in dispersal, or both.

sterol Type of lipid that has a rigid backbone of four fused carbon rings.

stoma, plural **stomata** The gap between two plumped guard cells in leaf or stem epidermis; allows the diffusion of water vapor and gases across the epidermis and prevents it when closed.

stroma The semifluid matrix between the thylakoid membrane system and two outer membranes of a chloroplast; where sucrose, starch, cellulose, and other end products of photosynthesis are built.

substrate A reactant molecule that is specifically acted upon by an enzyme.

substrate-level phosphorylation The direct, enzyme-mediated transfer of a phosphate group from a substrate to another molecule.

surface-to-volume ratio Physical relationship in which volume increases with the cube of the diameter, but surface area increases with the square; constrains increases in cell size.

syndrome A set of symptoms that characterize a disease or a genetic abnormality or disorder.

tandem repeat One of many copies of short base sequences positioned one after the other along a chromosome; used in DNA fingerprinting.

telophase The last stage of meiosis I; two haploid clusters of chromosomes have formed. The last stage of mitosis and of meiosis II, haploid clusters of chromosomes that decondense into threadlike form, and a nuclear envelope forms around each.

temperature A measure of molecular motion.

test, scientific Any standardized or innovative means by which a prediction based on a hypothesis might be disproved; may require designing and conducting experiments, making observations, or developing models.

testcross A cross that might reveal the (unknown) genotype of an individual showing dominance for a trait; the individual is crossed with a known homozygous recessive individual.

thylakoid membrane A chloroplast's inner membrane system, often folded into flattened sacs, that forms a single continuous compartment in the stroma. Arrays of pigments and enzymes in the membrane function in the formation of ATP and NADPH in the first stage of photosynthesis.

tracer Any substance with a radioisotope attached; researchers can track it after delivering it into a cell, a multicelled body, ecosystem, or some other system.

transcription First stage of protein synthesis, in which a strand of RNA is assembled on a DNA template (a gene).

transfer RNA tRNA. One of a class of small RNA molecules that deliver amino acids to a ribosome; an anticodon that is part of each pairs with an mRNA codon during translation.

transition state The fleeting point when a chemical reaction can run either to product or back to reactant.

translation Second stage of protein synthesis; information encoded in an mRNA transcript guides the synthesis of a new polypeptide chain from amino acids; takes place at ribosomes.

translocation Of cells, the attachment of part of a broken chromosome to another chromosome.

transport protein Membrane protein that passively or actively assists specific molecules or ions across the membrane's lipid bilayer. The solutes cross inside a channel through the protein's interior.

transposon Transposable element. A stretch of DNA that jumps spontaneously and randomly to a different location in the genome; may cause mutation.

triglyceride A lipid with three fatty acid tails attached to a glycerol backbone.

tumor Tissue mass with cells dividing at an abnormally high rate. When benign, cells stay in the home tissue; When malignant, cells metastasize, or move to form tumors in new places in the body.

uracil One four nitrogen-containing nucleotide bases that occur in RNA; can base-pair with adenine in DNA.

variable A specific aspect of an object or event that may differ over time and among individuals. In an experimental test, a single variable is manipulated in carefully controlled ways that might disprove or support a prediction based on a hypothesis.

vesicle One of a great variety of small, membrane-bound sacs in cytoplasm that function in the transport, storage, or digestion of substances.

wavelength Of wavelike forms of energy in motion, the distance between the crests of two successive waves.

wax A type of lipid with long-chain fatty acids that are attached to long-chain alcohols or carbon rings.

X chromosome inactivation Gene control mechanism; the programmed shutdown of many genes on one of the two X chromosomes in somatic cells of a mammalian female.

xanthophyll One of a class of accessory pigments in photosynthesis that reflect yellow to orange light.

xenotransplantation Surgical transfer of an organ from one species to another.

x-ray diffraction image Pattern formed when x-rays that have been directed at a molecule are scattered; the resulting pattern of streaks and dots is used to calculate positions of atoms in the molecule.

Art Credits and Acknowledgments

This page constitutes an extension of the book copyright page. We have made every effort to trace the ownership of all copyrighted material and secure permission from copyright holders. In the event of any question arising as to the use of any material, we will be pleased to make the necessary corrections in future printings. Thanks are due to the following authors, publishers, and agents for permission to use the material indicated.

page i, iii, Russ Lowgren

TABLE OF CONTENTS **Page v** Raychel Ciemma. **Page vi** right, Larry West/FPG/Getty Images. **Page vii** top, © Professors P. Motta and T. Naguro/SPL/Photo Researchers, Inc.; bottom, © Jennifer W. Shuler/Science Source/Photo Researchers, Inc. **Page viii** top from left, © George Lepp/CORBIS; (both) Courtesy of Carl Zeiss MicroImaging, Thornwood, NY; © McLeod Murdo/Corbis Sygma. **Page ix** top from left, Model, courtesy of Thomas A. Setitz from *Science*; Courtesy of Joseph DeRisa from *Science*, 1997 Oct. 24; 278 (5338) 680–686.

INTRODUCTION NASA Space Flight Center

CHAPTER 1 **1.1** John McColgan/Bureau of Land Management, Alaska Fire Service; (inset) © Peter Turnley/CORBIS. **1.2** (a) Rendered with Atom In A Box, copyright Dauger Research, Inc.; (b, above left) PDB file courtesy of Dr. Christina A. Bailey, Department of Chemistry & Biochemistry, California Polytechnic State University, San Luis Obispo, CA.; (b, above center) PDB ID: 1BBB; Silva, M. M., Rogers, P. H., Arnone, A.; A third quaternary structure of human hemoglobin A at 1.7-A resolution; J Biol Chem 267 pp. 17248 (1992); (b, above right) PDB file from Klotho Biochemical Compounds Declarative Database; Photographs: (d) © Science Photo Library/Photo Researchers, Inc.; (e) © Bill Varie/CORBIS; (f) © Jeffrey L. Rotman/CORBIS; (g) © Jeffrey L. Rotman/CORBIS; (h) © Jeffrey L. Rotman/CORBIS; (i) Peter Scoones; (j–k) NASA. **1.3** David Neal Parks. **1.4** © Y. Arthus-Bertrand/Peter Arnold, Inc. **1.6** Photographs by Jack de Coningh. **1.8** Page 8, Clockwise from above left, © P. Hawtin, University of Southampton/SPL/Photo Researchers, Inc.; CNRI/SPL/Photo Researchers, Inc.; © Dr. Harald Huber, Dr. Michael Hohn, Prof. Dr. K. O. Stetter, University of Regensburg, Germany; R. Robinson/Visuals Unlimited; Page 9, Clockwise from above left, © Lewis Trusty/Animals Animals; John Lotter Gurling/Tom Stack & Associates; Edward S. Ross; © Stephen Dalton/Photo Researchers, Inc.; Edward S. Ross; Robert C. Simpson/

Nature Stock; © Oliver Meckes/Photo Researchers, Inc.; Courtesy © James Evarts; Emiliania Huxleyi. Photograph by Vita Pariente. Scanning electron micrograph taken on a Jeol T330A instrument at the Texas A & M University Electron Microscopy Center; Carolina Biological Supply Company. **1.9** Left, Photographs courtesy Derrell Fowler, Tecumseh, Oklahoma; right, © Nick Brent. **Page 11** © LWA-Stephen Welstead/CORBIS. **Page 12** © SuperStock. **1.11** (a–b) © Chris D. Jiggins; (c) Background, © Kevin Schafer/Peter Arnold, Inc.; (d) © Martin Reid. **1.12** © Gary Head. **Page 16** © Digital Vision/PictureQuest. **Page 17 Unit I** © Wim van Egmond, Micropolitan Museum

CHAPTER 2 **2.1** © Owaki-Kulla/CORBIS. **2.2** (c) Rendered with Atom In A Box, copyright Dauger Research, Inc. **2.4** (a) © CC Studio/Photo Researchers, Inc.; (d) Harry T. Chugani, M.D., UCLA School of Medicine. **Page 22** © Michael S. Yamashita/CORBIS. **2.5** Rendered with Atom In A Box, copyright Dauger Research, Inc. **2.8** Photographs (a) above, Gary Head; below, Micrograph © Bruce Iverson; (c, page 25) PDB ID:IBNA; H .R. Drew, R. M. Wing, T. Takano, C. Broka, S. Tanaka, K. Itakura, R. E. Dickerson; Structure of a B-DNA Dodecamer. Conformation and Dynamics, PNAS. **2.9** (a,b,c, left) PDB file from NYU Scientific Visualization Lab; (b, right) © Steve Lissau/Rainbow; (c, right) © Mark Newman/Bruce Coleman USA. **2.11** (a) © Lester Lefkowitz/CORBIS. **2.13** Michael Grecco/Picture Group. **2.14** © National Gallery Collection; by kind permission of the Trustees of the National Gallery, London/CORBIS. **2.15** © H. Eisenbeiss/Frank Lane Picture Agency.

CHAPTER 3 **3.1** Left, © 2002 Charlie Wait/Stone/Getty Images; right, © Dr. W. Michaelis/Universitat Hamburg. **Page 33** © John Collier. Great Britain, 1850–1934, *Priestess of Delphi*, 1891, London, oil on canvas, 160.0 x 80.0 cm. Gift of the Rt. Honourable, the Earl of Kintore, 1893. **Page 34** (right, top) PDB file from NYU Scientific Visualization Lab; (right, center), PDB file from NYU Scientific Visualization Lab; (right bottom), PDB file from Klotho Biochemical Compounds Declarative Database. **3.5** Photograph, Tim Davis/Photo Researchers, Inc. **3.9** © Steve Chenn/CORBIS. **3.10** © David Scharf/Peter Arnold, Inc. **3.11** (a) PDB file courtesy of Dr. Christina A. Bailey, Department of Chemistry and Biochemistry, California Polytechnic State University, San Luis Obispo, CA. **3.12** Left, © Kevin Schafer/CORBIS. **3.13** (a) PDB file courtesy of Dr. Christina A. Bailey, Department of Chemistry.

3.14 (a) © Scott Camazine/Photo Researchers, Inc. **3.15** (b–e) PDB files from NYU Scientific Visualization Lab. **3.16** (b, right) After: *Introduction to Protein Structure*, 2nd ed., Branden & Tooze, Garland Publishing, Inc.; (c, left) PDB ID: 1BBB; Silva, M. M., Rogers, P. H., Arnone, A.; A third quaternary structure of human hemoglobin A at 1.7-Å resolution; J Biol Chem 267 pp. 17248 (1992); (c, right) After: *Introduction to Protein Structure*, 2nd ed., Branden & Tooze, Garland Publishing, Inc. **3.17** PDB ID: 1BBB; Silva, M. M., Rogers, P. H., Arnone, A.; A third quaternary structure of human hemoglobin A at 1.7-Å resolution; J Biol Chem 267 pp. 17248 (1992). **3.18** (a,b) PDB files from New York University Scientific Visualization Center; (c) © Dr. Gopal Murti/SPL/Photo Researchers, Inc.; (d) Courtesy of Melba Moore. **3.19** PDB files from Klotho Biochemical Compounds Declarative Database. **3.21** PDB ID:1BNA; H. R. Drew, R. M. Wing, T. Takano, C. Broka, S. Tanaka, K. Itakura, R. E. Dickerson; Structure of a B-DNA Dodecamer. Conformation and Dynamics; PNAS V. 78 2179, 1981. **3.22** © ThinkStock/SuperStock. **3.23** Left, PDB ID: 1AKJ; Gao, G. F., Tormo, J., Gerth, U. C., Wyer, J. R., McMichael, A. J., Stuart, D. I., Bell, J. I., Jones, E. Y., Jakobsen, B. K.; Crystal structure of the complex between human CD8alpha(alpha) and HLA-A2; Nature 387 pp. 630 (1997); right, Al Giddings/Images Unlimited.

CHAPTER 4 **4.1** © Tony Brian and David Parker/SPL/Photo Researchers, Inc. **4.2** Left, National Library of Medicine; right, Armed Forces Institute of Pathology. **4.7** Leica Microsystems, Inc., Deerfield, IL. **4.8** © Geoff Tompkinson/SPL/Photo Researchers, Inc. **4.9** Photographs: (hummingbird) © Robert A. Tyrrell; (human) © Pete Saloutos/CORBIS; (redwood) © Sally A. Morgan, Ecoscene/CORBIS. **4.10** Photographs by Jeremy Pickett-Heaps, School of Botany, University of Melbourne. **4.11** (a) K. G. Murti/Visuals Unlimited; (b) R. Calentine/Visuals Unlimited; (c) Gary Gaard and Arthur Kelman. **4.12** (a) © University of California Museum of Paleontology; (b) © University of California Museum of Paleontology; (c) © Russell Kightley/SPL/Photo Researchers, Inc. **4.13** M. C. Ledbetter, Brookhaven National Laboratory. **4.14** Micrograph, G. L. Decker. **4.16** Stephen L. Wolfe. **4.17** Left micrograph, Don W. Fawcett/Visuals Unlimited; center micrograph, A. C. Faberge, *Cell and Tissue Research*, 151: 403–415, 1974. **4.18** Micrographs: (a) Stephen L. Wolfe; (c,d) Don W. Fawcett/Visuals Unlimited, computer enhanced; (e) Gary Grimes, computer enhanced. **4.19** Right micrograph, Keith R.

www.microimaging.ca. **10.14** © Lisa O'Connor/ZUMA/CORBIS.

CHAPTER 11 **11.1** Left, Robert E. Basye, Rose Breeding and Genetics Research Program, Department of Horticultural Sciences, Texas A&M University; right, Department of Horticultural Sciences, Texas A&M University. **Page 169** Bob Cerasoli. **11.2** Moravian Museum, Brno. **11.3** (a) Jean M. Labat/Ardea, London. **11.10** © David Scharf/Peter Arnold, Inc. **11.11** Photographs: above, William E. Ferguson; below, © Francesc Muntada/CORBIS. **11.12** Tedd Somes. **11.13** (a–b) Michael Stuckey/Comstock, Inc.; (c) Bosco Broyer, photograph by Gary Head. **11.14** © Bettmann/CORBIS. **11.17** Left, © Pamela Harper/Harper Horticultural Slide Library; right, from Prof. Otto Wilhelm Thomé, Flora von Deutschland Österreich und der Schweiz. 1885, Gera, Germany. **11.18** Left (Down from top), Frank Cezus/FPG/Getty Images; Frank Cezus/FPG/Getty Images; Ted Beaudin/FPG/Getty Images; © Michael Prince/CORBIS; right, © Lisa Starr. **11.20** Courtesy of Ray Carson, University of Florida News and Public Affairs. **11.21** Left, © Tom and Pat Leeson/Photo Researchers, Inc.; right, © Rick Guidotti, Positive Exposure. **11.22** (a) Courtesy of Wayside Gardens/www.waysidegardens.com; (b) © Gene Ahrens/SuperStock; (c) © Karen Tweedy-Holmes/CORBIS; (d) © Clay Perry/CORBIS. **11.23** Leslie Falteisek/Clacritter Manx. **Page 185** Tedd Somes. **11.24** © Maximilian Stock Ltd./Foodpix.

CHAPTER 12 **12.1** Left, © Reuters/CORBIS; center, Gene Griessman/www.presidentlincoln.com; right, © Hulton-Deutsch Collection/CORBIS. **12.2** © Andrew Syred/Photo Researchers, Inc. **12.3** (b) right, © Charles D. Winters/Photo Researchers, Inc.; (f) © Omikron/Photo Researchers, Inc. **12.4** From *Multicolor Spectral Karyotyping of Human Chromosomes*, by E. Schrock, T. Ried, et al, *Science*, 26 July 1996, 273:495. Used by permission of E. Schrock, T. Reid and the American Association for the Advancement of Science. **12.5** Above, © Frank Trapper/CORBIS Sygma. **12.6** © Lois Ellen Frank/CORBIS. **12.7** Eddie Adams/AP Wide World Photos. **Page 192** Above, © 2001 PhotoDisc, Inc.; below, © 2001 EyeWire. **12.8** (b) from M. Cummings, *Human Heredity: Principles and Issues*, 3rd Edition, p. 126. © 1994 by Brooks/Cole. All rights reserved; (c) after Patten, Carlson & others. **Page 193** Top right, Carolina Biological Supply Company. **12.9** Left, © Carolina Biological/Visuals Unlimited; right, © Terry Gleason/Visuals Unlimited. **12.11** Art, After V. A. McKusick, *Human Genetics*, 2nd Ed., © 1969. Reprinted by permission of author; photograph

Bettmann/CORBIS. **12.12** Left, photos by Gary L. Friedman, www.FriedmanArchives.com. **Page 195** © Russ Schleipman/CORBIS. **12.13** Courtesy G. H. Valentine. **12.14** From *Multicolor Spectral Karyotyping of Human Chromosomes*, by E. Schrock, T. Ried, et al, *Science*, 26 July 1996, 273:496. Used by permission of E. Schrock and T. Reid and the American Association for the Advancement of Science. **12.16** (a) © CNRI/Photo Researchers, Inc. **Page 199** UNC Medical Illustration and Photography. **12.18** © Stapleton Collection/CORBIS. **12.19** Dr. Victor A. McKusick. **12.20** Steve Uzzell. **12.21** © Saturn Stills/SPL/Photo Researchers, Inc. **12.22** From Lennart Nilsson, *A Child is Born*, © 1966, 1977 Dell Publishing Company, Inc. **12.23** © Matthew Alan/CORBIS; Inset, Fran Heyl Associates © Jacques Cohen, computer-enhanced by © Pix Elation. **12.24** Stefan Schwarz.

CHAPTER 13 **13.1** PA News Photo Library. **13.2** A. C. Barrington Brown © 1968 J. D. Watson. **13.4** (c) photograph, Eye of Science/Photo Researchers, Inc. **13.6** PDB ID: 1BBB; Silva, M. M., Rogers, P. H., Arnone, A.: A third quaternary structure of human hemoglobin A at 1.7-Å resolution. J Biol Chem 267 pp. 17248 (1992). **13.9** (1–3) © James King-Holmes/SPL/Photo Researchers, Inc.; (4) © McLeod Murdo/CORBIS Sygma. **13.10** Photos by Victor Fisher, courtesy Genetic Savings & Clone. **13.12** © SPL/Photo Researchers, Inc. **13.13** Right, Shahbaz A. Janjua, MD/Dermatlas; http://www.dermatlas.org.

CHAPTER 14 **14.1** Left, © Vaughan Fleming/SPL/Photo Researchers, Inc.; right, PDB ID: 2AAI; Rutenber, E., Katzin, B.J., Ernst, S., Collins, E.J., Mlsna, D., Ready, M. P., Robertus, J. D., Crystallographic Refinement of ricin to 2.5 A. Proteins 10pp.240 (1991). **14.7** Above, tRNA model by Dr. David B. Goodin, The Scripps Research Institute. **14.8** (a) Courtesy of Thomas A. Steitz from *Science*. **14.11** Left, Nik Kleinberg; right, P. J. Maughan. **14.12** © John W. Gofman and Arthur R. Tamplin. From *Poisoned Power: The Case Against Nuclear Power Plants Before and After Three Mile Island*, Rodale Press, PA, 1979. **14.14** © Dr. M. A. Ansary/SPL/Photo Researchers, Inc.

CHAPTER 15 **15.1** From the archives of www.breastpath.com, courtesy of J. B. Askew, Jr., M.D., P. A. Reprinted with permission, copyright 2004 Breastpath.com. **Page 231** Courtesy of Robin Shoulla and Young Survival Coalition. **15.2** (b) From the collection of Jamos Werner and John T. Lis. **15.4** (a) Dr. Karen Dyer Montgomery. **15.5** Jack Carey. **15.6** (a) Above, Juergen Berger, Max Planck Institute for

Developmental Biology–Tuebingen, Germany; (a) below, © Jose Luis Riechmann; (b) below, © Jose Luis Riechmann. **15.7** (a) Left, © Visuals Unlimited; (a) right, UCSF Computer Graphics Laboratory, National Institutes, NCRR Grant 01081; (c) from left, © Walter J. Ghering/University of Basel, Switzerland; © Carolina Biological/Visuals Unlimited; © Carolina Biological/Visuals Unlimited; © Carolina Biological/Visuals Unlimited; Courtesy of Edward B. Lewis, California Institute of Technology **15.8** (a) Palay/Beaubois after Robert F. Weaver and Philip W. Hedrick, *Genetics*. © 1989 W. C. Brown Publishers; (b–c) © Jim Langeland, Jim Williams, Julie Gates, Kathy Vorwerk, Steve Paddock, and Sean Carroll, HHMI, University of Wisconsin-Madison. **15.9** PDB ID: 1CJG; Spronk, C. A. E. M., Bonvin, A. M. J. J., Radha, P. K., Melacini, G., Boelens, R., Kaptein, R.: The Solution Structure of Lac Repressor Headpiece 62 Complexed to a Symmetrical Lac Operator. Structure (London) 7 pp. 1483 (1999). Also PDB ID: 1LBI; Lewis, M., Chang, G., Horton, N. C., Kercher, M. A., Pace, H. C., Schumacher, M. A., Brennan, R. G., Lu, P.: Crystal structure of the lactose operon repressor and its complexes with DNA and inducer. *Science* 271 pp. 1247 (1996); lactose pdb files from the Hetero-Compound Information Centre - Uppsala (HIC-Up). **15.11** © Jim Langeland, Jim Williams, Julie Gates, Kathy Vorwerk, Steve Paddock, and Sean Carroll, HHMI, University of Wisconsin-Madison.

CHAPTER 16 **16.1** (a) Per Hardestam/www.hardestam.se; (b) Dr. Jorge Mayer, Golden Rice Project. **Page 243** ScienceUV/Visuals Unlimited. **16.3** (a) Dr. Huntington Potter and Dr. David Dressler; (b) with permission of © QIAGEN, Inc. **Page 248** © TEK IMAGE/Photo Researchers, Inc. **16.9** Left, © David Parker/SPL/Photo Researchers, Inc.; right, Cellmark Diagnostics, Abingdon, UK. **16.10** Right, © Volker Steger/SPL/Photo Researchers, Inc. **16.11** Courtesy of Joseph DeRisa. From *Science*, 1997 Oct. 24; 278 (5338) 680–686. **Page 252** © Professor Stanley Cohen/SPL/Photo Researchers, Inc. **16.12** (a) Courtesy Calgene, LLC; (b) Dr. Vincent Chiang, School of Forestry and Wood Products, Michigan Technology University. **16.13** (d) © Lowell Georgis/CORBIS; (e) Keith V. Wood. **16.14** (a) © Adi Nes, Dvir Gallery Ltd.; (b) Transgenic goat produced using nuclear transfer at GTC Biotherapeutics. Photo used with permission; (c) © Matt Gentry/Roanoke Times. **16.14** R. Brinster, R. E. Hammer, School of Veterinary Medicine, University of Pennsylvania. **Page 256** © Jeans for Gene Appeal.

Index

The letter i *designates illustration;* t *designates table;* **bold** *designates defined term;* ▪ *highlights the location of applications contained in text.*